Franz Pichler · Heinz Schwärtzel (Eds.)

CAST
Methods in Modelling

Computer Aided Systems Theory for the Design of Intelligent Machines

With 181 Figures

Springer-Verlag
Berlin Heidelberg NewYork
London Paris Tokyo
HongKong Barcelona Budapest

Dr. phil. Franz Pichler

Professor, Institute for Systems Sciences
Johannes-Kepler-University, Linz / Austria

Dr. techn. Heinz Schwärtzel

Corporate Research and Development
Systems Technologies
Siemens AG, Munich / Germany

ISBN-13: 978-3-642-95682-9 e-ISBN-13: 978-3-642-95680-5
DOI: 10.1007/978-3-642-95680-5

Library of Congress Cataloging-in-Publication Data

CAST: methods in modelling:
computer aided systems theory for the design of intelligent machines
Franz Pichler, Heinz Schwärtzel (eds)
Includes bibliographical references and index
ISBN-13: 978-3-642-95682-9
1. Systems engineering--Mathematical models
2. Computer-aided engineering
3. Computer-aided design
4. Computer vision
I. Pichler, Franz, Ing., Dr. phil.
II. Schwärtzel, H.
TA 168. C37 1992
620'.001'5118--dc20 92-8524

Softcover reprint of the hardcover 1st edition 1992

Typesetting: Camera-ready by authors

62/3020-5 4 3 2 1 0 - Printed on acid-free paper

Preface

The merits of Systems Theory for the education of students, particularly in communications and control engineering, have been recognized for a long time and are demonstrated even today by numerous dedicated courses. Before CAD tools came into common use, systems-theoretical methods could be applied in practical engineering only as long as the size of the problems was sufficiently small. Since about 1975, the appearance of complex building blocks, such as microprocessors, signal processors, and other highly integrated circuits – together with the corresponding CAD tools – has changed the situation in engineering. Modelling techniques supplied by Applied Mathematics and Systems Theory lost much of their earlier importance. With the new powerful building blocks and the new CAD tools for software and hardware development, engineers were now able to solve design and analysis tasks well enough for practical application. The use of theoretical means for optimization could be avoided in most cases and the seemingly unlimited computational power of the new technologies covered most of the deficiencies caused by poor modelling.

Today we encounter a different situation. "Micro-technologies" (i.e., microelectronics, micromechanics, and other basic technologies for manufacturing microsystems) and their corresponding CAD tools have meanwhile reached a level of sophistication that requires the application of theoretical means on all modelling levels of design and analysis. Also, there is a growing need for a scientific approach in modelling again. Many concepts provided by Systems Theory again turn out to be of major importance. This is especially valid for the design of "machines with intelligent behavior".

When dealing with complex systems, the engineering design has to be supported by CAD tools. Consequently, the methods of Systems Theory must also get computerized. The newly established field of "Computer Aided Systems Theory" (CAST) is a first effort in this direction. The goal of CAST research and development is to provide "Systems Theory Method Banks" which can be used in education and to provide a platform for the migration of CAST methods into existing CAD tools. This book is based on different research and development projects in CAST, undertaken at the Institute of Systems Science at the Johannes Kepler University in Linz. Of special importance for our work was a steady contact with industry experts. Especially our cooperation with Siemens AG Munich, Corporate Research and Development, deserves to be mentioned. A short overview on the contents of this book and a motivation for the selection of the topics reads as follows:

After a general introduction to problem solving – as seen from the systems science point of view – in Chapter 1, we deal with the construction of CAST method bank systems in Chapter 2. As an example we show CAST.FSM, a method bank system for Finite State Machines, and its use for CAD in microelectronics. In Chapter 2 we have problem areas in mind, where classical Systems Theory still works for both phases of modelling: for model building and for model analysis. Chapter 3 (Modelling and Simulation) deals with modelling tasks which are more complex. We introduce advanced modelling methods, which use Systems Theory for model building, but depend on simulation for model analysis. Finally, in Chapter 4 (Machine Vision), we consider a problem area, where systems-theoretical methods are of restricted value for both model building and model analysis and have to be supported by methods from Artificial Intelligence. The book concludes with a discussion of programming techniques which are useful for the implementation of CAST tools.

The book is written for engineers who are interested in using and developing CAST systems, particularly in the field of Information- and Systems Engineering. The editors take the opportunity here to give their thanks to all members of the CAST research group at the University of Linz who contributed to this book. Mrs. Erika Draxler's help in typing the manuscript and the cooperative efforts of the Springer Verlag are highly appreciated.

Linz and Munich, May 1992 Franz Pichler, Heinz Schwärtzel

Contributors

Wilhelm Burger, University of Linz	Ch. 4, 5
Martin Geiger, Siemens AG Munich	Ch. 2
Hermann Hellwagner, Siemens AG Munich	Ch. 2
Rudolf Mittelmann, University of Linz	Ch. 2, 5
Thomas Müller-Wipperfürth, University of Linz	Ch. 2, 5
Franz Pichler, University of Linz	Ch. 1, 2
Herbert Praehofer, University of Linz	Ch. 3, 5
Heinz Schwärtzel, Siemens AG Munich	Ch. 1

Contents

1 INTRODUCTION

1.1 PROBLEM-SOLVING BY MODELS

Problems arise contextually in connection with open answers to questions concerning a given reality. A reality can already exist or be planned to exist. Existing realities are typical for problems of science, planned realities are the main objects in engineering.

Problem-solving is realized by a process of taking different steps towards the goal of reaching a solution. In science and engineering the most important approach is the use of models which allow for the application of scientific methods to realize the problem-solving process. A possible structuring of the problem-solving process (Figure 1.1) can be made by considering the following three steps (see for example [Pichler 91]):

1. The first step consists of the phase of *model building*. It includes the task of defining the problem in an appropriate form and stating the goal which has to be reached in order to solve the problem. The concept of a model serves as a vehicle and desired means for reaching a solution . A model should contain all relevant knowledge concerning the addressed reality which is useful in realizing the problem-solving process.
2. The second step realizes the phase of *model application* (*model analysis*). It consists of the different steps through which the general algorithm for the "computation" of the solution is derived by means of the underlying model. The result of this step is a "blue print" in which the procedure for reaching the solution is already expressed in technological terms relevant for the actual implementation.

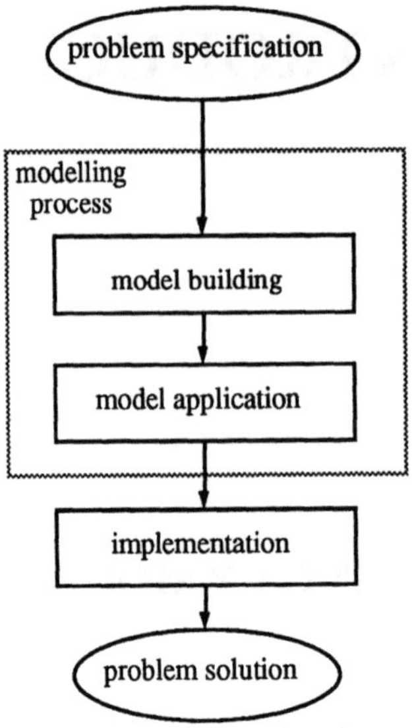

Figure 1.1 Phases in problem-solving.

3. The third and last step in our structuring of the problem-solving process is the step of *implementation*. It establishes the desired solution in reality using the "blue print" which results from the phase of model application. Depending on the kind of the solution, a certain technology is applied to "physically" realize the solution. As examples we mention CMOS technology to manufacture VLSI circuits for computer- and automation engineering applications – being a "hardware technology" – and certain specific programming techniques to realize programs in a language which can be executed on a certain computer – being considered as a "software technology".

By "modelling" we generally understand here the activity of establishing the first two steps, namely the phases "model building" and "model application" of the problem-solving process.

1.2 SYSTEMS THEORY INSTRUMENTED MODELLING

Systems Theory, as developed in the classical sense in the fields of Electrical Engineering and Control Engineering, has proven to be of important value for many applications in modelling. For certain areas of engineering applications the modelling process can successfully be supported by Systems Theory methods such that a high quality "blue print" (the result of the modelling process) is reached. The term "high quality" depends on differences in the properties of modelling results such as the degrees of correctness, testability, robustness concerning variations of the implementation technology and others.

In complex engineering tasks, such as, for example, the design of intelligent machines, it is generally necessary to optimize modelling. Systems Theory methods applied to modelling can support such engineering tasks. In model building, the conceptual framework, as it is available in Systems Theory, may help to construct a model which best represents the stated problem and which allows for efficient model-refinement towards the solution as determined by the stated goals. Of importance in this respect is the availability of transformations as part of Systems Theory which allow multi-level representations of models [Mesarovic, Macko, Takahara 70].

For the reason of completeness and in order to more clearly point out the philosophical aspects of our approach in this book, we give a short introduction to the general systems framework which we use in modelling [Pichler, Schwärtzel 90].

We want to distinguish between five different types of models, namely the

- *black box* type,
- *generator* type,
- *dynamics* type,
- *algorithm* type,
- *network* type.

A black box describes the model by its interaction with the environment. In the most simple case a black box model consists of the set of variables which describes the attributes of the model which are essential for interaction with the environment. These variables may depend on the kind of space or time set in which the model is embedded (space-variables, time-variables). More specific examples of black box models are reached in cases where the variables have a well-defined value-set (analogue variables, digital variables) or a direction of their interaction with the environment (input-variables, output-variables).

A generator reflects local laws which are valid for the state-variables of a model. Specific generators are given, for example, by systems of differential equations, which reflect fundamental physical laws of nature. For models with discrete-time variables difference equations and finite state machines are well-known models of the generator type. Another generator-type model is given by discrete event systems or by marked Petri nets.

Dynamics (or dynamical systems) describe the global change of states. Since state reflects the most important properties of a model, global changes of states are very often an essential part of the desired solution. This type of model is, therefore, of great interest in model building. In many cases a dynamical system is reached from a generator type model by "integration".

Algorithms consists of operations of computation (the processing part of an algorithm) and of an organizational scheme (the control part) which determines at which step a certain operation has to be applied. Algorithms must possess a well-defined set of initial input data to specify the starting position and they must have a well defined stop rule to end the computation. Algorithms obviously have a wide field of application in science and engineering. Besides computing the value for a function, they are equally important for giving a scheme for the cooperation of different component models in an overall system. In this sense they determine the control structure for a network model (to be discussed next) which uses a discrete event type of coupling.

Networks are model-types which consist of components (determined by the model specifications which we have introduced above) together with a set of coupling rules which interface the components. Networks therefore define a model type which possesses a topology. A network type of model is also called "structured". Different specializations of networks are possible. Important are, for example, recursive networks, where each component is again allowed to be of network type. Recursive networks give cause to hierarchical multi-level models. The counterpart of recursive networks are in some respect flat networks, where each component is considered to be atomic (not decomposable).

The different types of models which are available constitute a conceptual framework which is of great help in model building. The general types of models, which we have introduced, together with the known specializations of them, allow us, as part of the problem-definition phase, to construct a model (of the first kind). Usually this model will not be suited for application in the sense that the computation of the wanted solution is directly derivable from it. Generally it will be necessary to take more steps and "process" the model to other representations.

This idea leads to the concept of a *model-transformation*. If **M** denotes a model, then a model-transformation (which fits to **M**) is an operation T which maps **M** to another model **M'** = T(**M**). **M'** is assumed here to have the same ability for problem-solving as **M**. However **M'** should already be somewhat better suited for the computation of the solution than **M**. In general it will be necessary to apply a sequence of model-transformations $T_1, T_2, \ldots, T_n$ (starting from the initial constructed model **M**) to reach after n steps with $\mathbf{M}' = T_n(T_{n-1} \ldots (T_1(\mathbf{M})) \ldots)$ a model from which an algorithm to compute the desired solution can directly be derived. The finding of the right type of model-transformations T_i is rather an "art" than a science. Furthermore even when a large set of such transformations is known to the modeler it needs expert-knowledge to select the proper one to form the right sequence.

In our discussion of modelling we have now reached a point where Systems Theory can be brought into discussion. In Systems Theory, formal mathematical models (i.e., models which are abstract and expressed in mathematical language) are considered and model-transformations are defined in precise mathematical terms.

It is quite obvious that systems theoretical models have certain advantages in problem solving. The mathematical theory which comes along with such models offers many ways for model representation and model transformation. It is therefore most often easier to reach a desired goal in a systematic way. On the other hand such a mathematical approach may also have some disadvantages. The modeller may be misguided in the sense that he selects models in favour of available efficient methods for model application while neglecting essential features of the given problem by abstraction. Balancing the modelling activity between "problem adequacy" and "method efficiency" is an essential art part of applying Systems Theory to modelling.

Systems Theory can be considered as a field which offers *instruments* (in the form of mathematical concepts and methods) for problem solving to the modeller [Nagel 61]. The proper selection of the right type of instruments in the process of problem solving is done by the expert in the specific domain. This is what we mean by the terms "Systems Theory Instrumented Problem Solving (STIPS)" and "Systems Theory Instrumented Modelling".

1.3 CAST – COMPUTER AIDED SYSTEMS THEORY

Today the computer is used as the main tool in realizing engineering problem solving environments. Computer aided design tools (CAD tools) exist for many different engineering problem solving tasks. They support an expert in certain specific modelling activities such as model building (e.g. drawing a specific diagram of electronic circuitry which uses predefined building blocks) and simulation as part of model application (e.g. computation of the I/O behavior of a specific electronic circuit). CAD tools are usually tailored to a specific domain of application (such as for example "digital electronic circuits" or "mechanical machinery components") and automate classical engineering approaches such as engineering drawing and investigation of a design by models of reduced size and reduced functionality (e.g. a wooden-model of a bridge). CAD tools are supplemented by CAM tools, which are tools for computer aided manufacturing. Such tools help to automate implementation.

The field of CAD tools is still expanding very fast. Besides going into new technological areas, such as, for example microsystems technology, there is the tendency to develop CAD tools for early steps of modelling such as problem definition (building a model of the first kind) and model application (which are based on theoretical results) available for specific model types.

In current developments of CAD tools in engineering, Systems Theory instruments are need to support modelling. Computer Aided Systems Theory (CAST) has as its goal the supplementation of current and future CAD tools by Systems Theory software (CAST tools) which can be applied in model building and in model application (Figure 1.2). CAST tools allow the design engineer to apply different theoretically based model transformations as part of the problem solving process. They allow for the application of theoretical knowledge needed to reach optimal results in

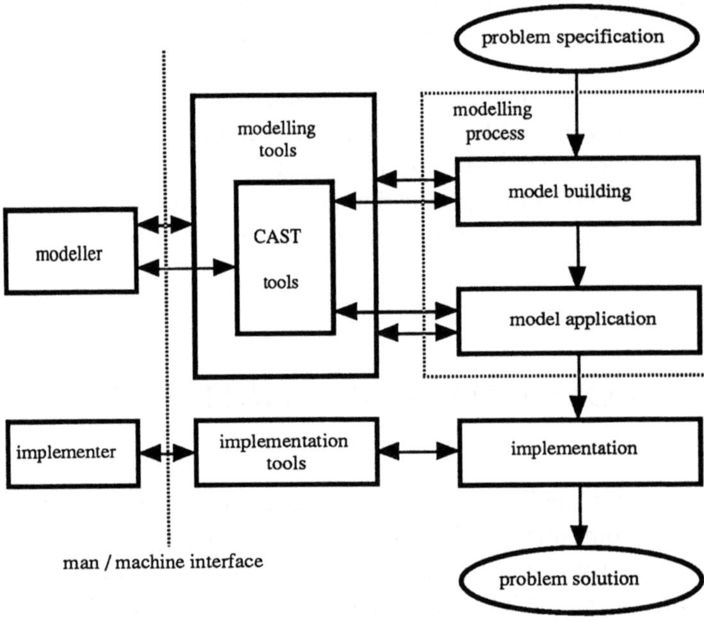

Figure 1.2 CAST tools in modelling.

modelling complex systems. They are the proper means of realizing a Systems Theory instrumented modelling philosophy [Pichler 88].

1.4 CAST METHODS IN MODELLING: AN OVERVIEW

In the following chapters of this book we will cover different topics of CAST research and development of CAST tools. In Chapter 2 we will outline some of the major concerns in the making of CAST tools. Of central interest is the building of CAST method bank systems which can be integrated in existing CAD environments.

As a conceptual framework for the development of CAST methods in modelling we will introduce a state machine (Systems Theory Instrumented Problem Solving machine; STIPS machine). The states of this (non-deterministic) machine are formed by different formal model types which are available. State transitions are realized by the different formal model transformations of Systems Theory [Pichler 86; Pichler 88; Pichler, Schwärtzel 90].

In a CAST method bank system the STIPS machine is the basic concept for the realization of the application layer. It interfaces by its inputs (which select the sequence of model-transformations) and outputs (which indicate the reached state-set in modelling) to the user.

By special implemented examples of CAST method bank systems (CAST.FSM, a method bank system for the application of finite state machines in modelling, CAST.FOURIER for the application of general spectral techniques, CAST.LISAS for finite state machine models with a cellular architecture and CAST.PN for modelling with Petri-nets) we are able to show how different kinds of CAST method bank systems are constructed and implemented in software. Specific examples of applications will be discussed in order to indicate the manner in which these method bank systems are applied.

The emphasis in Chapter 2 is on the use of CAST tools in rather classical situations in modelling. As the subtitle of this book (Computer Aided Systems Theory for the Design of Intelligent Machines) promises, we will cover also non-classical applications of Systems Theory in modelling. One of the most important approaches in modelling complex systems is without any doubt simulation. Chapter 3 (Modelling and Simulation) introduces the systems theoretical framework to support a simulation modelling. The original work on this topic comes from [Zeigler 84].

However, by extending to coupled multi-formalism models, we are able to cover a very large range of problems related to the modelling intelligent machines.

Chapter 4 (Machine Vision) addresses questions which are on the limit of problem-solving by models. It is obvious that modelling of human intelligent behavior demands the application of knowledge which is outside of classical Systems Theory. Then, dynamical systems, reflecting "Newton" model-types, are no longer the dominating mean for problem-solving. New approaches in model building and model applications have to be used when dealing with human intelligence. The field of Artificial Intelligence, especially the topic of knowledge based systems, has to be merged with the systems approach. As a result models of the "Minsky type" are used [Zeigler 89].

Finally, in Chapter 5 of this book ("Implementation Issues"), we will deal with some of the important technical questions, which have to be answered in CAST research and development. CAST software demands the use of the most advanced software design and programming techniques. A CAST method bank system should undoubtedly offer more to the user than a traditional program-package with an additional glued graphic interface can. Programming styles and techniques which have their origins in the field of Artificial Intelligence support the goals of CAST method bank systems design. The resulting software has a fair chance to fulfill the high expectations of CAST applications in the task of modelling intelligent machines.

REFERENCES – CHAPTER 1

[Mesarovic, Macko, Takahara 70]
M. D. Mesarovic, D. Macko, Y. Takahara, *Theory of Hierarchical, Multilevel Systems*, Academic Press, New York 1970.

[Nagel 61]
F. Nagel, *The Structure of Science*, Harcourt, Brace & World, New York, 1961.

[Pichler, Schwärtzel 90]
F. Pichler, H. Schwärtzel, *CAST: Computerunterstützte Systemtheorie*, Springer Verlag, Berlin, 1990.

[Pichler 86]
F. Pichler, "Model Components for Symbolic Processing by Knowledge Based Systems: The STIPS Framework," in *Modelling and Simulation Methodology in the Artificial Intelligence Era*, Elzas, Ören, Zeigler (eds.), North Holland, Amsterdam, 1986.

[Pichler 88]
F. Pichler, "CAST - Computer Aided Systems Theory: A Framework for Interactive Method Banks", *Cybernetics and Systems*, pp. 731–736, 1988.

[Pichler 91]
F. Pichler, "Modellbildung und Modellanwendung in der Informationstechnik – Systemtheoretische Erfordernisse," in *Angewandte Informatik und Software*, H. Schwärtzel (Hrsg./Ed.), pp. 175–185, 1991.

[Zeigler 84]
B. P. Zeigler, *Multifacetted Modelling and Discrete Event Simulation*, Academic Press, London, 1984.

[Zeigler 89]
B. P. Zeigler, "Knowledge Representation from Newton to Minsky and Beyond," *Applied Artificial Intelligence*, vol. 1, pp. 87–107, 1989.

2 CAST METHOD BANK SYSTEMS

2.1 CAST METHOD BANK SYSTEM DESIGN

2.1.1 Systems Theory Method Banks

In Chapter 1 we introduced the reader to the world of the "systems approach" for modelling purposes from a general point of view. The concept of *system types*, specific examples, and the concept of *system transformations* and their linking to form a *system algorithm* were the main objects of concern. In the following, we will introduce the concepts and computer-aided means that make Systems Theory operational for practical applications. The concept of a "method bank", a special kind of database system which enables the user to apply the knowledge of Systems Theory, is put forward.

The architecture of a method bank can be viewed from three different points. In the case that a method bank is described by its function to the user we speak of the *logical architecture* of a method bank. Here the focus is on the description of the individual types of system specifications and their representations. I.e., the corresponding system transformations and additional functions which are relevant from the user's point of view, stand in the foreground. Logical architecture has the most important meaning for our purpose and it will thus be of central concern.

Apart from the logical architecture, we view the *software architecture* and *hardware architecture* of a method bank. Software architecture describes from a high level point of view how the individual functions of a method bank are realized by programs. While the establishment of the logical architecture stands at the beginning of the design of a method bank and is interesting, above all, for the user, software architecture provides the necessary framework for the implementation of a method bank.

The hardware architecture describes the apparatus-specific equipment of a method bank. In the past, the hardware architecture of method banks was determined mainly by already existing computers, and therefore could hardly be an object of extensive consideration when constructing a method bank. Today, however, a change has occurred. The development of personal computers and high performance workstations and the existing possibilities for the construction of distributive computer systems by means of networking technology have allowed for the adaptation of hardware architecture to special tasks done by method banks.

Method banks are, principally, special database systems. The same kind of auxiliary functions as available in conventional database systems are needed, in particular for:

- storing system types and system transformations;
- locating previously stored system types and system transformations;
- changing stored data.

In addition, there are specific functions which are used in a method bank, such as

- functions for the construction of system types, for the definition of problems and for model building on system level;
- functions for the application of system transformations called by the method bank to be executed;
- functions which support problem definition and solution interpretation (e.g. for the use of graphics and result representation).

Figure 2.1.1 shows the different basic functions of a *Systems Theory Method Bank* as a block diagram.

When using a Systems Theory Method Bank, we assume that a CAD system is already made available for problem solving. In consequence that means that computer support is already offered for the phases of "problem definition", "model building", and "model application". Then the typical procedure for problem solving using a Systems Theory Method Bank (which we also call a CAST *system*) can be described by the following three steps:

1. Representation of a model problem by a system problem.
2. Development of a system algorithm to solve the system problem.
3. Interpretation of the reached solution on the model.

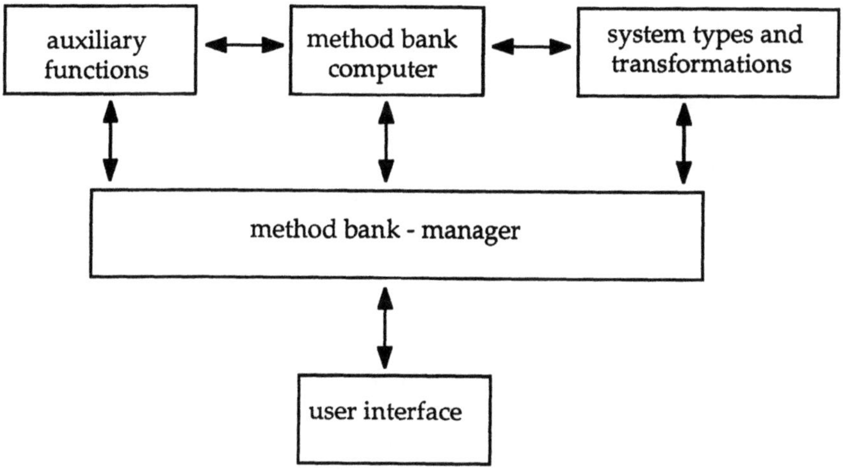

Figure 2.1.1 Block diagram of a Systems Theory Method Bank.

If a CAD system is complemented by a Systems Theory Method Bank (by a CAST system) in this way, then we speak of a *CAD/CAST environment*. While a CAST system is only thought to be for the computer-aided solution of system problems, a CAD/CAST environment provides a practical tool for a special technical discipline. A CAD/CAST environment can be considered as a specific type of a knowledge-based system. Each Systems Theory Method Bank represents a partial knowledge of Systems Theory. It is expected from the user to be capable of deriving the relevant system problems from model problems. Furthermore, he must know the various solution methods which can be aimed by the CAST system, and be capable of selecting the proper system transformations for developing system algorithms. But he has not to know how the algorithmical realization of the individual system transformations is achieved.

We have already used the term Systems Theory Method Bank without comparing it with method banks which exist in other application areas. We distinguish method banks by different classes, depending on how a method bank user performs as a problem solver. We can imagine four different situations *(a–d)* of method bank use:

(a) The method bank provides a set of programs for problem solving for a number of given problem specifications. The user only has to examine the problem situation until he is capable of choosing a suitable program and specifying its initial value. Computer programs used in operations research or in electrical network design are two typical examples for this kind of method bank. We call these types of method banks *program packages*.

(b) The method bank provides programs for important fundamental problems of a certain methodological discipline (e.g. statistics, Systems Theory, game theory). The appropriate auxiliary functions are available to the user for the application of the individual programs and for linking them interactively. Problem solving is done by the user through proper selection of individual programs and linking them together. Because of the importance of the interaction between the problem solver and the method bank, we call these types of method banks *interactive method banks.*

(c) The method bank is capable of completely supporting a certain mathematical computation through an appropriately installed program. The task of the user consists mainly in the exact specification of the problem in the language of the appropriate model. The derivation of the solution follows automatically. Typical for these types of method banks are those used for certain mathematical computation, e.g. for mathematical logic, algebra or analysis. We call them *deductive method banks.*

(d) The user is provided with a high-level programming language for the formulation of algorithms for problem solving. The user must develop the algorithm to solve a given problem situation and program it in order to reach a solution. This type of tool is not normally called a method bank, but a *programming environment.*

After this discussion it should be clear to the reader that Systems Theory Method Banks, as introduced by us, belong to the class of interactive method banks.

In the framework of computer-aided problem solving it would be advantageous to have a method bank available which more or less combines all the four classes introduced above. This leads us to a presentation of a *hierarchically structured* method bank in the form of a multi-level system. Figure 2.1.2 shows the architecture of a four-level method bank which is constructed from classes according to our previous division.

In the practical case of a Systems Theory Method Bank it is advisable to have the multi-level model of Figure 2.1.2 in mind. The most important part of such a method bank would undoubtedly be the interactive part consisting of an implementation of system types and system transformations. In addition, the method bank will also contain programs for special applications which result from the implementation of developed system algorithms. In the same way one will make use of the possibility of the means of a deductive method bank for certain well-defined problem situations of a mathematical nature in a Systems Theory Method Bank. Moreover, the availability of a programming environment for the development of additional programs in the process of system algorithm construction is necessary for the implementation of

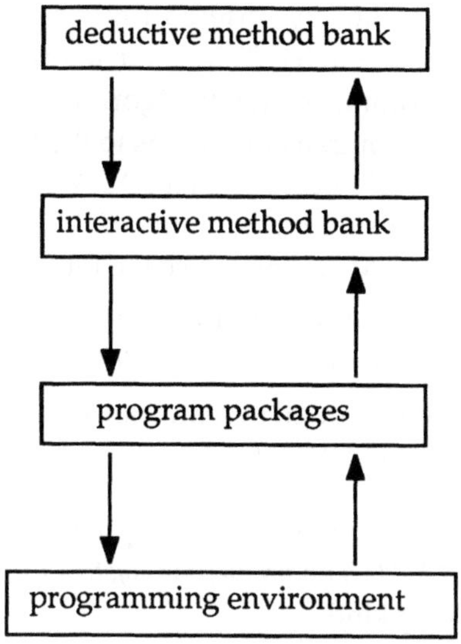

Figure 2.1.2 Hierarchical multi-level method bank.

new system types and new system transformations. We will clearly recognize these features of a practical CAST system later, when we discuss CAST.FSM, a method bank for the theory of finite state machines.

2.1.2 The STIPS Machine STIPS.M

We come now to the actual concern of our matter, namely the concrete construction of a Systems Theory Method Bank which uses the general systems concepts introduced in Chapter 1. In order to describe the logical architecture of such a method bank, we use a state machine concept in the form of the *STIPS machine* (Systems Theory *I*nstrumented *P*roblem *S*olving machine) which will be described here. Starting from an initial state S_0, which defines the initial system specification of the problem, this machine's task is to reach a final state S_n after n-steps, which is a system specification of a solution, through the directed application of system transformations. A STIPS machine describes system algorithms by its state trajectories. It presents a conceptually simple frame for the organization of system types and system transformations for the purpose of problem solving. This type of representation of the problem solving process corresponds in the area of Artificial Intelligence to the concept of "problem solving in a state space" [Newell, Simon 72].

States are represented in STIPS.M by ordered finite sets of specific system types (system specifications). Each state in STIPS.M is the intermediate result in the course of the development of a deterministic system algorithm. State transformation is defined by the application of system transformations to the individual system specifications as they are contained in the states of STIPS.M. With this picture in mind, system transformations can be considered as inputs of STIPS.M. In order to come to a more formal definition of STIPS.M we introduce the following sets:

1. Let $SYST$ denote the set of all specific system types which are available to STIPS.M. We call $SYST$ the *system bank* of STIPS.M.
2. Let $TRANS$ denote the sets of all system transformations of STIPS.M. $TRANS$ is the *transformation bank* of STIPS.M.

Next we define the relation which describes the application of transformations from $TRANS$ to the elements of $SYST$. The *production system* P(STIPS.M) of STIPS.M contains elements of the following kind:

(P1) $(S,T) \rightarrow S^*$

(P2) $(U,T) \rightarrow S^*$

where $S,S^* \in SYST$, $T \in TRANS$ and $U \subset SYST$.

By (P1) a specific system type S will be assigned to an other specific system type $S^* = T(S)$ by application of the transformation T. If T is a deductive transformation then S^* is, in a certain sense, a *homomorphic* (in special cases also an *isomorphic*) image of S. By an inductive transformation T, S is, in contrast, a homomorphic (isomorphic) image of S^*.

The rules of type (P2) realize a coupling of the system types $S \in U$ to a new system type S^* by means of the transformation T.

Each production system P(STIPS.M) can be naturally represented by a bipartite graph G(STIPS.M) in the following way: The nodes of the first kind are given by the system types as given in $SYST$. The nodes of the second type are given by the transformation from $TRANS$. The edges of G(STIPS.M) are defined by the production rules of P(STIPS.M) as follows: If the rule $(S,T) \rightarrow S^*$ is valid, then an edge goes from S to T and from T to S^*. If the rule $(U,T) \rightarrow S^*$ is valid than an edge from each system $S \in U$ goes to T and, further, we have an edge from T to S^*. We call the graph G(STIPS.M) constructed in this way the *generator graph* of STIPS.M. Next to the production system P(STIPS.M) the generator graph G(STIPS.M) is a convenient concept for the representation of a STIPS machine.

We give a simple example to demonstrate our STIPS.M representation method. A production system for the STIPS machine M_0 is given by the following list:

$$(S_1; T_1) \rightarrow S_2$$
$$(S_2; T_2) \rightarrow S_3$$
$$(S_4; T_4) \rightarrow S_1$$
$$(S_1, S_2; T_3) \rightarrow S_4$$
$$(S_2, S_4; T_5) \rightarrow S_5$$

We see that the last two productions are of (P2) type; all others are of (P1) type. The corresponding generator graph is shown in Figure 2.1.3.

The STIPS machine STIPS.M which we have introduced is of purely symbolic character. This means that the corresponding system specifications from *SYST* and the transformations from *TRANS* are symbolic and the accompanying variables and parameter values have still to be determined. Consequently, the system algorithms are of symbolic nature also.

A numerical representation will come from concretely (i.e., numerically) defined system specifications and transformations. In this case a STIPS machine "starts" from a system specification $S_0 := s_0$. This value assignment also determines the numerical value of the subsequent transformations and system specifications. We would like to differentiate between system specifications and transformations determined in this

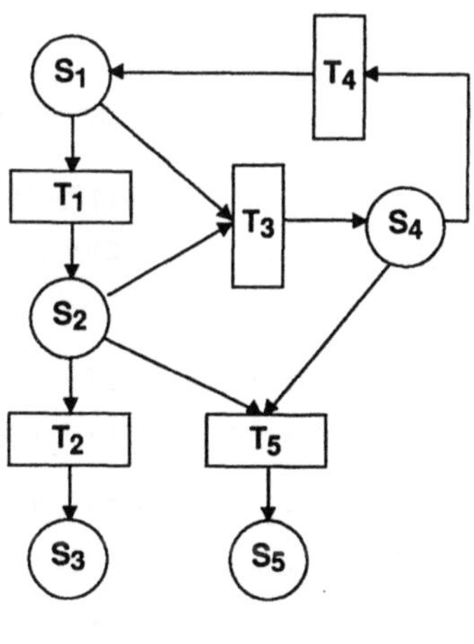

Figure 2.1.3 Generator graph for the STIPS machine M_0.

way by using small letters. Each of the system algorithms ALG developed with STIPS.M which symbolically computes a system solution $S_n = ALG(S_0)$, starting from S_0, solves a special numerical system problem by $s_n = ALG(S_0: = s_0)$.

One can generally proceed from the view that the designer of a CAST method bank has to deal mainly with the STIPS machine STIPS.M and its implementation on a suitable computer system. The problem solver uses this tool to solve a special system problem which is determined by $S_0 := s_0$. One can also say that the problem solver develops the solution by system algorithms with the help of system specifications which are created through value assignment of variables and parameters. Readers interested in further details on the STIPS approach should consult [Pichler 86; Pichler 88a; Pichler, Schwärtzel 90].

2.1.3 Special STIPS Machines

After we stated and viewed the general construction of STIPS.M and how the development of symbolic system algorithms and their evaluations by the problem solver take place in concrete cases, we would like to concern ourselves even closer with the individual specific system specifications and system transformations which are used in practical cases.

When defining special system types and transformations we principally follow the general classification scheme already introduced in Chapter 1. However, it does not make sense to do this in a universally applicable formalized way. It seems to be more purposeful to construct suitable STIPS machines for special system classes and related methods which are already considered as being important in certain applications. Of special interest is the construction of the following three specific STIPS machines:

(FSM) the construction of a STIPS machine STIPS.FSM to be used as the basic concept for the implementation of a method bank CAST.FSM for the application of the theory of finite state machines;

(LM) the construction of a STIPS machine STIPS.LM as the basis for the implementation of a method bank CAST.LM for Linear Systems Theory;

(GS) the construction of a STIPS machine STIPS.GS for the implementation of a method bank CAST.GS for General Systems Theory.

The construction of STIPS machines of these types and their implementation as CAST systems on advanced workstation computers appear to be an important matter for the education of engineers today. Students can investigate the span of accompanying theories and their associated methods more practically with such computerized tools. However, tools of this kind must also be considered useful to the working design engineer when dealing with problem areas which are open to theoretical methods. We would like to examine in the following more closely the design of STIPS.FSM, the STIPS machine for problem solving using methods of finite state machine theory.

CAST.FSM should provide the user with all important methods of finite automata theory including the results which are known for sequential switching circuits. Since sequential switching circuits are realized by networks of binary memory devices and memoryless combinatorial circuits, we propose the following three-level model for STIPS.FSM:

- The top level EA represents the abstract finite automata and accompanying methods for the solution of automata-theoretic problems. Mathematical concepts for the construction of system specifications and transformations are, e.g., set theory and parts of general algebra (lattices, finite fields, linear spaces over finite fields, and others).
- The middle level SW of STIPS.FSM, which is of greatest importance for the practical application of CAST.FSM in the area of microelectronics, represents the system specifications and transformations which will be needed for binary-coded finite state machines. Additional mathematical concepts related to the finite field GF(2) and linear spaces over GF(2) are also considered essential.
- The bottom level (GS) of STIPS.FSM is reserved for finite state machines which are realized by logic gates. This level forms the interface to CAD design tools for physical (electronic) realization. Certain algebraic methods, which have been developed for gate-level design, especially methods of Boolean algebra, can be used for this purpose. Figure 2.1.4 shows the three levels of STIPS.FSM.

Our following task is to describe individual system specifications and system transformations of STIPS.FSM. Naturally, we cannot do this here in detail, and we have to limit ourselves to the basics. However, the acquisition of additional details should not be difficult for the reader by consulting the corresponding existing literature [Pichler, Praehofer 88; Pichler, Schwärtzel 90].

For each of the three levels, a systematically ordered collection of special system types should exist. To avoid an overlapping of levels, we want to divide the set *SYST* of system types of STIPS.FSM according to their assignment in the three levels disjunctively into the sets *SYST*(EA), *SYST*(SW), *SYST*(GS). An analogous partition of system transformations *TRANS* of STIPS.FSM gives the disjunctive sets *TRANS*(EA), *TRANS*(SW) and *TRANS*(GS), of transformations for the corresponding levels EA, SW and GS. In this context we speak also of *horizontal system transformations*. In addition to the horizontal system transformations, *vertical system transformations* also exist in STIPS.FSM. These represent the interfaces between neighboring levels. We denote them as *TRANS*(EA/SW), *TRANS*(SW/EA), *TRANS*(SW/GS), and *TRANS*(GS/SW), respectively. Figure 2.1.5 shows the partition of system types and system transformations, with respect to the introduced hierarchical architecture of STIPS.FSM. Further details on the systems-theoretical nature of the system types and system transformations of STIPS.FSM are postponed to Section 2.2 where we discuss CAST.FSM.

After this discussion of STIPS.FSM and its three level architecture we will proceed to investigate STIPS.LM and STIPS.GS [Pichler, Schwärtzel 90].

Linear Systems Theory belongs, without any doubt, to the best developed and most well-known areas of Systems Theory when it comes to applications in communications engineering and control engineering [Zadeh, Desoer 69; Casti 77]. It is a well accepted part of engineering education today and its applicability for practical problems has frequently been demonstrated in the past. Therefore, it appears especially desirable to have a CAST tool available for the application of Linear Systems Theory to concrete engineering problems.

It cannot be our goal here to go into the details of STIPS machine construction for linear systems. We will limit ourselves to the general discussion of important examples for the considered system types and system transformations. For a more extensive look into the construction of a method bank for Linear Systems Theory we recommend consulting the existing technical literature and looking at examples of installations that exist.

Finite Automata	Level EA
Binary-Coded Finite Automata	Level SW
Logic Gate Circuits	Level GS

Figure 2.1.4 STIPS.FSM as a three level system.

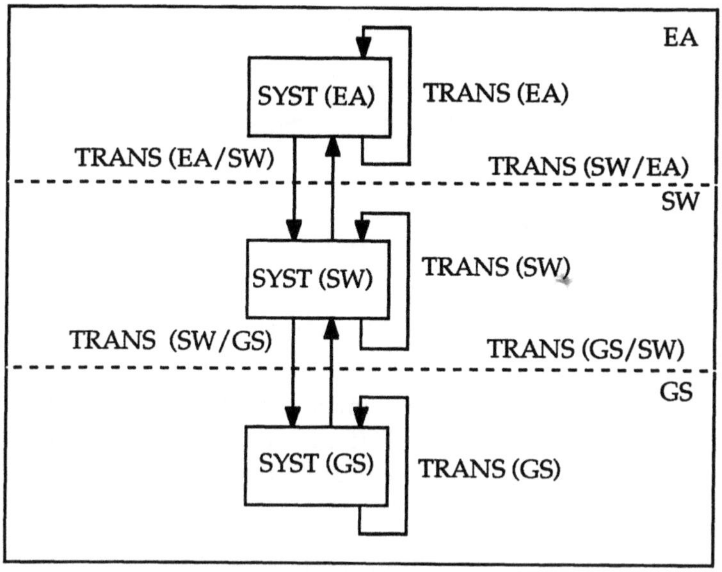

Figure 2.1.5 System types and system transformations in STIPS.FSM.

A classification of Linear Systems Theory can be done by distinguishing the following parts:

(ALS) algebraic theory of linear systems;

(LDTS) linear differential systems;

(LDCS) linear difference systems;

(LFSM) linear finite automata.

Although Linear Systems Theory also exists in the case of time-variable systems, we want to assume for reasons of importance and simplicity that all of the linear system specifications to be considered here possess a time invariant I/O-behavior and a time-invariant state process.

The algebraic theory of linear systems (ALS), which has been established mainly by the work of Kalman [Kalman 69], makes use of an extension of the theory of linear spaces, the R-module theory. It thereby succeeds in defining important system types of Linear Systems Theory, such as the transfer function or the I/O-behavior so generally that the corresponding objects in (LDTS), (LDCS), and (LFSM) are special cases of it. Analogous statements are valid for important system transformations, such as

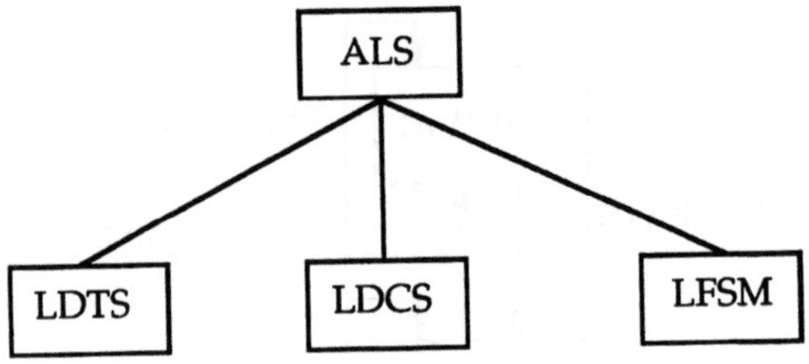

Figure 2.1.6 Hierarchical structure of Linear Systems Theory.

the realization transformation, the transformation in a spectral description or the decomposition into the "Kalman-canonical" form. It therefore appears appropriate that system types and system transformations are ordered according to the hierarchy shown in Figure 2.1.6.

The differences in system types and system transformations in (LDTS), (LDCS) and (LFSM) are mainly caused by differences in the individually used time scale as well as in the different value sets for signals and state variables. Linear differential systems have, for example, a continuous time scale (modelled by the real numbers **R** or $\mathbf{R}_0$) and the local state transition is described by the differential quotients of state trajectories. Linear difference systems, on the contrary, possess a discrete time scale (modelled by the integers **Z** or the natural numbers $\mathbf{N}_0$). The local state transition of a state at time t is given by assigning a new state at time t+1.

Linear automata differ from linear difference systems only in the value sets of signals and state variables. Due to the requirement of finiteness, the value sets must be finite fields GF(q).

Examples of system types, which come into consideration for STIPS.LM, are the following:

BLM (black boxes): functions (given by tables), linear functions (represented as matrices);

GLM (generators): linear differential systems (A,B,C), linear difference systems (A,B,C), linear automata (A,B,C);

DLM (dynamics) given by the "solutions" of linear systems (A,B,C) that are: linear continuous dynamical systems, linear discrete dynamical systems, dynamical systems of linear automata;

ALM (algorithms): algorithms for the realization of signals, algorithms for the realization of dynamics;

NLM (networks): parallel coupling of systems (A,B,C), serial coupling of systems (A,B,C), cascade coupling of systems (A,B,C), feedback coupling of systems (A,B,C).

Many important system transformations for engineering applications exist in Linear Systems Theory. It is very desirable that the following be made available in STIPS.LM:

TRANS(BLM/GLM): Realization transformation, i.e., calculation of (A,B,C) from a given I/O-behavior described in the time or spectral domain. The corresponding algorithms have been developed by Ho/Kalman, Rissanen, and Massey/Berlekamp.

TRANS(GLM/BLM): computation of the impulse response of (A,B,C), computation of the transfer function of (A,B,C).

TRANS(BLM/BLM): Spectral transformation from time domain to spectral domain (impulse response to transfer function) by Fourier transform, Laplace transform, z-transform or D-transform.

TRANS(GLM/NLM): Kalman decomposition of (A,B,C), shift register realization of (A,B,C) by means of a coordinate change in the state space.

TRANS(BLM/NLM): serial decomposition of transfer functions (by means of factorizing), parallel decomposition of transfer functions (by means of partial fraction expansion).

If we structure STIPS.LM according to the introduced hierarchy of Linear Systems Theory we get a network of STIPS machines as shown in Figure 2.1.7. We see that in addition to the "vertical" transformations from and to STIPS.ALS, "horizontal"

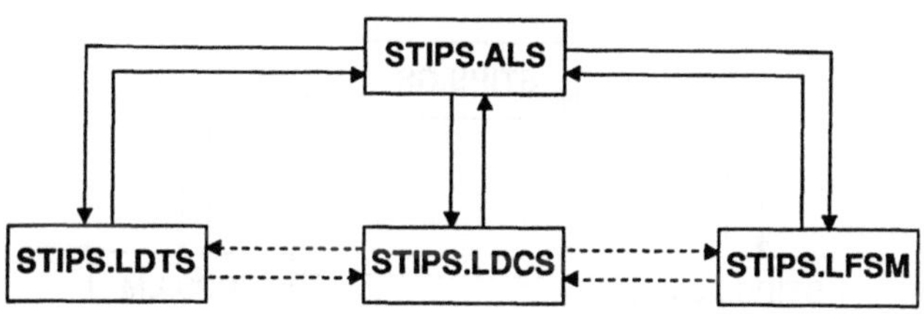

Figure 2.1.7 STIPS.LM as a network of special STIPS.L machines.

transformations (shown by dotted lines) exist between the STIPS machines of the lower level. Typical examples of such transformations compute a time discrete approximation in the form of a linear difference system for a time-continuous differential system. Examples of this are approximation operations in digital filter design or the quantization of continuous signals into a finite set of values (e.g., in the case of PCM coding).

Next, we will briefly consider the construction of STIPS.GS.

In General Systems Theory, concepts and methods are considered, which, in the context of model building and model application, are of very general nature and multidisciplinary use. One can distinguish between two different orientations in General Systems Theory. A philosophy of science-oriented General Systems Theory PGST and a mathematical-oriented General Systems Theory MGST. While PGST deals with basic formulations of modelling (for instance with the problem of "causality"), the mathematical-oriented direction MGST orients itself towards the study of system types and system transformations which have already been proven to be useful in special parts of Systems Theory and attempts to come to a unified view through their generalizations, and in general, to create applicable tools to solve analysis and synthesis problems, at higher levels of modelling.

STIPS.GS, as we consider it here, should more closely follow the mathematical-oriented direction of general systems research. The importance of the relation of STIPS.GS to STIPS.FSM and STIPS.LM is very natural in this direction and offers integration in an overall STIPS machine. Figure 2.1.8 shows such an integration of STIPS machines.

We do not intend to discuss individual system types and system transformations for STIPS.GS. Instead, we refer to the pioneering work of Mesarovic and Takahara

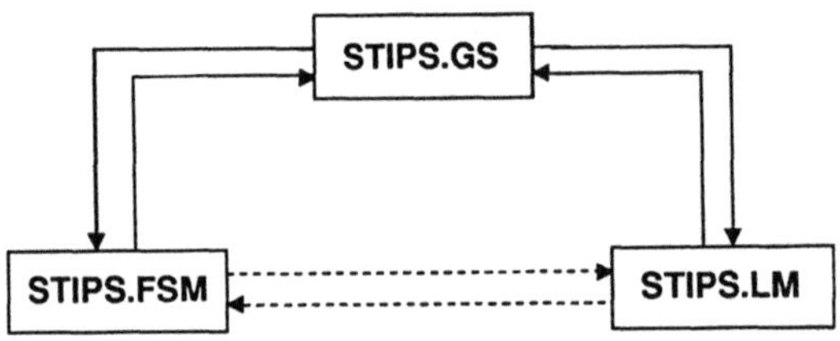

Figure 2.1.8 Integration of special STIPS machines.

[Mesarovic, Takahara 75; Mesarovic, Takahara 89], Wymore [Wymore 67], and especially to the work of Klir [Klir 85; Klir 91], where many results aiding the construction of a STIPS machine for General Systems Theory can be found. The *General Systems Problem Solver* (GSPS) created by Klir and his group illustrates an implementation which meets many of our requirements for CAST method banks.

2.1.4 Navigation in Systems Problem Solving

2.1.4.1 Systems Theory Problem Solving

"Systems Theory Instrumented Problem Solving" [Pichler 86] as it was introduced in Section 1.2 is the underlying concept of Computer Aided Systems Theory [Pichler, Schwärtzel 90]. It means to solve problems arising in the world of various engineering disciplines by defining the problem in systems-theoretical terms, then searching a solution to the problem by applying an appropriate sequence of system transformations (developing a system algorithm), and finally by re-transforming the result to the domain (Figure 2.1.9).

In the *first* step, the problem to be solved has to be defined. A system problem consists of one or more initial system specifications, a set of constraints, and a final sys-

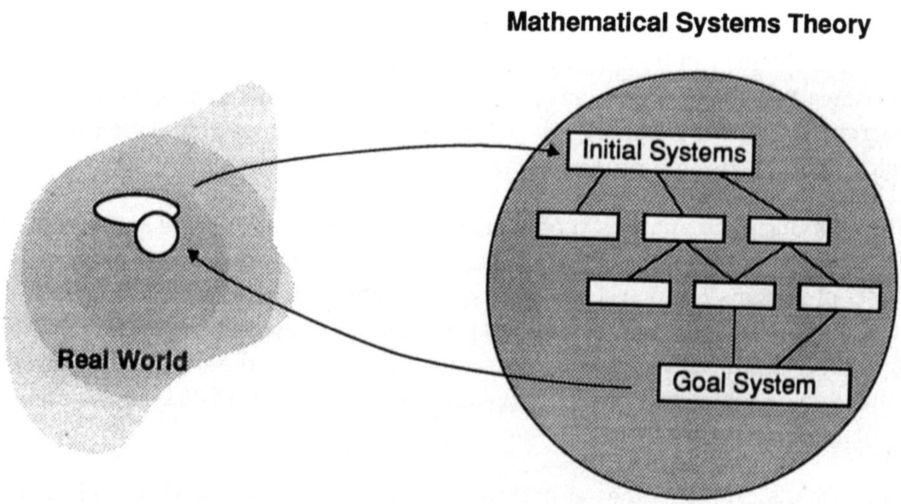

Figure 2.1.9 Systems Theory Instrumented Problem Solving.

tem specification [Mittelmann 87]. A system specification is either an instance of a special system type, i.e. a concrete system, or a general system specification. A general system specification in turn may be given by naming the type of the system or by defining the generic structure of the system in the form of its *entity structure.* In both cases additional constraints on the components of the system may be stated to complete the general system specification.

The *second* step is the development of a system algorithm. In some cases, an already available algorithm of the CAST knowledge base will fit. Otherwise, a new algorithm has to be constructed either by search techniques, or interactively by the user, or in a combination of these. The complete problem definition allows to use the search strategies developed by Artificial Intelligence researchers such as best-first search, bidirectional search, means-end analysis, and others [Nilsson 71; Nilsson 80; Kanal, Kumar 88; Pearl 84]. In some cases, the solution of the problem will be found in a way which is easily generalized and saved as a system algorithm in the knowledge base [Korf 85].

In the *third* and last step the result is interpreted in the original domain. Only in special integrated CAST applications it will be possible to automate this step, for example in closed CAD/CAST environments.

2.1.4.2 Realization Problems

System problems may be written as triples (*I*, *C*, *F*). Here *I* is a set of initial system specifications, *C* is a set of constraints and *F* is a final or goal system specification. System problems can be classified into two main categories: analysis and synthesis problems. For analysis problems, the set of initial system specifications is a set of concrete systems to be analyzed. The goal is expressed as a mathematical object, which states the desired property of the systems in *I*.

When solving such analysis problems, often a directly applicable analysis method will not exist. There are two possible strategies, the first one would be to transform the set of systems *I* into systems of a type for which an analysis method for the desired result exists. The task is then to apply a series of system transformations to an appropriate realization type. The second way is to solve the problem by means of simulation, in which case the set *I* has to be realized by a kind of system suitable for simulation.

For synthesis problems the set of initial systems defines the behavior and perhaps some details of the solution system (which is not known in the beginning) to be synthesized. The goal system *F* is given as a general system specification, describing known structure and other information. The set C contains predicates, which the solution system should fulfill. Solving of the problem is accomplished by successive transformation of the systems in I into other systems, until a state is reached where the computed systems form a realization of I, conform to F, and where all constraints in C are satisfied.

Although the two kinds of system problems are semantically rather different, the computation of the solution of both types can be seen as solving a kind of a general realization problem. A (general) realization problem is characterized as the task of transforming some initial systems into a goal system which realizes the initial systems. The following discussion therefore deals with Systems Theory instrumented realization problem solving.

2.1.4.3 Interactive CAST

Systems Theory Instrumented Problem Solving (STIPS) is the general approach to solve design, construction, analysis etc. tasks in various engineering disciplines by applying results of the systems sciences to those domains [Pichler 86].

In Section 2.2 we will discuss in detail the CAST method bank CAST.FSM. This CAST environment deals with finite state machines, Boolean functions, switching circuits, linear machines, Petri nets, and related topics. The user of the system, e.g. an engineer designing a digital circuit, is provided with the graphical representation of the implemented system types. The user selects a system type suitable for modelling his problem, creates an incarnation and fills in the values. The result is the initial sys-

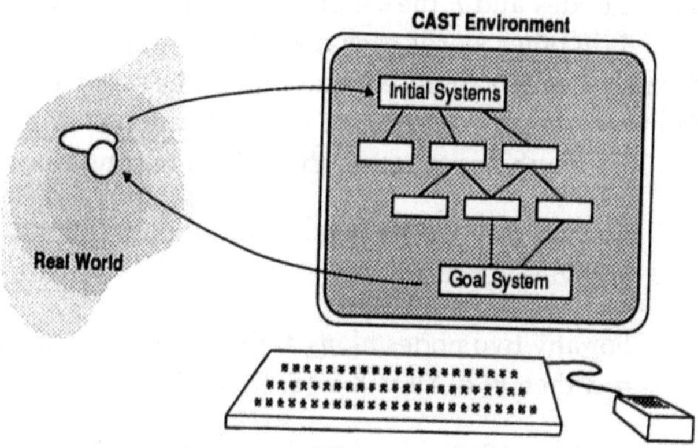

Figure 2.1.10 Using a CAST environment.

tem of the problem. Then he chooses an appropriate system transformation, analyzes the result, transforms it further, and so on, until the constructed realization tree is an acceptable solution of the problem. In CAST.FSM all this is as easy as moving the mouse and clicking. But the choice what to do next in each step is left to the user. Only the user knows the system problem specification. To enable a greater degree of automation in system problem solving as outlined in Section 2.1.2.1, in future CAST systems the problem has to be definable in a precise, but comfortable way.

2.1.4.4 Problem Definition and Recursive Entity Structure

As stated in Section 2.1.4.2, a formal problem definition consists mainly of initial and goal system specifications. In the general case, a system can be specified by describing its entity structure. The notion of entity structures is very useful in this context because it allows to describe several aspects of static decompositions of systems at the same time. For this purpose the so-called *recursive entity structure* (RES) [Mittelmann 90a] is used, which is a generalization of the system entity structure developed by Zeigler [Zeigler 84]. We distinguish between *composed systems* and *atomic systems*. The entity structure of an atomic system is simply the type of the system. A composed system consists of its components as well as an interconnection construction called its *coupling*. The decomposition is one possible aspect of the system. There are other kinds of aspects as well. By using the RES notion, it is possible for example to describe a system S as being a network of some subsystems, where some of the subsystems are of the same kind as S.

More formally, a RES may be defined as follows. A RES is a bipartite directed tree (E, A, F, e) with $F \subset E{\times}A \cup A{\times}E$, where E denotes the finite set of entity nodes, A denotes the finite set of aspect nodes and F the set of arcs; $e \in E$ is the root node of the RES. Each element of $E \cup A$ (in other words each node in the graph) has a name as well as other properties, which together represent one entity. The set of aspects A consists of three disjoint subsets D (decompositions), S (specializations), and M (multiple decompositions). Any RES graph has to satisfy the following conditions:

(bipartionedness) In each path in the tree an entity node follows an aspect node and vice versa.

(unique names) For any two nodes n_1, n_2, if they have the same name, then if $n_1 \in X$ then $n_2 \in X$ for $X \in \{E, D, S, M\}$.

(uniformity) Nodes with the same names have or represent the same (i.e. isomorphic) subtree. If two nodes lie in the same path and have the same names, both must be entity nodes.

(*recursive hierarchy*) If two (entity) nodes n_1, n_2 in the same path have the same name, then the node n_2 which has a greater distance to the root (we then speak of the "lower" node) must be a leave of the tree and must be a direct successor of a specialization $s \in S$, and s has to have at least one another successor (of type *entity*). The nodes n_1, n_2 will be called recursive entities.

(*brotherrule*) Nodes with the same direct predecessor (brothers) have different names and subtrees, they represent different structures.

(*decomposition*) A decomposition node has as direct successors those entities, which are the components of the decomposed system represented. The coupling of the decomposition is represented in the coupling property of the node.

(*multiple decomp*) A multiple decomposition node m represents the decomposition of a system into a number of systems of a single kind. The only successor of m is an entity e as a single representative of the component systems. The coupling of all components is again specified in m.

(*specialization*) A specialization has entities as successors which represent alternate specifications.

The rules guarantee that an interpretation of the RES graph may yield a meaningful system specification. An interpretation of a RES specification is obtained from the RES graph by the following process:

(*RES-interpr* 0) Expand the RES graph by the following: For any set of nodes with the same name except recursive entities, copy the subtree of the expanded node (if any) to all members of the set.

(*RES-interpr* 1) Put the root node in the OPEN list and create the root of the RES-i graph.

(*RES-interpr* 2) If OPEN is empty, terminate with step 5. Else select and delete one node n from OPEN.

(*RES-interpr* 3) If n has no successors, and is not a recursive entity, go to step 2. If n has no successors, and is a recursive entity, go to step 4. Else follow one arc from this node to an aspect node. Create a copy of the arc and the node in the RES-i graph. If this aspect is a (multiple) decomposition node, put all its successors into OPEN and copy them with their arcs to the RES-i graph. If it is a specialization node, select one of its successors, put it in OPEN and copy it with its arc to the RES-i graph. Continue with step 2.

(*RES-interpr* 4) *n* is a recursive leave, let its recursive partner be *m*. Replace n with a copy of *m*. Continue with step 3.

(*RES-interpr* 5) In the resulting graph, replace all aspect nodes and arcs with appropriate direct arcs from the respective entities to entities. Return the graph as the interpretation result.

As a consequence, a RES interpretation is represented by a subgraph of the RES graph consisting only of entity nodes and directed arcs. In fact, the interpretation is a tree with *e* as its root.

In general, there are many interpretations of a single RES specification.This fact stems from the arbitrary selection of aspects as well as from the arbitrary selection of specialization successors. The process of making an interpretation may never stop if there are recursive entities in the RES. But the recursive-hierarchy rule ensures the existence of finite interpretations for any RES [Mittelmann 90a].

In practical implementations of CAST systems on workstations with a graphical user interface the specification of initial and goal systems is done via a RES editor. The graphical RES editor is similar to browsers of object-oriented programming systems in that it allows editing the graphical representation of the RES under construction directly. In fact, it may be implemented as a specialization of an ordinary browser. The RES browser has different graphic representation methods for the four basic RES object types entity, decomposition, specialization and multiple decomposition. The latter three are special versions of the more general aspect type. That a RES is a bipartite graph of entity and aspect nodes, and several other constraints such as naming restrictions are enforced by the RES editor.

An example for using the RES specification method in the context of a CAST.FSM session is as follows: A finite state machine is to be defined. If this system is used as an initial system of a synthesis problem specification, the sets and functions of the state machine will be given, otherwise, if the system is to be used as the goal system specification, it will be defined as a tupel of variables. The example system to be specified is a network resulting from several decomposition steps applied to a finite state machine. The system's structure may be described by the RES diagram shown in Figure 2.1.11.

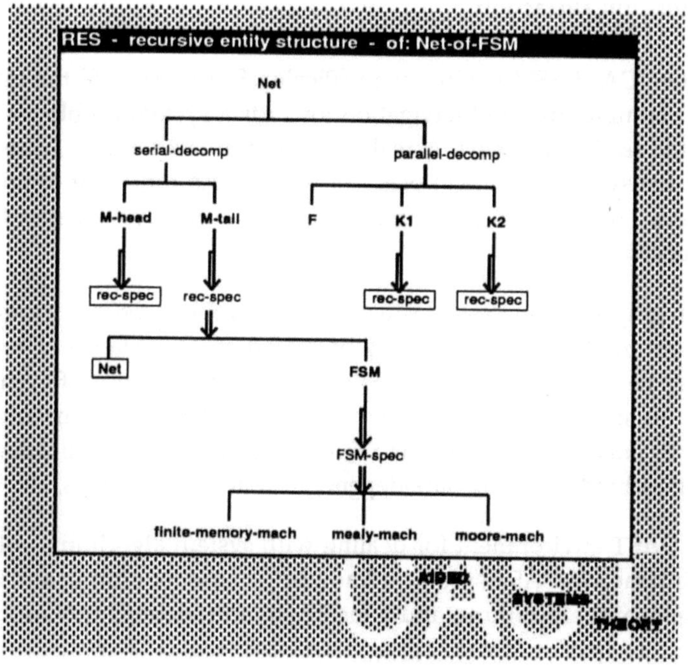

Figure 2.1.11 Example RES specification.

In CAST.FSM we have two decomposition methods, a parallel decomposition and a serial decomposition. By successive application of these methods a network with a characteristic structure is generated. The top node is a system of type *network*. This network is the result of either a serial or a parallel decomposition, shown as the two alternate aspects `serial-decomp` and `parallel-decomp`. The next lower level of the RES tree shows the components of each decomposition. A serial decomposed machine consists of the entities `M-head` and `M-tail`, whereas the parallel decomposed system has two components `K1` and `K2` plus an output coding function named `F`. Each system in this level except `F` may be described in the next lower level in the same way, shown as the specialization node `rec-spec`. The RES browser shows repetitive subtrees only once, so only one arbitrarily selected or user selected `rec-spec` node is expanded further. Any `rec-spec` is either a net itself, which is the recursive case, or it is an *atomic finite state machine*, which specializes into *Mealy*-type, *Moore*-type and others.

2.1.4.5 Algorithm Extraction

In the interactive CAST.FSM environment [Pichler, Schwärtzel 90] the intermediate results of the manually directed navigation towards a system problem solution are displayed as the *realization graph*. A realization graph is an acyclic, directed graph whose nodes are system specifications and whose arcs represent applied system transformations. The paths in the graph lead from initial system specifications to goal system specification.

The realization graph of a successful solved system problem defines a special case of a system algorithm. This algorithm may be stored as an instance of system algorithms and then be reused for similar problems. But in many cases of even very similar problems, this simple kind of algorithm will not lead to a problem solution. Then it is necessary to enhance the algorithm with control structures, which leads to general system algorithms including data-dependent conditionals, loops, etc.

An enhanced CAST environment for dealing with system algorithms will therefore consist of an interactive CAST system, a knowledge base for storing and associating system problems and algorithms, and a facility to edit, debug, and apply the algorithms.

Editing of system algorithms is supported in the CAST system by a *structure editor*. The formal language used to express the algorithms may be of a textual kind like the usual programming languages, but preferably an easy to use graphical language will be used, because the typical user will not be a programmer, but a specialist in some engineering discipline.

2.1.4.6 Search in CAST

The solving of a system problem may be described using the familiar state-space search description as developed in Artificial Intelligence research [Newell, Simon 72]. The state of the search process is definable as the set of concrete system specifications generated so far. The set of choices in each possible step of this problem space is the set of applicable system transformations. A general graph search procedure for solving a given systems problem is shown below:

```
general-search (initS constraints goalS)

   declare
      graph, open-set, closed-set, exp-set:     set
      result: list
   end-declare

   result := false
   graph := make-graph(initS)
   open-set := {initS}
   closed-set := {}
   while not(open-set = empty)
      node := select-node(open-set)
      open-set := remove(node open-set)
      closed-set := set-add(node closed-set)
      if satisfies(node constraints goalS) then
         result := list(node graph constraints goalS)
         return(result)
      end-if
      exp-set := expand(node rules heuristics)
      graph := add-nodes(graph node exp-set)
      open-set := set-union (open-set (set-difference(exp-set closed-set)))
      graph := set-pointers(graph node exp-set)
   end-while
   return(result)
end general-search
```

The function `add-nodes` expands the graph at the specified node *R* by those elements of the given node set not being ancestors of *R*, thus keeping the graph acyclic. The function `set-pointers` may be used to maintain a subgraph of optimal paths under a given cost function.

The crucial sub-functions of this rather general routine are `expand`, which decides how to generate the successor nodes of the last node, `select-node`, which selects one of the generated nodes for further expansion, and `satisfies`, which checks the node for constraints and solution properties. The definition of these sub-functions determine the kind of the search procedure. By choosing the appropriate functions, search methods as depth-first, breadth-first, best-first, A*, depth-first-iterative-deepening and others may be generated. The exact semantics of some functions used in the search algorithm depends on the intended usage. For example, for the application in CAST systems the function `expand` will not compute the successor node objects instantly, but determine which expansion steps are possible, and return those candidates unevaluated. Evaluation of the new nodes is delayed until select-node decided which candidate to choose.

The reason for this is that in the domain of CAST environments, the main problem is perhaps not the vast magnitude of the search space, but the high computational costs of a single expansion step, in contrast to typical AI problem domains. Realization problems arising in our research work are often solvable within less than ten (or perhaps a few ten) steps. Also the branching factor is fairly low for each node. But for realistic object size, the computation time for one step may be several hours or much more. Even worse, some basic system analysis methods may be not computable in any realistic amount of time and space in some cases, while in other cases they will finish almost instantly. This shows that another kind of (heuristic) background knowledge is needed for advanced CAST systems which estimates the computation time of node expansion and so helps to decide which step to go next. A different research topic arises here: how may sophisticated methods from AI can be adapted inside the computation of a single system transformation in order to get results in situations, where the usual algorithms will not terminate in any practical time limit?

2.1.5 Requirements for Applications in VLSI-Design

2.1.5.1 Introduction

This final section of our discussion of CAST method bank design deals with requirements for CAST method banks for the application in the design of VLSI circuits. First, relevant aspects of Systems Theory and VLSI design are considered (Sections 2.1.5.2–3). Then we derive potential fields of applications of Systems Theory software to the design process (Section 2.1.5.4) and discuss requirements this software has to match (Section 2.1.5.5).

In parallel to the development of highly integrated hardware technology, complex CAD tools are developed to support various design approaches. As a result, today there exists first class CAD software including highly sophisticated algorithms for design of VLSI-circuits running on powerful workstations. However, there is still important theoretical knowledge available from Systems Theory, which has not yet found applications in VLSI design.

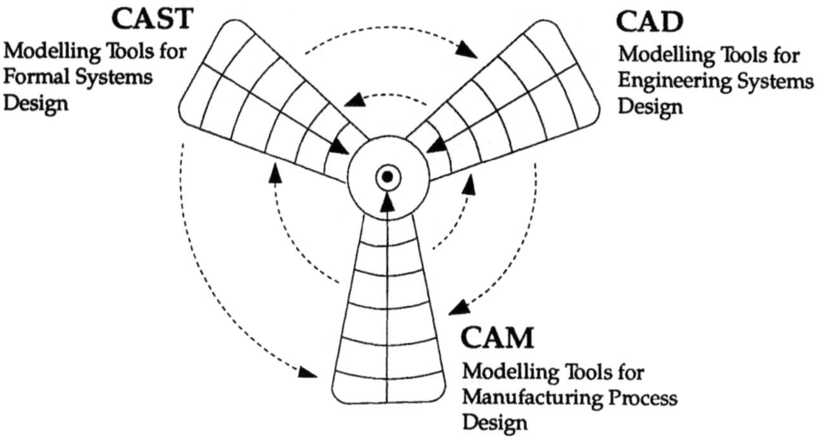

Figure 2.1.12 CAST/CAD/CAM - Design Propeller.

To show the relation of CAST with respect to CAD/CAM, the "Design Propeller" has been proposed [Pichler 90; Pichler, Schwärtzel 90] (Figure 2.1.12). We use it to point out the necessity of cooperation between CAST, CAD and CAM (Computer Aided Manufacturing) in order to improve the whole design process by "creating thrust at the center". It is a generalization of the *Y-chart* [Gajski, Kuhn 83; Walker, Thomas 85], which is used in VLSI design and which will be discussed in the following.

2.1.5.2 VLSI Design Process

A short overview of developments concerning the VLSI design process will help to see more closely potential fields of applications of CAST systems.

For ASIC (Application Specific Integrated Circuit) design – we will restrict the discussion to ASICs – the semi-custom approach (assembling of predesigned modules) gets more and more important. Much CAD software is available on the market which supports the design starting at the level of logic gates (first and second generation systems). However, emphasis is put on the development of more complex CAD systems starting from higher levels of modelling (third-generation systems).

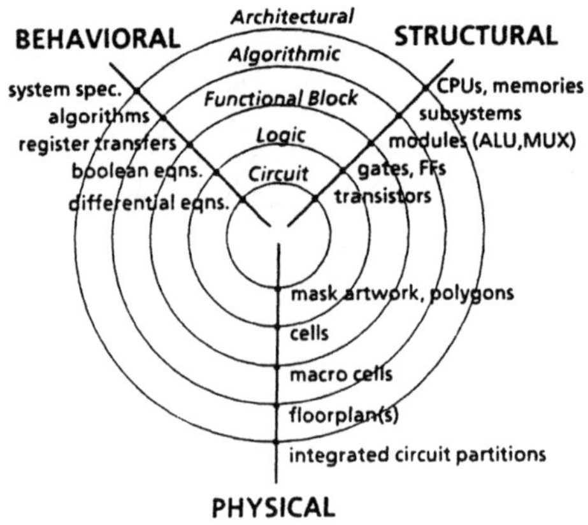

Figure 2.1.13 Y-chart.

The Y-chart (Figure 2.1.13), is used to illustrate the three domains of the design space and the different levels of data abstraction: The behavioral domain describes the function of the circuit on different levels. The structural domain comprises the technological building blocks and their interfaces and the physical domain defines the levels of physical implementations.

First and second generation CAD systems support the design process only by starting in the structural domain with schematic entry of the circuit structure on the switching and functional block levels. The transformation to the layout in the physical domain is done by assembling simple cells (gates, flipflops) and modules by placement and routing. Third-generation systems are often referred to as silicon compilers which start in the behavioral domain and transform a functional description into a final chip layout. By means of a hardware description language the behavior of a circuit is defined algorithmically. The automatic transformation into the structural domain is called *synthesis*. We distinguish *logic synthesis, register transfer synthesis* and *high level synthesis* corresponding to the input level.

Examples of industrial tools are VENUS and CALLAS [Hörbst et al. 87; März et al 89]. VENUS is a second generation system supporting various semi-custom approaches (standard cells, gate arrays, sea of gates), various technology specific cel libraries, module generators, and a test strategy. CALLAS is a high level synthesis system (Figure 2.1.14) that was designed as a front end of a silicon compiler. It can be linked to various state of the art CAD systems with different cell libraries (e.g. second generation systems).

The behavioral description specified by the user in a Pascal-like language is compiled into a flow-graph. The essential transformation from the behavioral description into a circuit structure as well as various optimizations are carried out by the synthesizer. The structure derived from the flow-graph consists of a data path

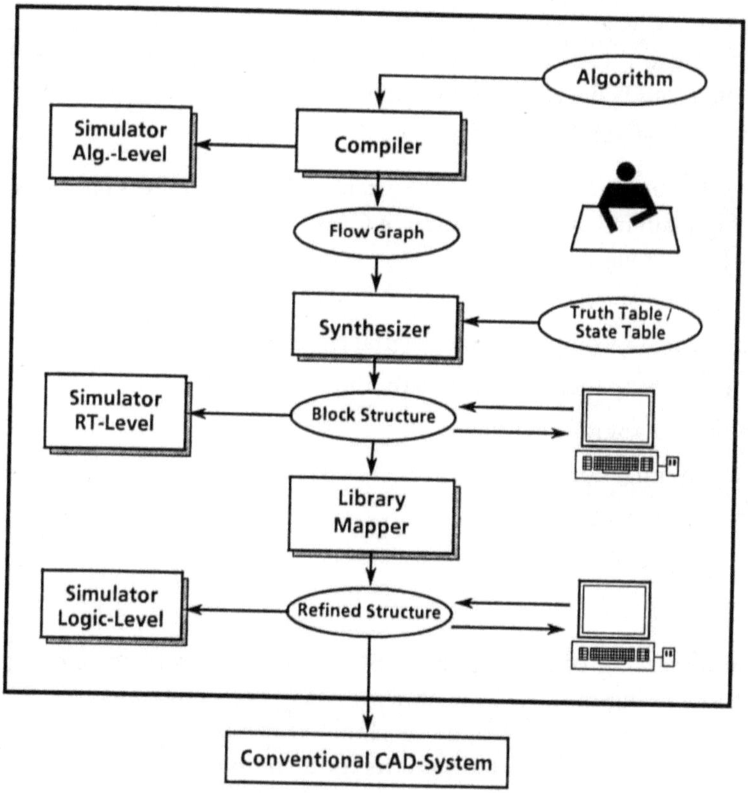

Figure 2.1.14 CALLAS architecture and design flow.

and a controller (i.e., a finite state machine). The states are introduced e.g. for loading registers or switching multiplexers of the data path. After optimizations (e.g. reducing parallelism, logic optimization) the controller is synthesized (e.g. single PLA, two PLAs with counter). The result of the data path and controller synthesis is a block structure containing library independent abstract modules. This structure is mapped to the target library and transformed into a refined structure by the library mapper. This final structure is input for a conventional CAD system for chip layout generation.

2.1.5.3 Potential Applications of CAST to VLSI Design

In general, Systems Theory methods have found many applications in engineering design. However, for use in VLSI design problems caused by algorithmic complexity have to be solved. Very often exact solutions cannot be calculated for realistic problem sizes because this would require enormous amounts of time. In these cases CAST systems could concentrate on specific problems and assist in finding heuristics to achieve computability.

Considering the various disciplines of Systems Theory and the VLSI design process, the following main fields of application of CAST systems can be seen from the VLSI-designer's point of view:

- design for testability,
- structuring (decomposition) of digital circuits,
- design of test pattern generators (built in self-test),
- synthesis of digital circuits,
- logic minimization,
- state assignment,
- controller synthesis,
- testing.

Due to the increasing complexity of VLSI circuits, the problems arising in circuit testing are increasing rapidly. Test strategies have been integrated considering only single stuck-at faults (i.e., one signal is clamped to power supply or ground). The main task here is to generate test patterns for a complete circuit test automatically. This is a very hard problem for combinational circuits and even much more difficult for sequential circuits. Design for testability includes various approaches to reduce these problems: insertion of scan paths (i.e. chaining latches to shift registers), reducing redundancies, enforcing design rules for testability, and hierarchical testing

[Hofestädt, Feiten 90]. Finite automata theory provides the theoretical foundations for work in this field, e.g., structure theory of finite automata [Hartmanis, Stearns 66], and methods for state identification and fault detection experiments [Kohavi 70]. The application of finite state machines is therefore promising [Booth 67; Pichler 88b] and should be supported by specific CAST systems to realize a CAD/CAST environment for solving VLSI testing problems.

For the automatic synthesis of digital circuits, again the theory of finite automata and sequential switching circuits provides possible theoretical foundations. Depending on the target architecture, synthesis may result in a data path and a corresponding controller (finite state machine). Logic optimization applied to the data path, state reduction and state assignment for controller synthesis purposes are promising fields of CAST systems applications. Further FSM transformations which are of interest are machine decomposition, shift register realization, linear realization, and the register flow machine representation [Bainbridge 75]. But the application of CAST seems not to be restricted to the controller of the synthesized circuit since much work has also to be done on the data-path/controller-interface and the data path itself. Aspects of design for testability apply as well and are closely related to state reduction and state assignment of FSMs [Devadas et al. 89]. Special care has to be taken of the complexity of the finite state machines in all cases.

In order to support the CAD user with a clear, easy to use, and flexible environment, new frameworks are to be investigated which comprise modelling of the VLSI design process as well as design policies and knowledge-based approaches [Bretschneider et al. 90].

2.1.5.4 Requirements for CAST Systems

From the experiences of applications of Systems Theory for the VLSI design process requirements for CAST systems can now be derived. However, more practical experience is still needed and the requirements mentioned may be topics for further discussions and investigations when designing CAST systems.

We distinguish two possibilities for the application of CAST systems: First, direct application of CAST systems to an actual circuit design task and second, indirect application by using CAST systems for designing CAD software. In this section emphasis is put on the indirect application.

Versatile tools and systems are necessary for the design of CAD software. In addition to conventional tools known from software engineering (e.g. full programming environments including graphic systems and window management) a theory tool is important for supporting the design of algorithms. The following requirements are essential for the development of CAST systems:

- functionality (algorithms, data interface),
- performance (complexity of algorithms and applications),
- user interface (interactions, design assistant),
- flexibility (portability, extensions).

Functionality is given by the available algorithms and data interfaces. Algorithms dealing with automata and switching circuits for the application to design for testability and synthesis of digital circuits are important here. They comprise structure theory of sequential switching circuits, state reduction and state assignment algorithms with respect to finite state machine realizations (PLA, multi level logic, shift register, register flow machine, etc.), logic optimization (two level, multi-level), and identification of redundancies. Petri nets should be supported for modelling applications, e.g. deadlock analysis.

But the mere implementation of algorithms is not sufficient – one must think about data interfaces as well. A finite state machine is often defined by the state table but for practical applications other definitions are used: state graph representations for small FSMs, algebraic representations (matrices for linear automata), compacted state tables or a special syntax for sparse or "regular" state tables (e.g. counter). Automata may also be given as binary coded sequential switching circuits. A specification by means of boolean equations and graphical representations of the logic gate circuitry (structural netlist) are necessary for this. The same is true for data interfaces using specific formats (e.g. EDIF). The support of coexistent descriptions on algorithmic or register transfer level and internal representations (e.g. flow-graph) as used in synthesis tools seems difficult due to different formats, but should be considered in the future. A graphical specification of Petri nets is required for modelling activities and may be used as an alternative to the flow-graph representation of algorithms.

The requirements concerning the performance of CAST systems are not to be underestimated. The complexity of the applied algorithms as well as of the size of the circuits to be worked on (e.g. the number of states of a FSM) define the application bandwidth. For interactive working the results have to be available within a "convenient" amount of time.

Another subject of great importance is the user interface. A graphic interface using a window standard (e.g. X-Windows) which supports interactive work by means of convenient dialogs, menus, help functions, etc. is state of the art. But there are further requirements: Especially a decision support system (e.g. knowledge-based) could help the user in deciding which of his realizations is the best to achieve a specific objective and which alternatives are possible. In synthesis one will expect estimations of area, number of terms, gates or states, etc. This functionality is referred to as design advisor or design assistant. Management of files and different versions have to be treated, too.

Flexibility comprises easy extendibility (e.g., integration of new algorithms by the user) and portability. Therefore state of the art workstations, standard operating systems, well suited programming languages, and software environments should be used for CAST systems implementations.

2.1.5.6 Summary

Computer Aided Systems Theory intends to make Systems Theory available to various fields of engineering by means of CAST method banks. The application of CAST systems to the VLSI design process is possible in many ways. Controller synthesis is an important topic in this aspect of design. Petri nets are used for framework activities. The main requirements for CAST system include functionality (algorithms, data interface), performance (complexity of algorithms and applications), user interface (interactions, design assistant), and flexibility (portability, open to extensions).

There is already some experience in the application of CAST system to VLSI design. Section 2.6.1 discusses the successful application of an algebraic decomposition method for the controller synthesis task and gives results on the MCNC benchmark FSMs. In Section 2.6.2 we will discuss the VLSI design of a one-way function for cryptographic applications using CAST.FSM and CAST.LISAS. Experience gained thereby will influence further designs and implementations of CAST method banks.

2.2 CAST.FSM – FINITE STATE MACHINE METHOD BANK SYSTEM

2.2.1 Introduction

This section outlines a CAST method bank which uses STIPS.FSM as the conceptual framework for the specification of the logical architecture. CAST.FSM is a software system implemented in InterLISP-D and LOOPS [Bobrow, Stefik 83]. The motivation for the development of CAST.FSM was a mainly pedagogical one: First of all, it is difficult today to get computer science students interested in Systems Theory if no computer support can be offered for it. Secondly, it seems to be especially important in the subject of Systems Theory, that the application of the theory to practical problem situations is done by examples. Otherwise the student remains on the level where he can principally recognize certain possible Systems Theory applications for practical problems, but has no chance to get close to concrete practical applications. Without a computer aided approach the student has to deal with too strongly idealized and too much simplified problem situations and is forced to use "paper and pencil" for their solutions. The situation is similar to teaching in electronic circuit design. There the student must definitely be confronted with real tools which are used in praxis in the form of a CAD system for electronics and must experience the practical development of a circuit, after having learned the fundamentals of electronic circuit design. The method bank system CAST.FSM has been used successfully in this respect in teaching.

Besides the pedagogically-oriented motivation, we also have the practical goal in mind of making the theory of finite state machines operational for engineers. Existing CAD systems for electronics usually do not offer theory-based tools for logic design in higher functional levels (from the register-transfer level upwards). In the future there is the hope that CAST tools will influence CAD systems and will bring existing theoretical knowledge to the workplace of the design engineer. This type of "CAST Knowledge Engineering" complements the present modern trend of providing the individual knowledge of a successful practitioner by expert systems to support engineering design activities [Pichler, Schwärtzel 90].

2.2.2 LOOPS Software Implementation of CAST.FSM

LOOPS [Bobrow 83] is an extension of the advanced programming environment InterLISP-D, running on personal workstations like the XEROX Dandelion and Dorado. The InterLISP Dialect of LISP was successful in the late seventies and early eighties in the development of very sophisticated programming tools like its DEdit structure editor, Programmer's Assistant, Spelling Corrector DWIM ("Do what I mean"), Masterscope, Inspector etc. But because of some oddities in the language itself, very few elements of InterLISP-D were acquired by the Common LISP standard committee for the standardized LISP language. The best features of InterLISP-D could not be included in a language promoting portability as one of its main goals, because they depended strongly on the display features of the D-workstations.

LOOPS was built on top of InterLISP-D and hence is also not portable. LOOPS provided several programming paradigms, each heavily supported by graphics oriented tools. The most successful paradigm introduced to LISP by LOOPS is the object-oriented one, therefore the authors of LOOPS proposed a portable version of the object concepts of LOOPS called PortableCommonLOOPS or PCL. The now standardized object-oriented enhancement of Common LISP (called CLOS) is a direct successor of PCL, hence a successor of LOOPS.

In LOOPS, the entities which are to be modelled are represented as objects. An object is a complex data structure of variables and other information, such as hierarchical dependencies to other objects, documentation records, etc. Each object can have local variables as well as common variables, which are shared with some other objects. Each object is an incarnation of a special kind of object, named class, which is used as a kind of blueprint to create new objects. Each object knows some set of methods, which are LISP functions with some syntactic conventions.

A LOOPS program executes by sending messages to objects. A message consists of the addressed object, the method selector, and the arguments for the method. The objects respond by applying their method for the received selector to the arguments and returning the result to the caller, which can send the next message.

To make these ideas reasonable, the additional concept of inheritance is needed. All classes in the system are arranged in an inheritance network. A subclass or specialization of a class has all variables and methods of its superclass by inheritance, but some variables and methods may be modified and some new ones may be added. There is one most general class, named *object*, from which all classes are specializations. In LOOPS, an object not being a class is called an instance, and a class, whose incarnations are classes is called a *metaclass*. An application of the metaclass concept and a graphical representation of the involved LOOPS classes is presented in Section 2.5.

LOOPS has shown to be well-suited for the development of CAST.FSM. The object-oriented programming style is very close to the systems-theoretical way of thinking. System types can be directly implemented as LOOPS classes. Methods applicable to systems are specified as LOOPS methods. The concept of multi-level class inheritance of LOOPS is largely equivalent to hierarchical requirements as they exist in classification of system types and system transformations. The browser built into LOOPS provides a convenient graphical representation of individual system classes of CAST.FSM (Figure 2.2.1). Another browser is the essential LOOPS tool for the development of system algorithms. It is thereby possible to develop a system algorithm graphically "on screen". The user has access to applicable methods by means of hierarchically organized menus. Figure 2.2.2 shows another browser illustrating the process of a system algorithm development for the reduction of a finite state machine.

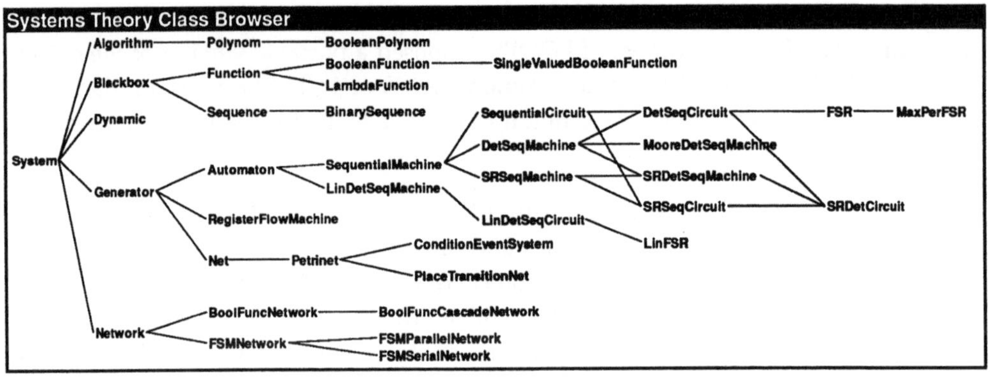

Figure 2.2.1 Browser of system types for CAST.FSM.

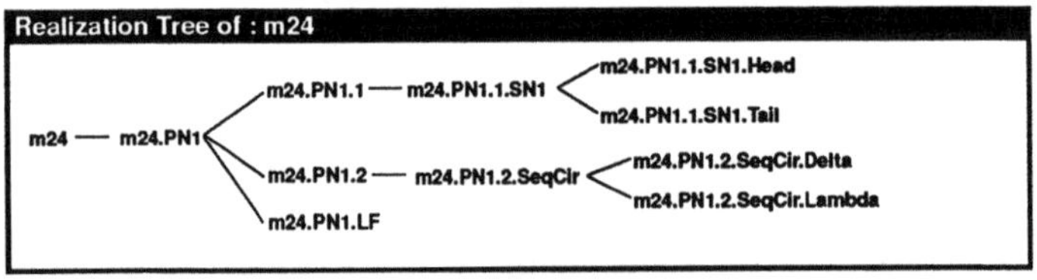

Figure 2.2.2 Browser for system algorithm developments.

The user interface of CAST.FSM offers the essential methods for manipulating finite machines available from classical text books. Aids offered by InterLISP-D and LOOPS, such as graphics, windows, mouse, and menu technology, are a great support. We follow the goal that all activities, which would otherwise be undertaken by "paper and pencil" using technical literature, could also be conveniently done on the machine. Various graphical representations of finite state machines and corresponding objects such as tables, graphs, and block diagrams are implemented in CAST.FSM. Additionally, the CAST.FSM user should be in the position to create individual tools for his workplace by drawing upon the means already available in CAST.FSM. Figure 2.2.3 shows a typical screen image of a CAST.FSM application.

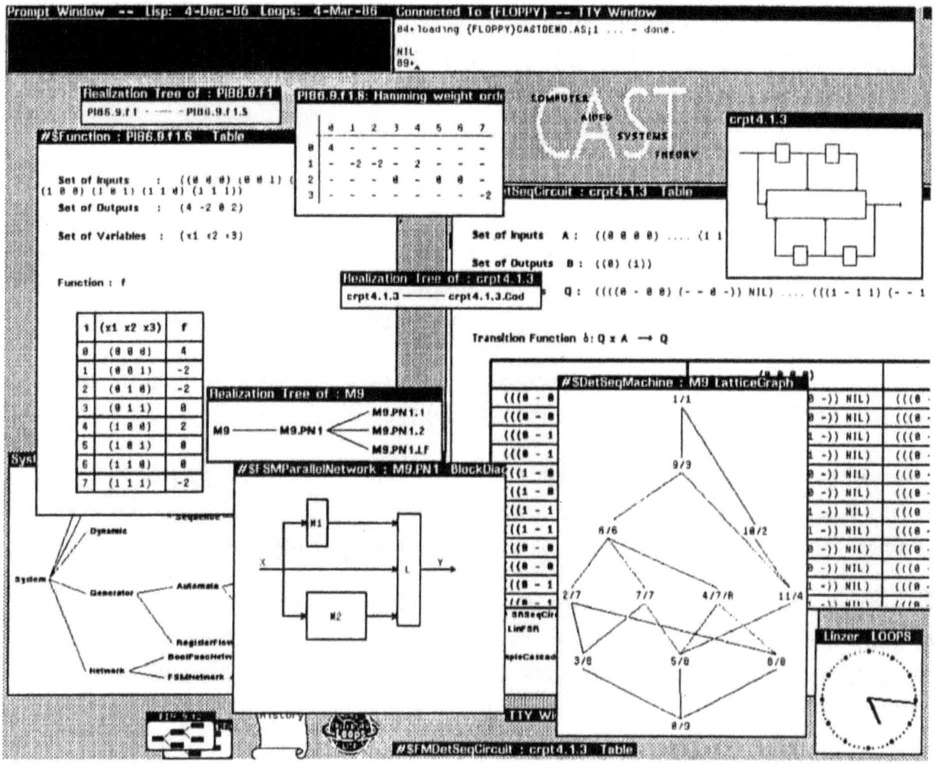

Figure 2.2.3 CAST.FSM screen.

2.2.3 STIPS.FSM Implementation

As already pointed out before, system types are associated with classes in LOOPS. This corresponds to a natural representation. Classes correspond to system types, incarnations of classes correspond to the actual system specifications of the respective types. A class determines which attributes (variables) an object will be characterized by and which method will be applicable to it.

The categorization of the systems-theoretical "world" proceeds from the general to the specific. Starting from the general type of a system, a specialization into five basic general system types (see Section 1.2) – which can be further specialized – is undertaken. This ordering is consistent with the concept of specialization and inheritance of the object-oriented programming paradigm. Instance variables and methods defined in general classes can be used in more specialized classes, to which additional attributes and methods can be attached.

In CAST.FSM a number of specific system types have been implemented. In the following, the most important implemented system types are described. As implementations of the general system type black box functions and their specializations, namely multi-output boolean functions and single-output boolean functions, exist in CAST.FSM. Figure 2.2.4 shows a table representation of a 2-output boolean function.

The system type *sequence* and its specialization, *binary sequence*, which are implemented in CAST.FSM are further examples of the type black box.

The most important general system type in CAST.FSM is the type *generator*. Its subtypes (*finite state machines, Petri nets, register flow machines, etc.*) build the core of CAST.FSM. Finite machines are further classified into *linear* and *non-linear* machines.

Non-linear machines are specified by the sets of inputs, outputs, and states and a delta/lambda-map which determines the state transition function and the output function of the machine. Figure 2.2.5 shows a machine in table representation, but it may also be represented by its state transition graph. There are a number of special-

#SBooleanFunction : BoolFunc1 Table

Set of Inputs : ((0 0 0) (0 0 1) (0 1 0) (0 1 1
) (1 0 0) (1 0 1) (1 1 0) (1 1 1))

Set of Outputs : ((0 0) (0 1) (1 0) (1 1))

Set of Variables : (x y z)

Function : BoolFunc1

i	(x y z)	BoolFunc1
0	(0 0 0)	(0 1)
1	(0 0 1)	(1 0)
2	(0 1 0)	(1 0)
3	(0 1 1)	(0 0)
4	(1 0 0)	(1 0)
5	(1 0 1)	(1 1)
6	(1 1 0)	(1 1)
7	(1 1 1)	(0 0)

Figure 2.2.4 Table representation of a function.

izations of non-linear finite machines with special applicable methods, e.g., *deterministic finite machines, Moore* machines, *sequential switching circuits* and *feedback shift regisers.*

Linear machines are defined by four matrices A, B, C, D over a Galois field GF(p^n), where p is prime (see Fig. 2.2.6). Since matrix calculus is important for this system type, a basic package of LISP functions for the most important matrix and vector operations is implemented. Important specializations of linear machines, which are also used in CAST.FSM, are *linear sequential switching circuits* and *linear feedback shift registers.*

Register flow machines (RFM) are implemented similarly to finite state machines. In place of the set of states, a set of *registers* is employed; the place of the state transition and output function is taken by the *flow function* and specific output variables. Each initial state of a finite state machine corresponds to an initial value of the flow function. RFMs may be represented in CAST.FSM as tables or flow-graphs (Fig. 2.2.7).

Petri nets are arranged in CAST.FSM in three classes: *General Petri Nets, Condition Event Systems,* and *Place Transition Nets.* In Condition Event Systems, a place can accept only one token. In Place Transition Nets the arrows can be weighted and the

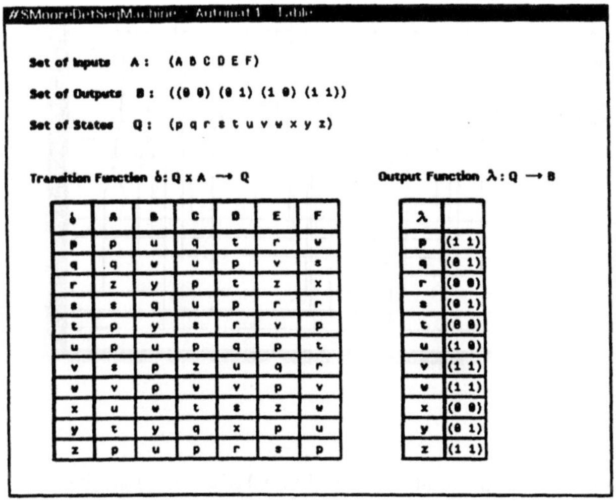
#SMooreDetSeqMachine : Automat 1 Table

Set of Inputs A : (A B C D E F)

Set of Outputs B : ((0 0) (0 1) (1 0) (1 1))

Set of States Q : (p q r s t u v w x y z)

Transition Function δ: Q x A → Q

δ	A	B	C	D	E	F
p	p	u	q	t	r	w
q	q	w	u	p	v	s
r	z	y	p	t	z	x
s	s	q	u	p	r	r
t	p	y	s	r	v	p
u	p	u	p	q	p	t
v	s	p	z	u	q	r
w	v	p	w	v	p	v
x	u	w	t	s	z	w
y	t	y	q	x	p	u
z	p	u	p	r	s	p

Output Function λ: Q → B

λ	
p	(1 1)
q	(0 1)
r	(0 0)
s	(0 1)
t	(0 0)
u	(1 0)
v	(1 1)
w	(1 1)
x	(0 0)
y	(0 1)
z	(1 1)

Figure 2.2.5 Table illustration of a finite state machine.

places can have limitations in capacity. All net types can be graphically edited with the mouse. I/O-experiments (in the form of token games) can be carried out.

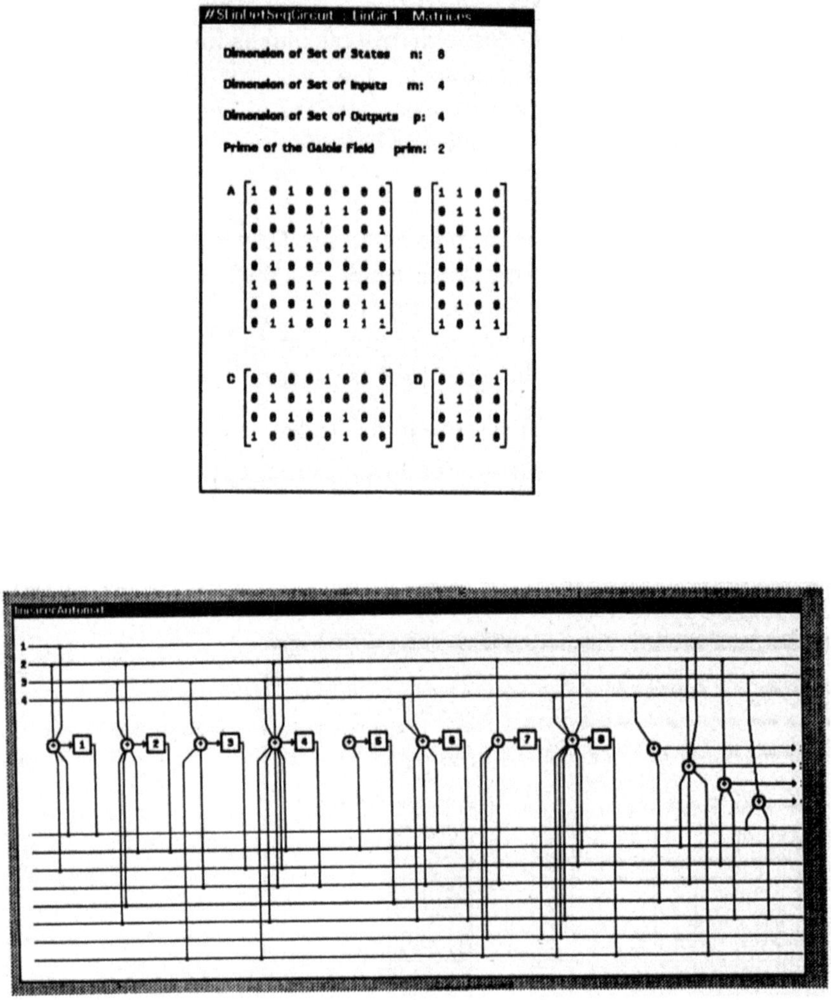

Figure 2.2.6 Matrix illustration (*top*) and circuit diagram (*bottom*) of a linear machine.

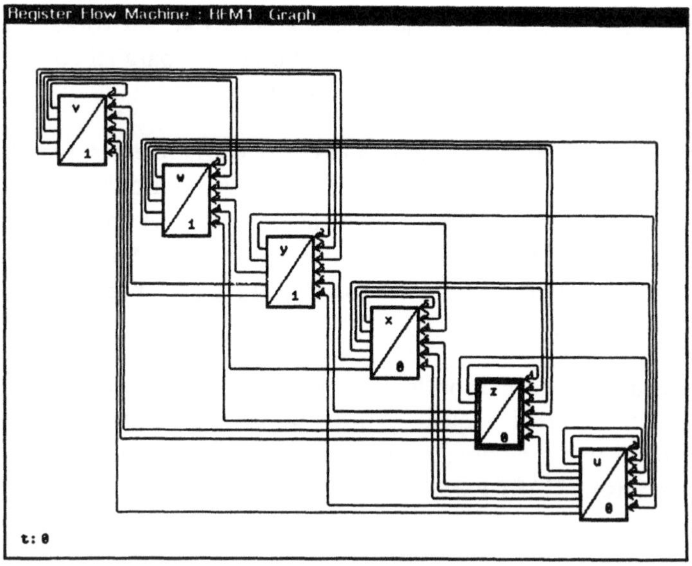

Figure 2.2.7 Flow-graph of a register flow machine.

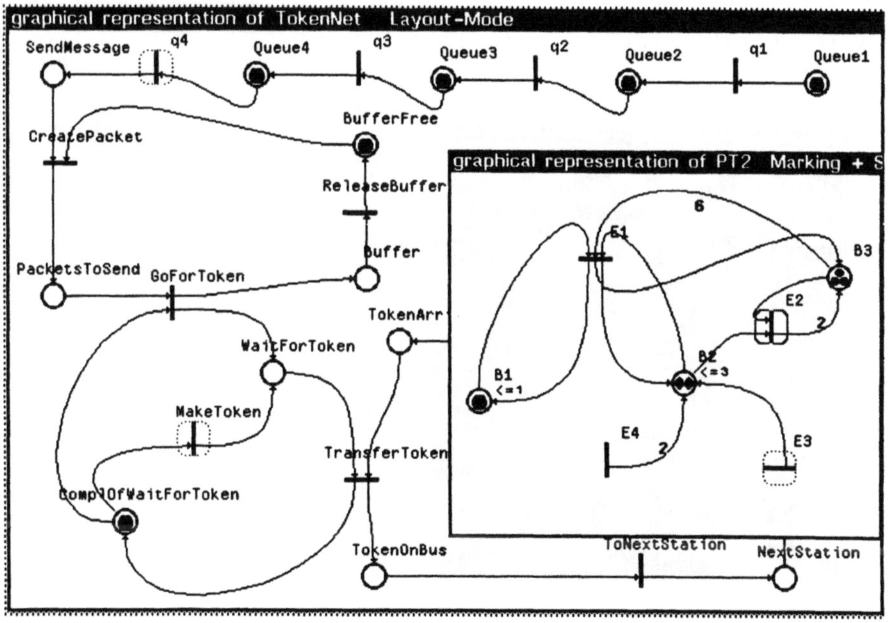

Figure 2.2.8 Petri nets.

A further important general system type is the network. Networks are defined by their *components* and by the *coupling structure.* Serial and parallel couplings of finite machines are important specializations of networks.

The system transformations implemented in CAST.FSM are extensive (details can be found in the Appendix of this book). At this point, we refer only to the most important examples and to some implementation characteristics. A specific transformation of a black box to a generator is given, for example, by the *Massey-Berlekamp* algorithm. It computes the shortest linear feedback shift register that can generate a given binary sequence (Fig. 2.2.9).

Another important transformation is performed by the *Quine-McCluskey* algorithm which computes a minimal boolean expression for a given single-output Boolean function. For boolean expressions a corresponding programmable logic array (PLA) can be computed using CAST.FSM (Fig. 2.2.10).

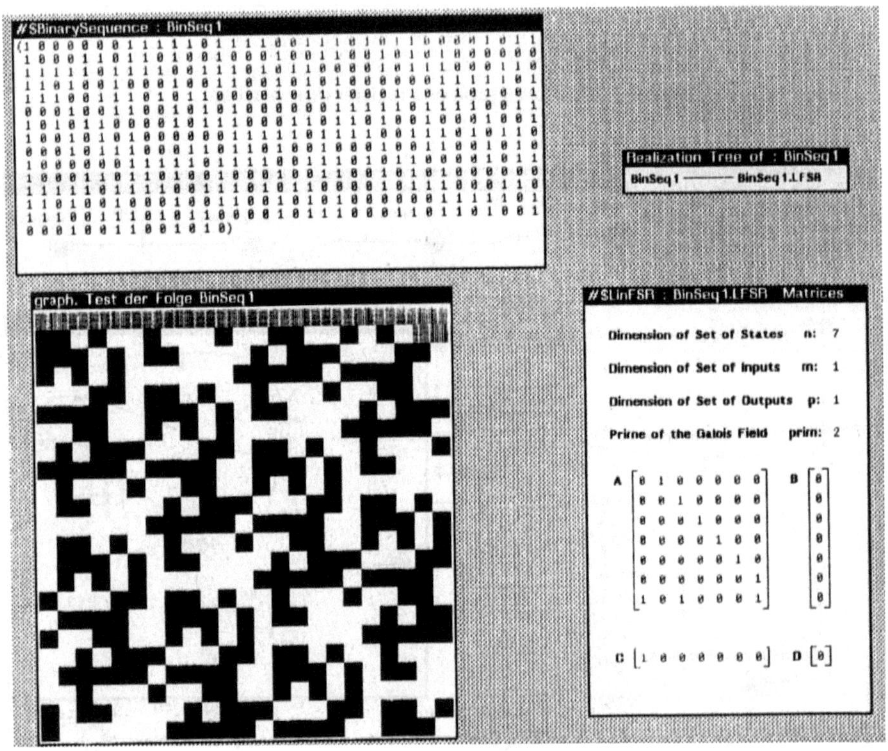

Figure 2.2.9 Massey-Berlekamp algorithm.

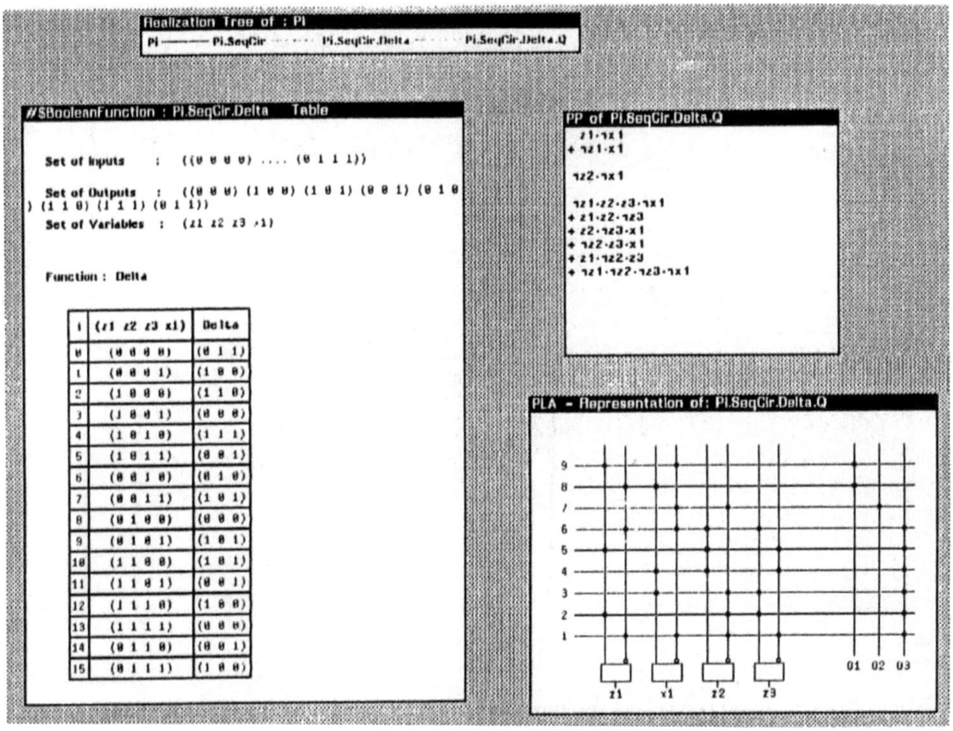

Figure 2.2.10 Quine-McCluskey algorithms and PLA realization.

Further important CAST.FSM transformations operate upon finite state machines. Examples are the computation of the *lattice* of the machine and the serial and parallel decompositions. In the implementation of these methods the computation of *closed partitions* on the set of states is essential. Figure 2.2.11 shows the lattice $L(M)$ of the finite state machine M.

Other important transformations for non-linear machines are the *linear realization* transformation, the computation of the *inverse machine*, the *shift register realization* of machines, and the computation of a register flow machine which simulates a given finite state machine. Furthermore, special methods for linear machines are also implemented in CAST.FSM. A special reduction algorithm for linear machines and the shift register decomposition of linear machines are two examples.

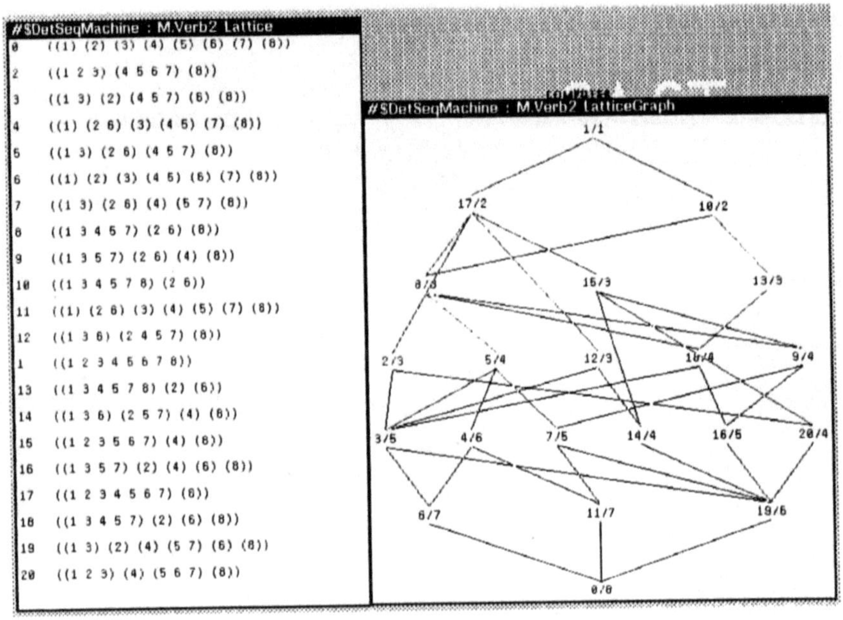

Figure 2.2.11 Lattice $L(M)$ for the structuring and reduction of a finite state machine M.

2.2.4 CAST.FSM Application Examples

The examples given in this section are intended to illustrate the use of CAST.FSM. The detailed steps with CAST.FSM will be described in the first example which shows the reduction of a finite machine. CAST.FSM is in our classification of Section 2.1.1 an interactive method bank using modern concepts of man/machine communication, such as graphics, multiple-window interface, mouse, menus, and browsers. The user interface is designed that the main interaction is accomplished with the mouse and pop-up menus. Further details with respect to the use of CAST.FSM can be found in [Mittelmann 89] and the Appendix of this book.

A CAST.FSM session begins with the creation of a class browser by selecting the corresponding menu entry in CAST.FSM's logo icon menu (Fig. 2.2.12).

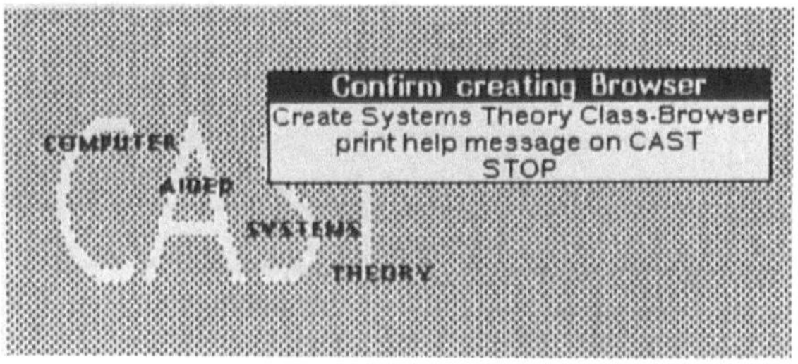

Figure 2.2.12 Starting a CAST.FSM session.

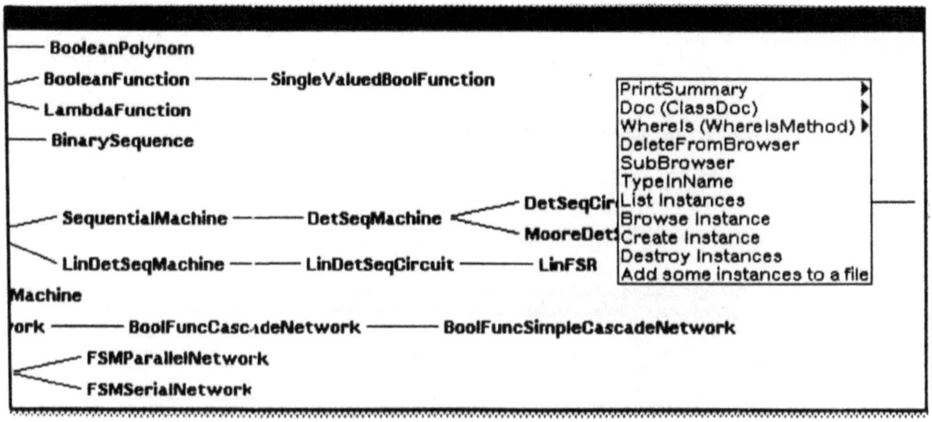

Figure 2.2.13 CAST.FSM class browser.

The class browser which was already shown in Figure 2.2.1 helps the user organize his object world. He can create or delete individual system specifications, store them in files, or represent them in browsers. As an example, a *Moore*-type machine can be created by selecting the class *MooreDetSeqMachine* and the menu entry *createInstance* followed by entering a name for the machine (Figure 2.2.13).

This machine can now be defined in another window (Figure 2.2.14). Furthermore, an incarnation browser is created for this object. This incarnation browser (Figure 2.2.15) will be used as a second command unit, where operations on the individual system specifications can be applied and the resulting system specifications are represented. This kind of an incarnation browser is called a *realization tree*.

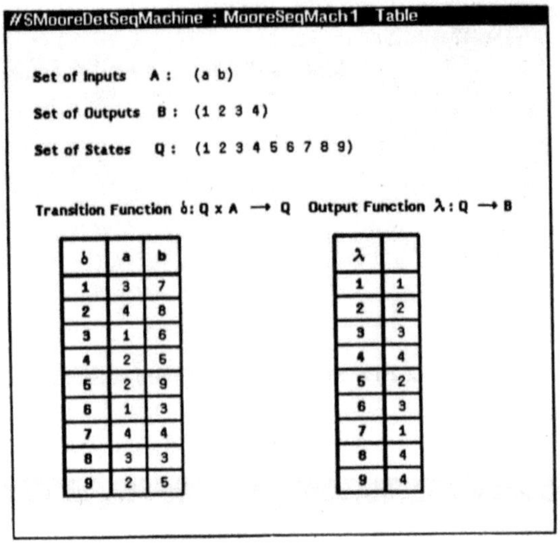
#SMooreDetSeqMachine : MooreSeqMach1 Table

Set of Inputs A : (a b)

Set of Outputs B : (1 2 3 4)

Set of States Q : (1 2 3 4 5 6 7 8 9)

Transition Function δ: Q x A → Q Output Function λ: Q → B

δ	a	b
1	3	7
2	4	8
3	1	6
4	2	5
5	2	9
6	1	3
7	4	4
8	3	3
9	2	5

λ	
1	1
2	2
3	3
4	4
5	2
6	3
7	1
8	4
9	4

Figure 2.2.14 Definition of a finite *Moore* machine *MooreSeqMach1*.

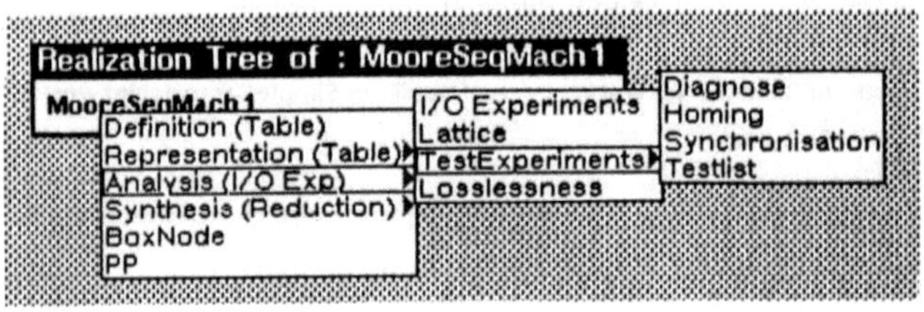

Figure 2.2.15 Incarnation browser for the machine *MooreSeqMach1*.

The lattice *L(MooreSeqMach1)* of congruence relations of the machine *MooreSeqMach1* is computed by the operation *Lattice* (Fig. 2.2.16). A user can see immediately from *L(MooreSeqMach1)* that the machine is not reduced (observe that the congruence relation number 4 is the reduction equivalence) and that *MooreSeqMach1* can be decomposed in many different ways.

The *quotient machine* is computed by the operation "reduction". It will be attached to the object as a realization.

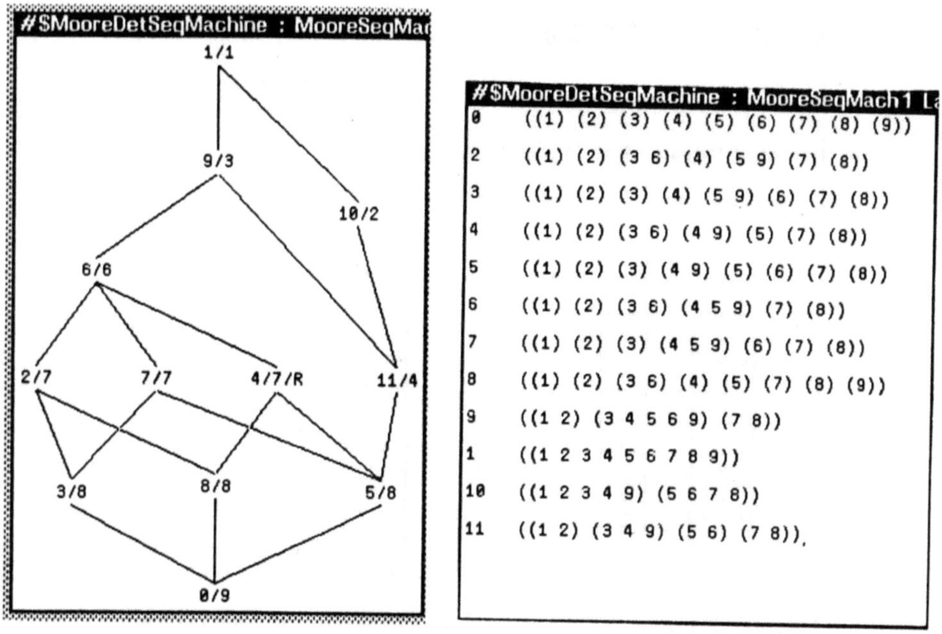

Figure 2.2.16 Lattice of congruence relations of the machine *MooreSeqMach1*.

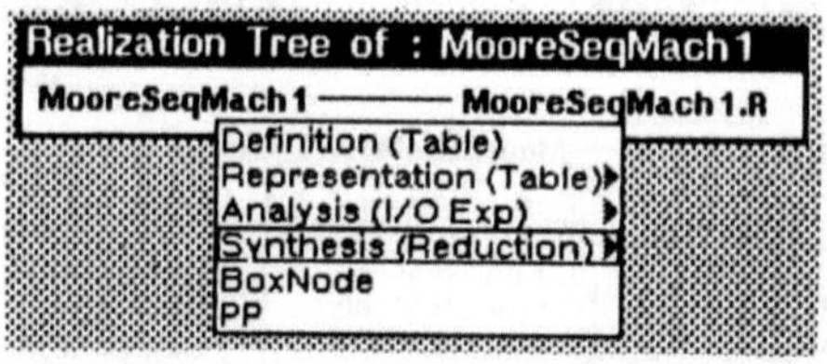

Figure 2.2.17 Computation of a quotient machine *MooreSeqMach1.R*.

Figure 2.2.18 shows the table illustration of the computed quotient machine *MooreSeqMach1.R*. The lattice *L*(*MooreSeqMach1*) shows that a parallel decomposition is possible. The task is to find two congruence relations whose intersection is empty. The computed networks and the components of the network are shown in the realization tree (Figure 2.2.19). Figure 2.2.20 shows the resulting network as a block diagram, whereby the size of the blocks *M1* and *M2* relates to the size of the respective state sets. Figures 2.2.21 and 2.2.22 show the representation of component machines and the output function by means of tables, respectively.

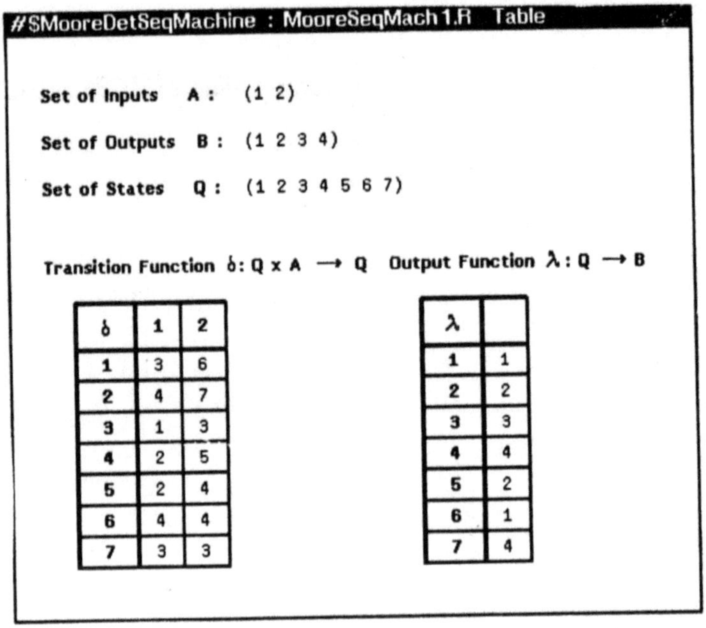

Figure 2.2.18 Table representation of a quotient machine.

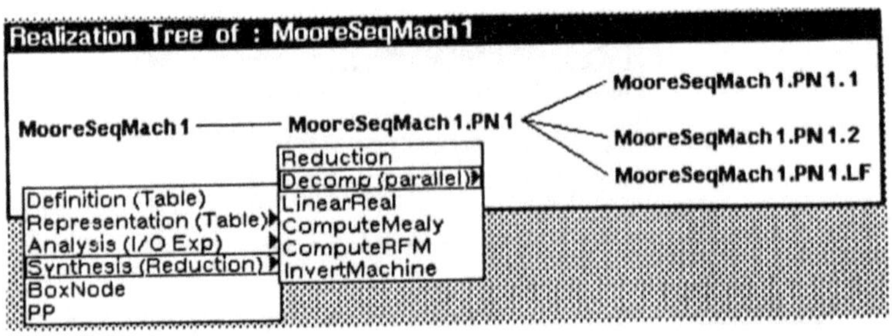

Figure 2.2.19 Parallel decomposition of the machine *MooreSeqMach1*.

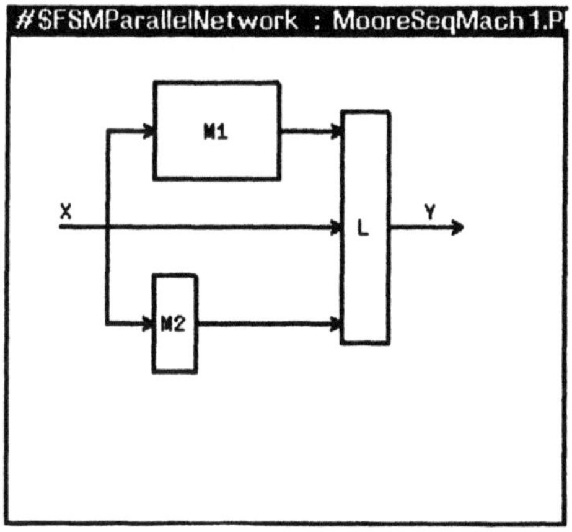

Figure 2.2.20 Block diagram of the parallel decomposition of *MooreSeqMach1*.

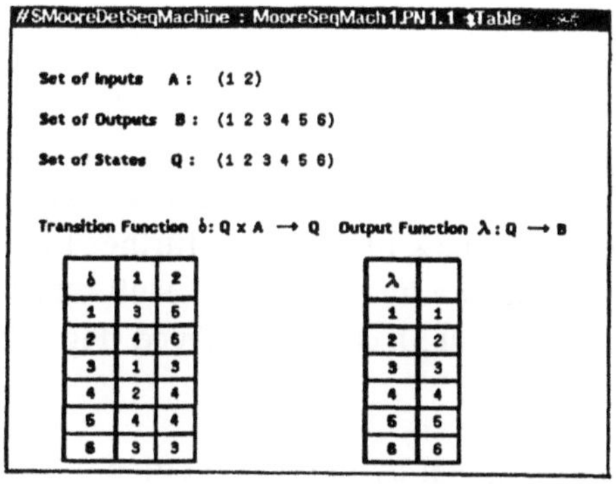

Set of Inputs A : (1 2)

Set of Outputs B : (1 2 3 4 5 6)

Set of States Q : (1 2 3 4 5 6)

Transition Function $\delta: Q \times A \rightarrow Q$ Output Function $\lambda: Q \rightarrow B$

δ	1	2
1	3	5
2	4	6
3	1	3
4	2	4
5	4	4
6	3	3

λ	
1	1
2	2
3	3
4	4
5	5
6	6

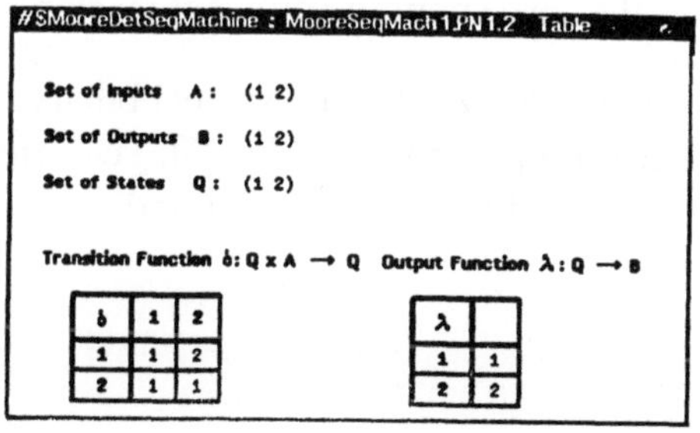

Set of Inputs A : (1 2)

Set of Outputs B : (1 2)

Set of States Q : (1 2)

Transition Function $\delta: Q \times A \rightarrow Q$ Output Function $\lambda: Q \rightarrow B$

δ	1	2
1	1	2
2	1	1

λ	
1	1
2	2

Figure 2.2.21 Component machines of *MooreSeqMach1*.

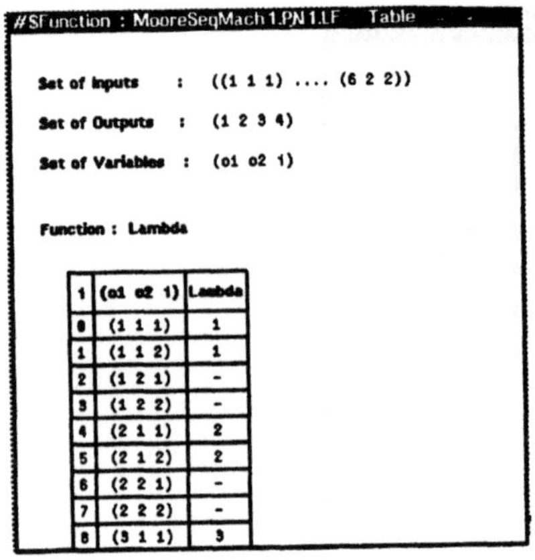
#SFunction : MooreSeqMach1.PN1.LF Table

Set of Inputs : ((1 1 1) .... (6 2 2))

Set of Outputs : (1 2 3 4)

Set of Variables : (o1 o2 1)

Function : Lambda

1	(o1 o2 1)	Lambda
0	(1 1 1)	1
1	(1 1 2)	1
2	(1 2 1)	-
3	(1 2 2)	-
4	(2 1 1)	2
5	(2 1 2)	2
6	(2 2 1)	-
7	(2 2 2)	-
8	(3 1 1)	3

Figure 2.2.22 Output function *L* of *MooreSeqMach1*.

In the second example, an autonomous finite state machine *a1* serves as initial system specification for the development of a system algorithm (Fig. 2.2.23). A suitable boolean realization should be found for this machine. Two different methods are available for this; the first is the realization of the machine by a linear sequential switching circuit and the second is the realization of the machine by a sequential switching circuit implemented by PLA.

The following figures show the results of a linear realization computation. Figure 2.2.24 shows the matrix representation, Figure 2.2.25 the circuit diagram of the resulting linear machine *a1.L*. One sees that the linear machine *a1.L* is a binary linear feedback shift register in which each cell has a feedback line. With a further transformation, a parallel reduction *a1.L.RKF* consisting of a linear feedback shift register of length 4 and a shift register of length 1 (storage cell 5), can be achieved (Fig. 2.2.26–2.2.28).

An alternative to this linear realization of *a1* is given by the computation of a corresponding optimally structured sequential switching circuit. The computation of the state assignment is done by means of the machine lattice *L(a1)*, shown in Figure 2.2.29. One sees that machine *a1* is not reduced (the congruence relation 19 is the reduction equivalence). Figure 2.2.30 shows the corresponding quotient machine.

By means of the lattice of the quotient machine (Fig. 2.2.31), an optimal binary state assignment is computed according to the method given in [Pichler 75]. The result is shown in Figure 2.2.32. Finally, Figure 2.2.33 shows the corresponding circuit *a1.R.seqCir* in table form.

#$DetSeqMachine : a1 Table

Set of Inputs A : (0)

Set of Outputs B : (0 1)

Set of States Q : (1 2 3 4 5 6 7 8 9 10 11 12)

Transition Function δ: Q x A → Q

δ	0
1	12
2	8
3	2
4	1
5	9
6	4
7	6
8	10
9	3
10	5
11	7
12	11

Output Function λ: Q x A → B

λ	0
1	0
2	0
3	1
4	0
5	0
6	1
7	0
8	0
9	1
10	0
11	0
12	1

Figure 2.2.23 Finite machine *a1*.

#$LinDetSeqCircuit : a1.L Matrices

Dimension of Set of States n: 5

Dimension of Set of Inputs m: 1

Dimension of Set of Outputs p: 1

Prime of the Galois Field prim: 2

$$A \begin{bmatrix} 1 & 1 & 1 & 1 & 1 \\ 1 & 0 & 0 & 0 & 0 \\ 0 & 1 & 0 & 0 & 0 \\ 0 & 0 & 1 & 0 & 0 \\ 0 & 0 & 0 & 1 & 0 \end{bmatrix} \quad B \begin{bmatrix} 0 \\ 0 \\ 0 \\ 0 \\ 0 \end{bmatrix}$$

$$C \begin{bmatrix} 0 & 0 & 0 & 0 & 1 \end{bmatrix} \quad D \begin{bmatrix} 0 \end{bmatrix}$$

Figure 2.2.24 Linear realization of *a1.L*.

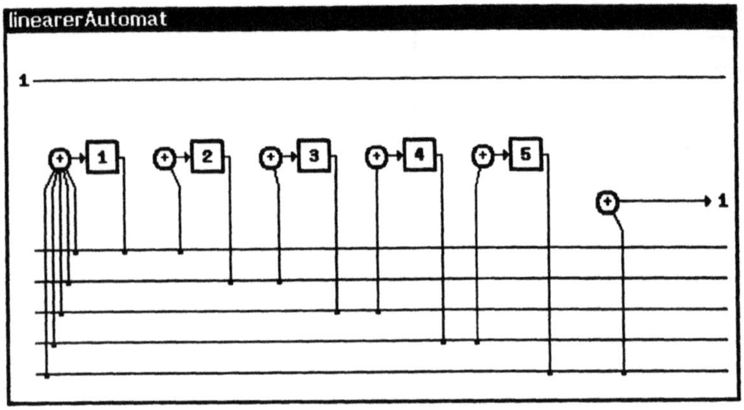

Figure 2.2.25 Circuit diagram of *a1.L*.

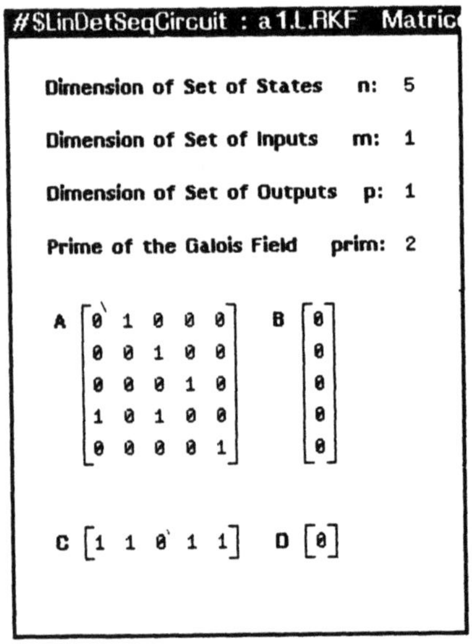

Figure 2.2.26 Linear machine *a1.LRKF*.

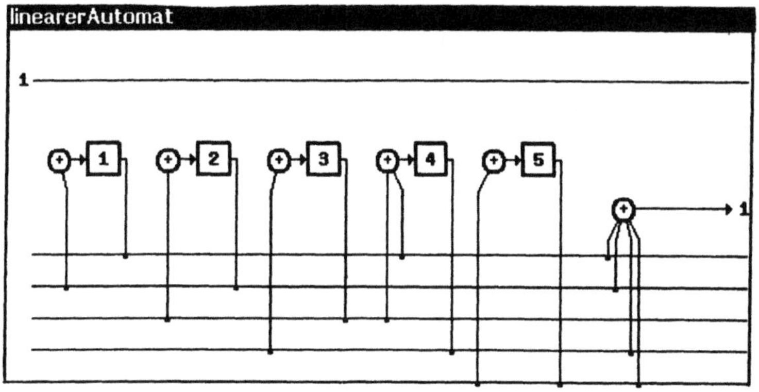

Figure 2.2.27 Circuit diagram of linear machine *a1.LRKF*.

Figure 2.2.28 Realization tree for the computation of *a1.L* and *a1.L.RKF*.

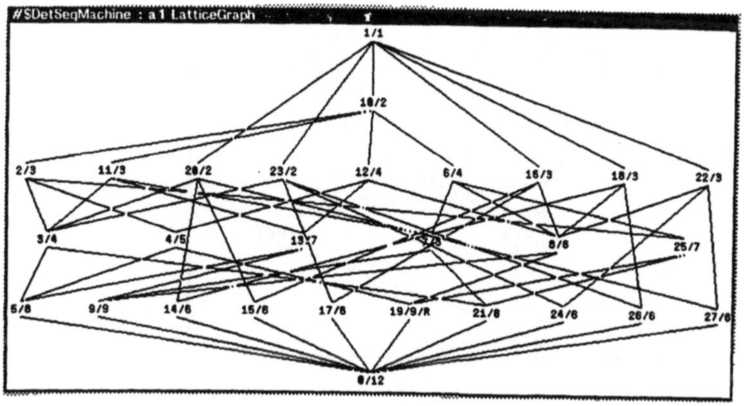

Figure 2.2.29 Lattice of *a1*.

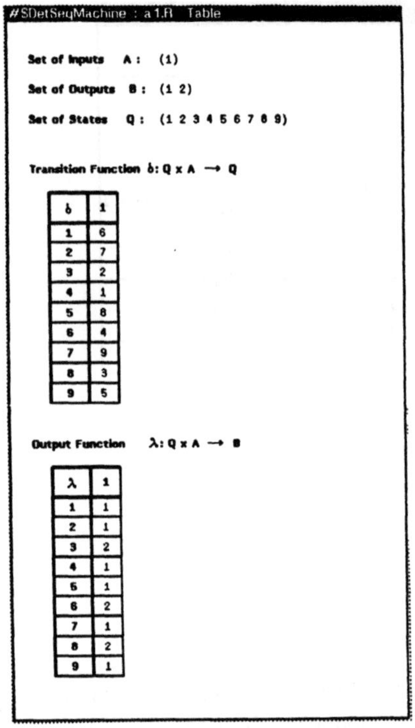
#$DetSeqMachine : a1.R Table

Set of Inputs A : (1)

Set of Outputs B : (1 2)

Set of States Q : (1 2 3 4 5 6 7 8 9)

Transition Function $\delta: Q \times A \rightarrow Q$

δ	1
1	6
2	7
3	2
4	1
5	8
6	4
7	9
8	3
9	5

Output Function $\lambda: Q \times A \rightarrow B$

λ	1
1	1
2	1
3	2
4	1
5	1
6	2
7	1
8	2
9	1

Figure 2.2.30 Quotient machine *a1.R.*

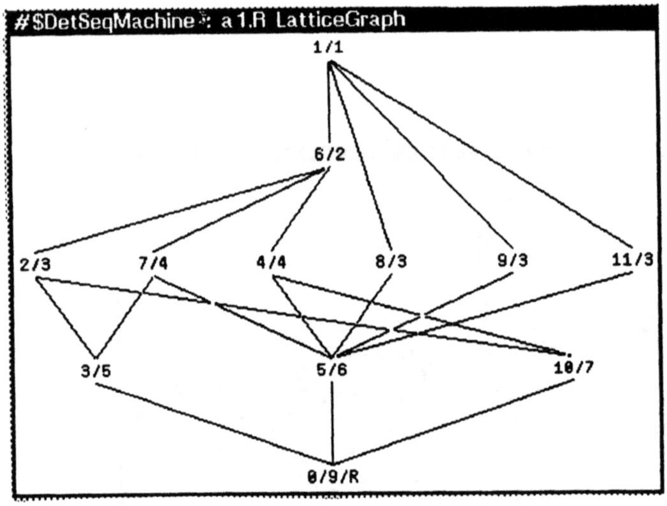

Figure 2.2.31 Lattice of quotient machine *a1.R.*

#$Function : NIL Table

Set of Inputs : (1 2 3 4 5 6 7 8 9)

Set of Outputs : ((0 0 0 0) .... (0 1 0 1))

Set of Variables : (x)

Function : Gamma

i	(x)	Gamma
0	1	(0 0 0 0)
1	2	(0 0 0 1)
2	3	(0 1 1 0)
3	4	(0 1 0 0)
4	5	(0 0 1 0)
5	6	(1 0 0 0)
6	7	(1 0 1 0)
7	8	(1 0 0 1)
8	9	(0 1 0 1)

Figure 2.2.32 State assignment of *a1.R.*

#$DetSeqCircuit : a1.R.SeqCir Table

Set of Inputs A : ((0))

Set of Outputs B : ((0) (1))

Set of States Q : ((0 0 0 0) .... (0 1 0 1))

Transition Function $\delta: Q \times A \longrightarrow Q$

δ	(0)
(0 0 0 0)	(1 0 0 0)
(0 0 0 1)	(1 0 1 0)
(0 1 1 0)	(0 0 0 1)
(0 1 0 0)	(0.0 0 0)
(0 0 1 0)	(1 0 0 1)
(1 0 0 0)	(0 1 0 0)
(1 0 1 0)	(0 1 0 1)
(1 0 0 1)	(0 1 1 0)
(0 1 0 1)	(0 0 1'0)

Output Function $\lambda: Q \times A \longrightarrow B$

λ	(0)
(0 0 0 0)	(0)
(0 0 0 1)	(0)
(0 1 1 0)	(1)
(0 1 0 0)	(0)
(0 0 1 0)	(0)
(1 0 0 0)	(1)
(1 0 1 0)	(0)
(1 0 0 1)	(1)
(0 1 0 1)	(0)

Figure 2.2.33 Sequential Switching Circuit *a1.R.SeqCir.*

2.2.5 CAST.FSM Extension and Perspectives

CAST.FSM is not a closed system, but is open to extensions and subject to continuous further development. New specific system types can be defined by the user and new methods can be added. The advantages of structured object-oriented programming in LOOPS facilitates the progressive extension of CAST.FSM with comparatively little effort. A convenient reuse of methods and structures from other classes is made possible by a consequent use of the inheritance mechanism. The creation of new classes and methods takes place dynamically and interactively and can be immediately tested by the developer within the whole CAST developmental environment. A cumbersome translation and binding of the overall system (as common in conventional programming environments) is not necessary. Extending CAST.FSM is, thereby, an incremental process which may proceed in parallel to its application.

In spite of the practical improvements of CAST.FSM, certain weaknesses of this prototypical implementation are apparent. One major limitation is insufficient computing power. The compiled CAST.FSM code currently has the size of 1 MByte which, together with 5 MBytes of LOOPS code, leads to performance problems on workstations with only 1.5 MByte of physical main memory. The object-oriented programming paradigm has not been applied with enough consequence in all cases, leading to problems and additional work. For example, a finite machine consists of three sets (input, output and state) and two functions (state transition and output). Since these two functions are not implemented as objects of the system type function in CAST.FSM , but as LISP data types, the methods of system type function (e.g. binary assignment, Quine-McCluskey minimization) can not be directly applied to the state transition and output function of a machine. The operations must be initiated in CAST.FSM by an additional explicit conversion into a function object.

As a further example, the relatively time-consuming methods for the interactive definition of system specifications for new classes must always be written again, in order to enter into the specific class characteristics. They differ only in the details leading to multiple versions of program parts and to the known disadvantages, such as poor maintenance. With a uniform representation of sets and functions, multiple implementation is eliminated. The inheritance lattice of implemented classes is also given partially arbitrary and leads to additional expenditure in the methods. The careful formation of inheritance hierarchies appears to be the key to good object-oriented programming. Finally, CAST.FSM suffers from the inefficiency of InterLISP-D and LOOPS. These software systems are better suited for the rapid exploration of new ideas (rapid prototyping, explorative programming) as for the development of run-time efficient systems.

In addition to extending CAST.FSM, the design of a new, run-time optimal, strictly object-oriented implementation based on object-oriented Common LISP (CLOS) is under way. Besides providing a better run-time behavior, a strong emphasis lies on very clean design. The inheritance graph will not directly reproduce the system type hierarchies, but will be strictly laid out according to the rules of object-oriented programming. Double implementation and inheritance of non-relevant details will then be avoided.

A larger step into the future of CAST systems will be reached through the inclusion of complete system algorithms. Let us first repeat the discussion of system algorithms development. In STIPS.M (as well as in its implementation CAST.FSM) system algorithms are executed by making the user manually choose appropriate system transformations from a set of system-specific methods. The realization tree, which is constructed by continued selection and application of system transformations shows the intermediate results and the final result of the applied system algorithm. Through the interactive selection of transformations the CAST.FSM user creates a trajectory of the underlying abstract STIPS machine. This process offers every possible freedom, but the user must own the proper systems-theoretical knowledge in order to come to reasonable results.

Consequently, a natural extension would be to keep complete system algorithms as methods, including the initial system specifications and on-line documentation about the system algorithms´ characteristics and goals. Among the prerequisites for the realization of such a CAST.FSM successor are *(a)* the possibility to apply system transformations in a non-interactive manner and *(b)* to supply criteria for selecting alternative branches.

A further evolutionary step in CAST technology will be to introduce "intelligent" search mechanisms capable of developing system algorithms autonomously. The user defines the system problem by means of initial system specifications, goals and constraints. Such an advanced CAST system – let us call it STEXP (*Systems Theory Expert*) – would derive a system algorithm from its knowledge base plus heuristic rules and subsequently executes it. STEXP (Figure 2.2.34) can thereby expand its knowledge base by memorizing successful algorithms and heuristics.

Interaction facilities are provided in this system to request additional support and advice from the human user. STEXP consists of three main components: an *inference engine* which navigates and executes the derivation process, a *database* which stores the initial problem specification and the intermediate results in the form of generated

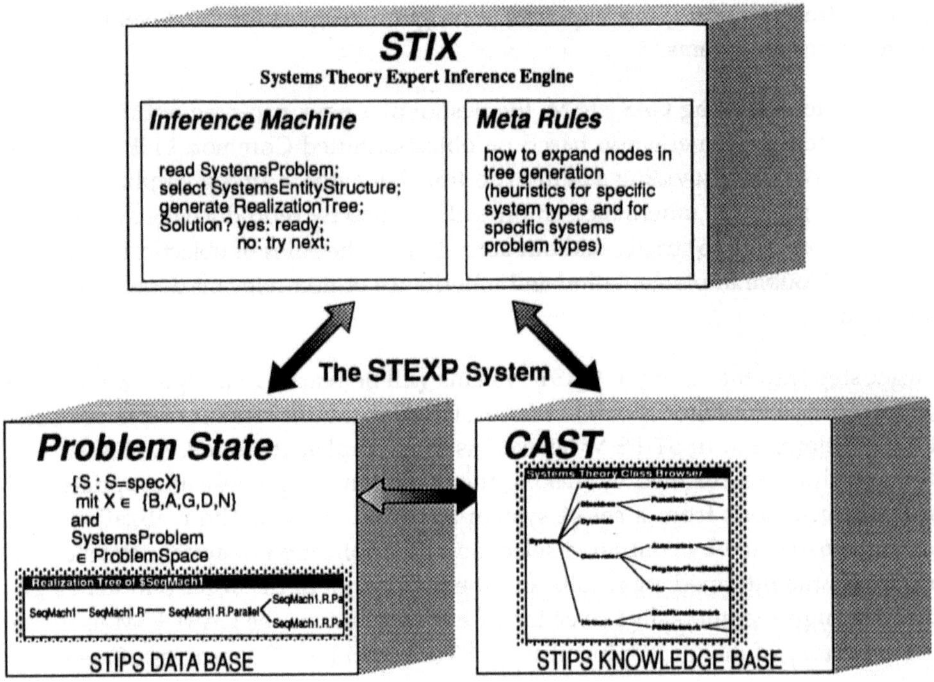

Figure 2.2.34 Systems Theory Expert (STEXP).

realizations, and the *knowledge base* which corresponds to a STIPS machine [Mittelmann 90a, 90b].

2.3 CAST.FOURIER – METHOD BANK FOR GENERALIZED SPECTRAL TECHNIQUES

2.3.1 Introduction

Computer Aided Systems Theory (CAST) denotes, as we have pointed out earlier, the effort of systems scientists to provide various engineering disciplines with software implementations of Systems Theory based concepts and methods and, thus, to make the results of Systems Theory applicable and beneficial for a broader range of application domains. This attempt also aims at raising the level of acceptance that Systems Theory has found in the engineering community. To this end, Systems Theory approaches and techniques are best being offered as convenient and versatile interactive method banks, preferably implemented on high-performance graphics workstations.

These CAST systems have to support the user in modelling his problem (in terms of systems theory concepts), to offer a collection of methods and to enable him to interactively and, to some extent, exploratively navigate through the problem space, i.e., apply and concatenate (synthesis and analysis) methods until an acceptable solution has been found. In this setting, it is up to the user to decide which methods are applied and how they are combined to finally yield a solution for the problem. This problem-solving process is called, as we know by now, Systems Theory Instrumented Problem Solving (STIPS).

An example of a prototype CAST tool, which we have discussed in Section 2.2, is CAST.FSM. Application areas which can benefit from this system are digital circuit design, in particular design for testability, cryptology and Systems Theory education.

In this section, a CAST system intended for use in a quite different application environment will be outlined. This system, called CAST.FOURIER, will serve as a tool for investigating and using (generalized) spectral techniques in the areas of (1-D) digital signal processing (DSP) and, in particular, digital image processing (DIP). The underlying motivation, goals, concepts, state of implementation, and future directions of the project will be sketched.

2.3.2 Motivation

There are two principal reasons that motivate the development of a CAST tool for the domains mentioned above:

First, in DIP, many applications require a highly interactive and experimental problem-solving strategy [Gonzalez, Wintz 87]. DIP problems and techniques can be divided into two broad classes. In the first category, pictorial information has to be improved for human inspection and interpretation. Clearly, human guidance is essential in selecting the appropriate enhancement and restoration techniques to obtain images of satisfying quality. Problems of the second type deal with processing scene data for autonomous machine perception (computer vision). The results of this early step may be crucial for the success of the entire vision task, so the designer of a pre-processing algorithm will carefully select, analyze and combine suitable methods. In this process, he may have to go through a number of experiments and iterations until the right computational subtasks are found. This interactive style of tackling problems calls for powerful and versatile software support. It also corresponds to the STIPS methodology indicated above.

Secondly, DSP and DIP have long made use of concepts which are considered to be important parts of Systems Theory. Examples are design and analysis of digital filters, spectral transforms, and related frequency-domain techniques. There is also a number of tools which make results of these well-established areas available for practical applications. Generalized spectral techniques, however, can be considered a new branch of Systems Theory (or applied mathematics) which has not yet generally been accepted by people working on DIP problems. This is partially due to the lack of convincing demonstrations of the usefulness of these approaches, but certainly also due to the lack of appropriate software tools that make these techniques work in practice. CAST.FOURIER is designated to eventually fill this gap.

It should be evident now that DIP and, more specifically, the role and use of (generalized) spectral techniques in DIP is an interesting field to demonstrate the feasibility and usefulness of the CAST approach.

It should be noted that our system is not only intended to enhance the toolbox of an engineer faced with DIP problems, i.e., to "instrument" problem-solving, but also to serve as a *research* tool which will help to gain insight into and experience about (generalized) spectral methods. As will be discussed in more detail below, knowledge about these novel approaches is limited, and also a researcher exploring possible applications must have an appropriate tool at hand.

2.3.3 Generalized Spectral Techniques in Digital Image Processing

Most of the spectral techniques developed in DIP (and in 1-D DSP, of course) are based on the "classical" discrete Fourier transform (DFT). Due to the well-known theoretical background, the obvious interpretation of the spectral domain in terms of spatial frequencies and the existence of the fast Fourier transform (FFT), frequency-domain methods are well-established in DIP. Examples include filter design and filtering, image restoration and image encoding (data compression) methods [Gonzalez, Wintz 87].

Another significant transform is the Walsh-Hadamard transform (WHT) which is perhaps the most well-known of the non-sinusoidal, orthogonal transforms. Many theoretical results have been derived, fast algorithms have been proposed, and the concept of *sequency* has been suggested for interpreting signals in the spectral domain [Harmuth 69; Ahmed, Rao 75]. Since the WHT can essentially be computed using additions and subtractions only and is therefore more desirable than the DFT from a computational point of view, it has gained some prominence in various applications. Examples include image compression, image enhancement as well as pattern recognition tasks [Beauchamp 84]. Various modifications of the WHT are known.

Besides these two notable classes of spectral techniques, a variety of more specific transforms and related spectral-domain methods has been developed. Some of these orthogonal transforms are listed in the sequel. For details, the interested reader is referred to, e.g., [Ahmed, Rao 75; Beauchamp 84; Gonzalez, Wintz 87; Pratt 78].

- Discrete *Karhunen-Loève* transform (KLT), also referred to as *Hotelling* or *eigenvector* transform or as principal component analysis. It is most often employed in data compression applications. Since the KLT is based on the statistical properties of an image, it yields the highest data reduction ratios. However, fast algorithms do not exist in general.
- Discrete *cosine* transform (DCT). The DCT closely approximates the KLT in a statistical sense but can be computed with fast algorithms. Consequently, the DCT is often used in place of the KLT for practical applications.
- Discrete *sine* transform (DST).
- *Haar* transform (HT).
- *Slant* transform (ST).

  These two transforms are, to some extent, related to the WHT.
- Discrete *Hartley* transform (DHT), which has been proposed as a substitute for the DFT for some purposes [Bracewell 84].

Furthermore, the DFT has been generalized in different respects. This effort has yielded, for instance, number-theoretic transforms and polynomial transforms [Nussbaumer 82] and various versions of the so-called *generalized* discrete Fourier transform (GDFT). As an example which will be referred to repeatedly below, the GDFT as used by [Hellwagner 89; Kunz 77; Pichler, Schwärtzel 90] is introduced in more detail.

The basic notion for the GDFT is to consider signals $\mathbf{x}$ defined on an algebraic structure, namely a finite Abelian group $\mathbf{G}$, that is,

$$\mathbf{x}: \mathbf{G} \to \mathbb{C} \quad (\text{or } \mathbb{R})$$

It is well-known from algebra that any finite Abelian group can be represented by a direct product of cyclic groups $\mathbf{Z}_{N_k}$ of prime order power $N_k = p_k^{a_k}$. Hence, the discussion can be restricted to the following case:

$$\mathbf{G} \cong \mathbf{Z}_{N_1} \times \mathbf{Z}_{N_2} \times \ldots \times \mathbf{Z}_{N_m}, \quad N_1 N_2 \ldots N_m = N = |\mathbf{G}|.$$

Then, the GDFT, in the 1-D case, is a linear, orthogonal transform defined by the matrix operation

$$\mathbf{y} = \mathbf{V}_{N_1 N_2 \ldots N_m} \cdot \mathbf{x}$$

where $\mathbf{x}$ and $\mathbf{y}$ denote the signal and spectral vectors of order N, respectively, and the $N \times N$ GDFT-matrix $\mathbf{V}_{N_1 N_2 \ldots N_m}$ is obtained as the *Kronecker* product of the individual Fourier matrices $\mathbf{U}_{N_k}$ belonging to the cyclic groups $\mathbf{Z}_{N_k}$:

$$\mathbf{V}_{N_1 N_2 \ldots N_m} = \mathbf{U}_{N_1} \otimes \mathbf{U}_{N_2} \otimes \ldots \otimes \mathbf{U}_{N_m}.$$

It is easily seen that the GDFT denotes a *family* of orthogonal transforms where an individual transform is determined by selecting a specific group **G**. There are two notable special cases. If the group **G** is chosen to be the cyclic group of order N, $\mathbf{G} = \mathbf{Z}_N$, the transform obtained coincides with the regular (cyclic) DFT. On the other hand, if $N = 2^n$ and the underlying group is the so-called *dyadic group* of order n, i.e., a direct product of n cyclic groups of order two, $\mathbf{G} = \mathbf{Z}_2 \times \ldots \times \mathbf{Z}_2$ (n times), we end up with the Walsh-Hadamard transform (in Hadamard order). It has been shown that, if $N = 2^n$, the GDFT comprises $N/2$ different transforms [Hellwagner 89] where the limiting cases are the DFT and WHT, respectively, and the other transforms enable a systematic transition from the DFT to the WHT. This situation is depicted in Figure 2.3.1.

Generalizations of the DFT have also been proposed by [Ahmed, Rao 75] and by [Bongiovanni et al. 76]. These approaches are quite different from the one outlined above. However, they also come up with a family or class of linear transforms where the DFT or both the DFT and the WHT occur as special cases.

A rigorous algebraic approach, even more general than the one adopted by the GDFT above, was first taken by [Nicholson 71]. Generalizations of the standard DFT with respect to the underlying group and to the number ring over which it is defined were developed. The underlying group **G** is assumed to be an arbitrary finite Abelian group. The number ring is no more restricted to the complex numbers $\mathbb{C}$, but is assumed to be a commutative integral domain. This direction was further explored by other researchers, notably by [Beth 84], who proposed the most general Fourier transforms in that, for example, also non-Abelian finite groups **G** are considered.

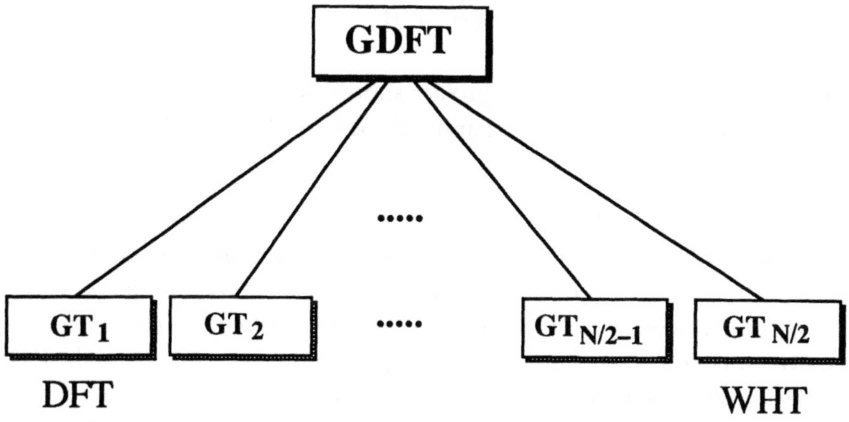

Figure 2.3.1 GDFT—a family of "general" orthogonal transforms.

In summary, there is a remarkable variety of (generalized) spectral techniques and, correspondingly, an extensive body of theoretical knowledge. However, it is our belief that know-how about the practical significance, benefits and drawbacks, and potential applications of these transforms is not equally well developed or, at least, not wide-spread.

As an example, consider the GDFT on finite Abelian groups in the sense outlined above. Many of the desirable properties of the DFT also hold for the other transforms of the class, e.g., linearity, (generalized) convolution property, straightforward inverse transform, and fast algorithm. In addition, a "general" transform can offer advantages over the DFT in that, for example, it may be computationally less expensive and, simultaneously, better approximate the optimum transform (KLT) [Kunz 77]. It was proposed, therefore, that a "general" transform can advantageously replace the DFT or WHT in the domains of image compression and generalized filtering [Pichler 80, 81]. However, this approach has, up to now, never been shown to operate in practice, nor have definite guidelines been worked out which help to select the transform best suited for a given problem.

It is the principal purpose of CAST.FOURIER to improve this situation, that is, to provide users as well as researchers with a tool which enables them to apply and investigate (generalized) spectral methods. In the above setting of the GDFT, the CAST system should, for instance, aid in defining criteria and in finding a systematic way to determine the "best" transform.

2.3.4 Goals and State of Development

Although many of the goals of the project should have become obvious from the previous discussion, they are briefly summarized here. They can be roughly divided into short-term goals and long-term goals.

The primary *short-term goal* is to thoroughly investigate the GDFT described above. First, a deeper understanding of the "general" transforms must be gained and reasonable interpretations be found. (The concept of "generalized frequency" that has been proposed [Ahmed, Rao 75] is very abstract and difficult to comprehend). Second, experiments will be performed to explore potential applications of the GDFT in DIP and (1-D) DSP. Comparison with established methods will reveal whether the above hypothesis holds that one or more "general" transforms can outperform the standard DFT (or WHT). If so, the final attempt will be to establish guidelines which direct the user in selecting a transform well suited for a given problem (application).

The *long-term goal* of the project is to make CAST.FOURIER a versatile software environment which enables the user/researcher to conveniently apply and investigate the most important (generalized) spectral techniques in DIP problems. The user should be in a position to learn about available techniques, to perform experiments, to explore new application areas, to develop new methods based upon available spectral transforms, and to solve certain DIP problems.

To this end, other generalizations of the DFT will have to be implemented and investigations similar to the ones described for the GDFT will have to be performed. For test, comparison, and evaluation purposes, standard frequency-domain as well as corresponding spatial-domain methods will have to be implemented as well. Furthermore, in order to increase the system's power, a component providing the user with some kind of assistance and guidance will have to be added. Clearly, it is desirable that a CAST system can offer advice and is able to support the user in selecting suitable methods. This demand is especially stringent in a domain like the role of generalized spectral techniques in DIP where the user is not expected to be an expert.

It is admitted that these goals are very ambitious; some of them may sound more like wishes than realistic objectives and probably will not be able to be met. In particular, it is an open question whether all this "knowledge" about generalized spectral transforms can be gathered and, if so, how it will be incorporated into CAST.FOURIER to make it a smart tool. Nevertheless, this represents a list of features which a CAST system for this domain is desired to offer.

Currently, as we are still in the early phases of the project, the emphasis is on the GDFT. The 2-D generalized fast Fourier transform (GFFT), a fast algorithm for the 2-D GDFT, and corresponding methods for scaling and displaying (generalized) spectral images have been implemented. Presently, three directions are being pursued:

1. The GDFT is investigated in a 1-D context. This work is to provide insight into and yield a comprehensible interpretation of "generalized spectra".
2. Case studies are performed which test the usefulness of the GDFT in specific DIP problems. Image compression seems to be a promising application area. Results obtained by means of the GDFT – more precisely, selected "general" transforms – are compared with and evaluated against established spectral techniques and/or corresponding spatial-domain methods. Since it is computationally infeasible to explore the entire class of transforms which the GDFT represents, the experiments are being performed with specific "general" transforms. Selecting representative transforms poses a major problem because, as stated above, our knowledge about the properties of "general" transforms is limited. Again, studying the principles of the GDFT is required.
3. In order to enlarge the method base, other spectral transforms are implemented simultaneously. These include the generalized transform as defined by [Ahmed, Rao 75], the DCT, DST, HT, DHT, as well as the KLT.

In the course of these developments, the system is permanently improved. So, first applications and enhancements of the tool go closely together.

2.3.5 Implementation

Fortunately, the implementation of CAST.FOURIER did not have to start from scratch. The system heavily relies on the Vision Kernel System (VKS), a research tool currently being developed at our department to support interactive machine vision programming [Burger 89]. VKS features basic functions for archiving and displaying images on a general-purpose color graphics workstation, for manipulating images and for performing primitive image operations. VKS is being implemented on Apollo Domain workstations. It provides a versatile programming interface based upon Common LISP (CL) and an object-oriented programming system (OOPS). Computationally intensive routines are coded in C and integrated into the CL environment. Images and various graphical representations are generated by means of Apollo's custom graphics routines. The underlying concepts and details of VKS are reported elsewhere [Burger 89] and are also the subject of Section 4.5 of this book.

CAST.FOURIER is a modular extension of VKS. It is based on the same concepts and makes use primarily of the VKS display and archiving facilities. It comprises all the methods which are usually associated with (generalized) spectral-domain approaches and which not necessarily have to be seen by the VKS users or developers. Examples are the GDFT (GFFT), various scale and display functions for (generalized) spectral images, miscellaneous other transforms, and the methods that are currently being developed to test the GDFT in an actual DIP task. The overall structure of the software is conceptually depicted in Figure 2.3.2.

It is argued in [Mittelmann 90a] that CAST systems be preferably implemented using the object-oriented programming paradigm. This statement holds for our system as well. System types and systems, which in our case are several sorts of images, for instance, naturally map to classes and instances, respectively, and system transformations, e.g., spectral transforms, map to methods that are activated by sending messages to image objects.

The CL-based OOPS represents a (nearly) persistent interactive programming environment that meets most of the requirements of an interactive Systems Theory Method Bank as proposed in section 2.1. Objects (e.g., images, filters, windows) are permanently available and constitute the database of the CAST system. A number of methods (e.g., spectral transforms) is at the user's disposal and can be applied and combined arbitrarily (as long as this makes sense from a mathematical or application point of view). In this way, the user can perform experiments and exploratively compose a solution for his problem. If required, object definitions and objects as well as functions (methods) can be added at any time and made part of the system for future use. In summary, the programming environment alone favorably matches to the experimental problem-solving strategy proposed by the STIPS framework and CAST methodology.

Figure 2.3.3 shows some of the facilities of VKS and the CAST system. Some images and the results of different "general" transforms of these images are displayed in several modes and windows. The text window represents the CL listener which, at present, is the primary interface to the user or programmer.

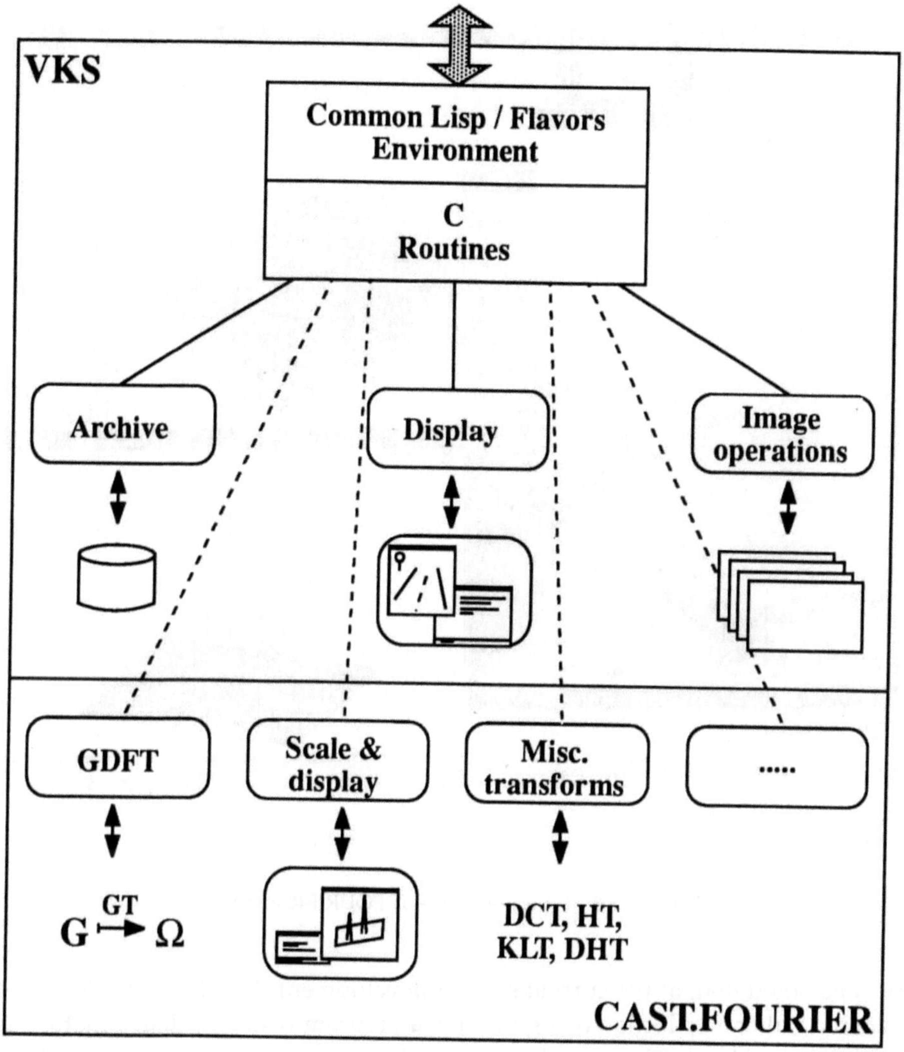

Figure 2.3.2 VKS / CAST.FOURIER structure (adapted from [Burger 88]).

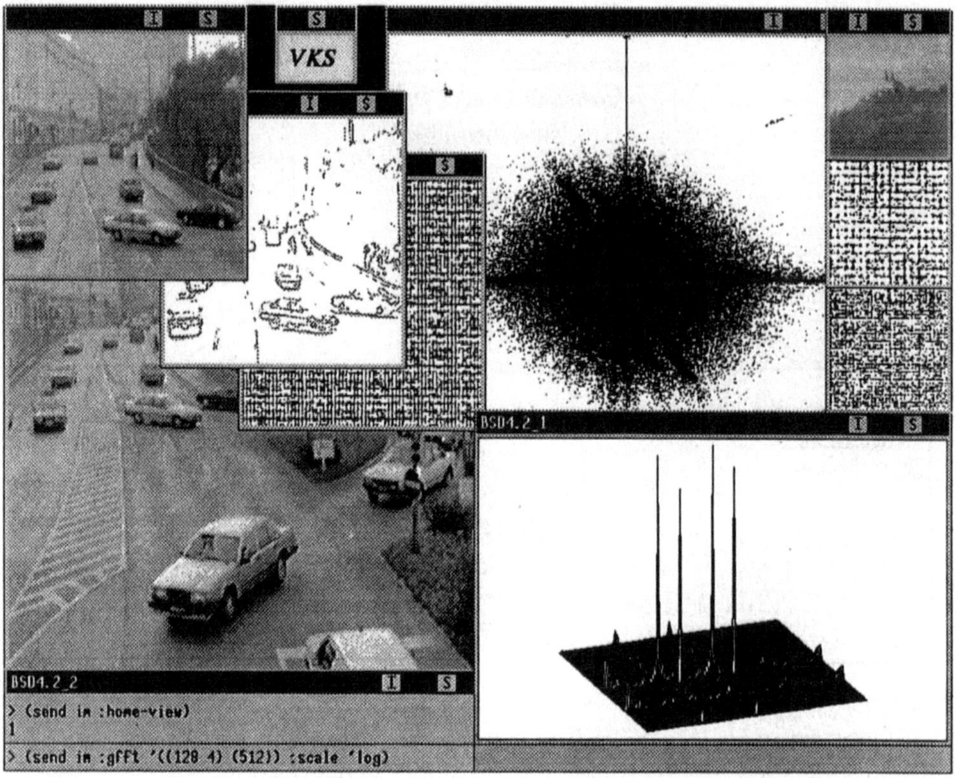

Figure 2.3.3 Typical VKS/CAST.FOURIER screen.

It must be noted that, at the current state of development, CAST.FOURIER must be considered rather a collection of algorithms (a program package) than a mature and coherent method bank. The textual man/machine interface is inconvenient and must be replaced by a state of the art graphical user interface as represented by the browsers in CAST.FSM, for instance. The system does not keep track of the system transformations and does not record the state of the problem-solving process as do the realization trees in CAST.FSM. Hence, it is the user's responsibility to bind and remember intermediate results so that they can be referenced in further processing, e.g., by typing

```
> (setq spectral-image (send image :fft)).
```

However, all the facilities to provide these convenient features are available in the CL and OOPS environment. There are of course plans to realize these important components of a CAST system in our tool since its acceptance and success may crucially depend on these features. In addition, the potential of LISP offers the opportunity to build a sort of "intelligent assistant" on top of the CAST system to make the knowledge gathered about (generalized) spectral techniques available for the average user.

2.3.6 Concluding Remarks

To some extent, this section should be seen as a proposal for the development of a CAST system rather than a report covering completed research. As we are in the early phases of the project, great achievements and significant results could not be reported.

However, it was attempted *(1)* to show that DIP and, specifically, the role and use of (generalized) spectral techniques in this application domain, is an interesting area to test and demonstrate the usefulness of the CAST approach; *(2)* to describe the basic concepts and state of the CAST system being developed for this purpose; *(3)* to reveal some of the directions that are intended to be explored in the further development of the tool; and *(4)* to make a contribution to the emerging field of CAST research by considering a more recent branch of Systems Theory which, from our point of view, can hardly be exploited without appropriate computer support.

2.4 CAST.LISAS – SIMULATION TOOL FOR REGULAR NETWORKS OF FINITE STATE MACHINES

2.4.1 Introduction

Large networks of regularly interconnected processing elements, operating in parallel, are hard to design and analyze due to massive parallelism, inherent in the computations. A "traditional" simulation method of moving transparencies over a network, performing computations at each cell and recording results before proceeding to the next time step, is tedious and error-prone. CAST.LISAS, a LOOPS implementation of a systolic array simulator has been successfully used to simulate complex networks, e.g., systolic architectures and cellular automata.

As systolic architectures and algorithms were the primary objectives for the design of CAST.LISAS, systolic systems are used for outlining the main ideas and facilities of the tool. Extensions and its usability for cellular automata are shown in the final example of this section.

At this place only a short introduction to systolic architectures will be given. Detailed information on systolic algorithms and architectures can be found in [Kung 82; Kung 80; Leiserson 81; Melhelm 85]; a definition of cellular automata and possible applications are presented in [Wolfram 86].

- A systolic system is a one or two-dimensional array of processing cells, operating in parallel (Figure 2.4.1).
- There are few different types of cells within an array. In an ideal systolic system all cells perform the same operation.
- The processing elements are acting synchronously and are driven by a global clock.
- The interconnection scheme between cells is highly regular and local. A cell communicates only with its up to eight neighbors.
- Data flow in a systolic system may be in multiple directions and at different speeds. Input-data, as well as partial results, are rhythmically passed along the communication lines. Figure 2.4.2 shows the flow of data in a linear network for the matrix-vector multiplication problem, introduced in [Kung 80].

Cellular networks are well suited for VLSI realizations because of their regular and local communication scheme which keeps routing costs low and leads to high densities. Systolic systems, however, are in most cases hard to understand and manipulate, due to multi-processing and pipelining. The difficulties for human design activ-

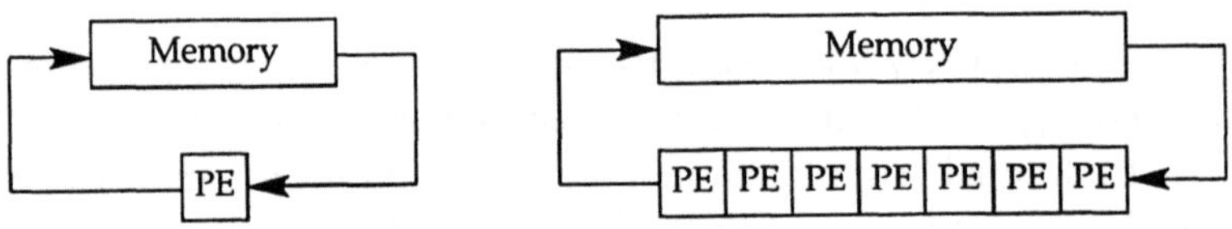

Figure 2.4.1 Basic principle of a systolic system [Kung 80].

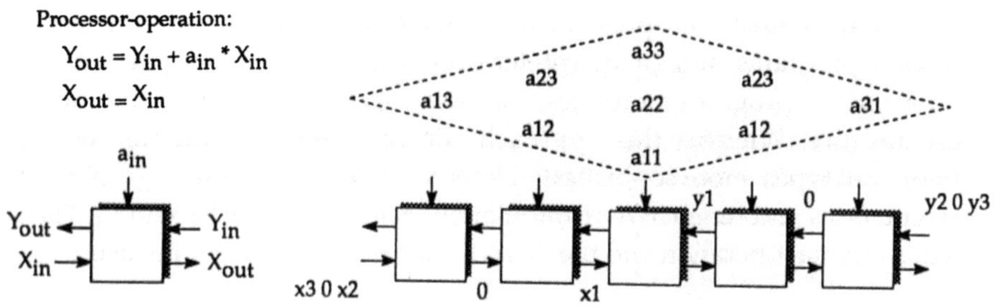

Figure 2.4.2 Linearly connected network for matrix-vector multiplication $\mathbf{y} = \mathbf{A} \cdot \mathbf{x}$ and streams of data.

ities originate in the specific and optimized use of data, vividly illustrated in [Kung 82]:

> *The function of the memory ... is analogous to that of the heart; it "pulses" data (instead of blood) through the array. The crux of this approach is to ensure that once an item is brought out from the memory it can be used effectively at each cell it passes while being "pumped" from cell to cell along the array.*

Therefore, simulation and visualization of data flow are desirable. CAST.LISAS visualizes the complex flow of data in networks of processing cells, where data elements have to be at the right place at the right time in the absence of global communication lines. It helps to design, analyze, verify, demonstrate, and understand highly parallel computations.

2.4.2 Main Features of CAST.LISAS

CAST.LISAS aims at combining the strengths of two published approaches on systolic array simulation:

- Sommer and Savage [Sommer, Savage 85] describe a simulator that provides a graphical environment in which systolic arrays can be monitored during simulation. This helps to understand the flow of data and the dynamical behavior of the investigated network. However, the simulator is restricted to arrays with completely homogeneous interconnection schemes, comprising only one cell type. These cells can be programmed to compute most arithmetic and logical functions. Symbolic simulation of systolic algorithms is not possible.

- Melhelm [Melhelm 85] presents a language for modelling systolic architectures as programs, and an interpreter to execute them. Running the interpreter on a certain program simulates the global (I/O) behavior of the modelled architecture. Whereas this approach allows to process complex networks (more cell types, more sophisticated interconnection structures), graphical representations and interactive manipulation facilities are not provided. The system's internal operation and the flow of data are hidden from the user.

CAST.LISAS has been designed to combine the following features: versatility, a graphical user interface, and performance.

Versatility.
One- and two-dimensional arrays of processing elements of nearly arbitrary size with linear, mesh, hexagonal, octagonal or mixed interconnection schemes can be simulated. Neither is the number of different cell types restricted, nor is the intercell communication limited to strict local and regular schemes. Non-local and torus-like links allow sophisticated networks to be modelled. Cells can be programmed to perform any desired operation, including string manipulation and symbolic computations. Using these building blocks, very general architectures can be modelled. CAST.LISAS is a powerful tool to model and simulate regular or partially regular networks of interconnected processing cells.

Graphical environment, convenient man-machine interface and high degree of interaction with the user.
CAST.LISAS provides a window-based graphical environment on the high-resolution display of the personal workstation Siemens 5815 (Xerox 1108). The simulator is implemented in InterLISP-D and LOOPS. CAST.LISAS exploits the extensive facilities of the InterLISP-D environment, yielding convenient means of communication between a user and the system. During simulation the user can view, control and manipulate a network's internal operation. He is able to monitor the parallel and pipelined flow of data and to verify that cells receive data as expected and produce correct outputs. This greatly simplifies the understanding of the overall network operation. Interactive modifications of data items let him imitate erroneous situations and analyze a system's reaction. Hence, CAST.LISAS is useful for finding design errors in systolic algorithms and systems and for demonstrating and teaching the systolic approach more effectively and vividly.

Performance.
Ease of use and wide applicability are of limited use, when the simulator is too slow. CAST.LISAS was designed and implemented to offer best performance possible on the Siemens (Xerox) hardware.

2.4.3 Cell Model and Activation Scheduling

During simulation, a network is driven by an imaginary clock. It should be noted that all cells operate synchronously since we adopted the processor model proposed in [Kung 80]. Actually, the clock is two-phased. In the first phase, the simulation scheduler lets all cells execute their programs virtually in parallel. The second phase is used by the scheduler to pass cell outputs to connected neighbors. Cells need not observe the availability of input data during execution phases as in asynchronous architectures: all values of transmission lines are set up and are fixed at the beginning of each clock tick and cannot be changed by cells during the calculation phase. Thereby, the cell's parallel operations are serialized for simulation in a deterministic way.

At first sight, this cell model conflicts with automata theory, where outputs of a finite state machine (Mealy-type) are supposed to be available at the same time the triggering inputs are provided. The two-phase clock of CAST.LISAS distributes calculation results during the second phase, when all calculations are finished. Outputs are propagated then and cannot be used by connected cells for calculations at the same clock-tick. However, this does not restrict the applicability of CAST.LISAS too much, since it is designed for the simulation of synchronous networks of processing elements, where connection lines are buffered anyhow.

2.4.4 Using CAST.LISAS

When CAST.LISAS is used to analyze or to develop a cellular network of finite state machines, three major tasks have to be performed:

1. specification,
2. simulation,
3. evaluation.

A simple, but illustrative systolic architecture for the matrix-vector multiplication problem is used to demonstrate both the complexity of the data flow and the usage of CAST.LISAS to analyze systolic architectures. The problem, the derived systolic algorithm, and architecture were introduced in [Kung 80].

The problem to be solved is the multiplication of a matrix $\mathbf{A} = [a_{ij}]$, $i, j = 0, 1, \ldots, N\text{-}1$, with a vector $\mathbf{x} = [x_0\, x_1 \ldots x_{N\text{-}1}]^T$, yielding a vector $\mathbf{y} = [y_0\, y_1 \ldots y_{N\text{-}1}]^T$:

$$\mathbf{y} = \mathbf{A} \cdot \mathbf{x}$$

The elements of the product **y** are describable as

$$y_j = \sum_{k=0}^{N-1} a_{jk} x_k,\ j = 0, 1, \ldots, N\text{-}1.$$

If the y_j are rewritten as

$$y_j = 0 + a_{j0}x_0 + a_{j1}x_1 + \ldots + a_{j(N-2)}x_{N-2} + a_{j(N-1)}x_{N-2},$$

the following recurrent formula can be used to specify a one dimensional systolic algorithm for the matrix-vector multiplication problem:

$$y_j^{(0)} = 0,$$

$$y_j^{(k+1)} = y_j^{(k)} + a_{jk}x_k,$$

$$y_j = y_j^{(N)},$$

$$j,k = 0, 1, \ldots, N\text{-}1.$$

Figure 2.4.2 (see earlier) shows a resulting systolic architecture for N=3.

2.4.4.1 Specification

During the specification phase, all static properties of cells and of a network are defined. Modelling a systolic architecture starts with the definition of all necessary cell types. This comprises the specification of unique names, the definition of communication ports, graphical appearances and cell operations.

A simple editor has been developed to handle the graphical aspects. Snapshots are given in Figure 2.4.3. After the shape and the size of a cell are selected, the local (regular) communication lines to/from up to its eight neighbors are specified. For that purpose, the user places the input/output ports at the cell's borderline. Thereby, the communication partner for each port is implicitly specified.

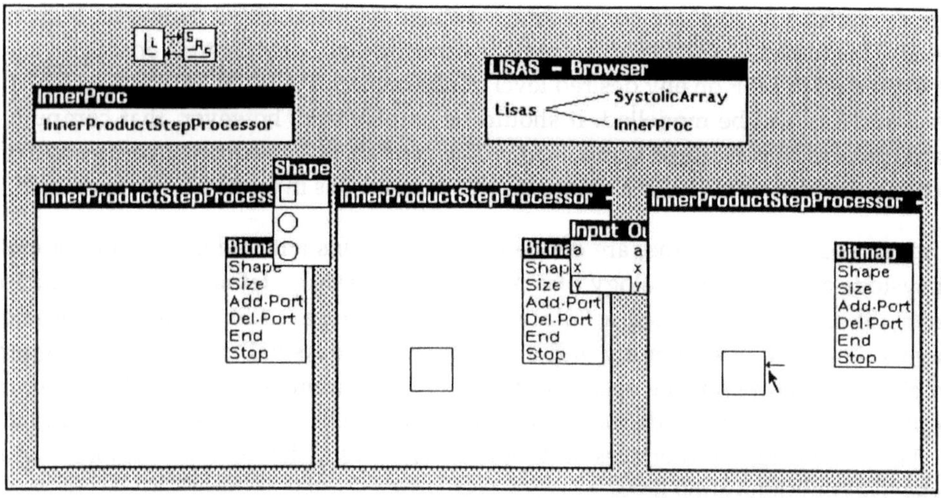

Figure 2.4.3 Definition of the graphical appearance and local interconnection scheme of *InnerProductStepProcessor*.

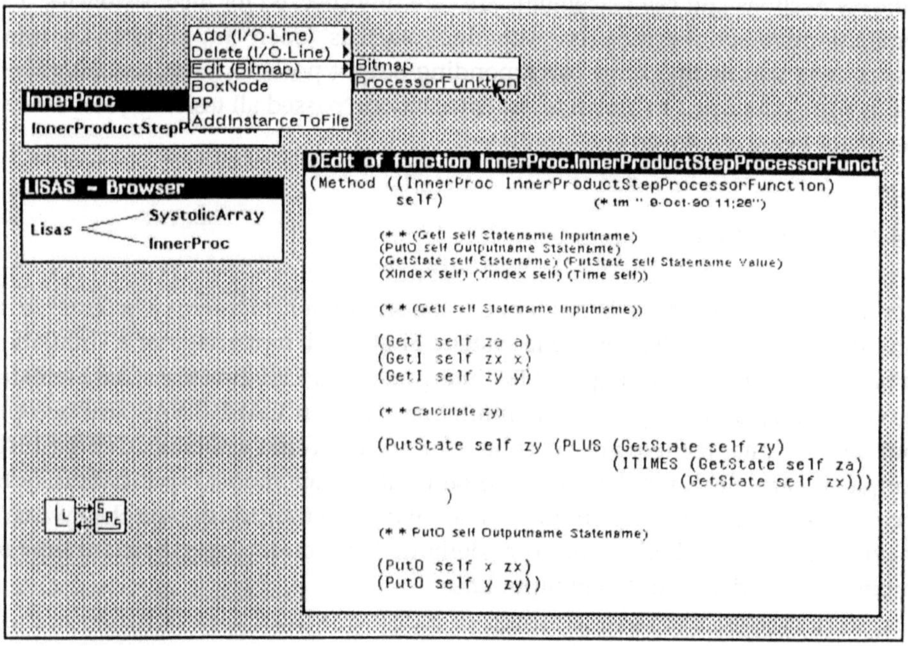

Figure 2.4.4 Definition of the cell operation of *InnerProductStepProcessor*.

The operations of cells are conveniently defined as InterLISP-D functions (Figure 2.4.4). This makes CAST.LISAS extremely versatile since any desired operation can be expressed in LISP on any desired level of abstraction. Typically, bit-level or word-level systems will be modelled. It should be emphasized, however, that computations can also be performed symbolically. This renders possible formal verification of systolic systems, since one can obtain output descriptions in terms of input symbols.

After all necessary cell-types are defined, they can be instantiated as building blocks of systolic arrays. The topology of a network is currently defined by means of an InterLISP-D function, called a "place-function" (Figure 2.4.5). Using mouse and menus would be user-friendly and straightforward for small systems. For bigger architectures, however, the specification would be a tremendous task. The current implementation offers predefined functions to place cells in a two-dimensional grid, to define the external interfaces and, optionally, to establish non-local (global) and *torus*-like communication lines.

Purely local and regular interconnections need not be specified since they are generated automatically from the definition of the individual cell and its actual neighbors within the system. When the array is actually set up according to this specification, all the connections and cell placements are carefully checked for inconsistencies. The user is informed about error-prone situations, such as "open" input ports, i.e. input ports of cells that neither have corresponding output ports in their local neighborhood nor are connected to global lines. After having passed all tests, a systolic array is ready for being displayed and simulated.

2.4.4.2 Simulation

After all static aspects have been defined, the simulation phase may be entered to monitor the dynamic behavior of the system. First of all, input data for a simulation run have to be provided. Using the LISP interpreter and LISP functions, data streams are comfortably defined prior to a simulation run. But CAST.LISAS also accepts input data on line. If there are no values for an input-line available, CAST.LISAS prompts at each clock tick for values to be used. In this mode, input data may be specified step by step during simulation, allowing the user to react on the system's behavior. Apparently, both modes (pre-definition and on-line specification) may be combined arbitrarily for different input lines.

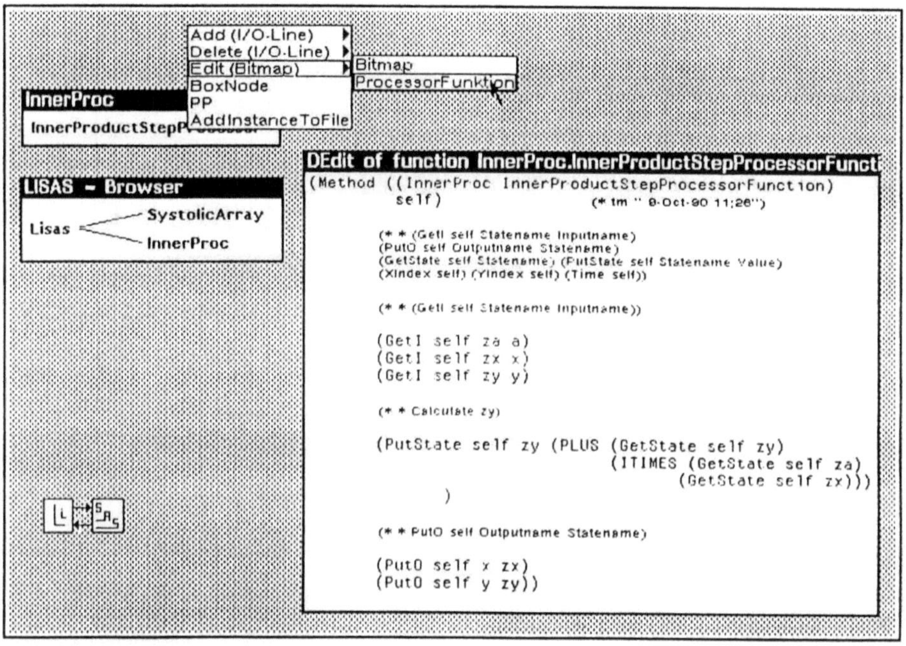

Figure 2.4.5 Placement function for the matrix-vector-multiplication system.

Systolic arrays can be displayed in different windows at different sizes. A scroll feature is implemented, enabling the user to display selected parts of a large network in a close-up view window. When several windows are used to monitor a simulation run, they can be utilized to display the state of the simulated array at different instances of time. Figure 2.4.6 shows two independent simulation windows, displaying two successive simulation ticks, thus illustrating the flow of data.

Time-connected windows are provided to display different regions of large architectures, when the whole system does not fit into a single window. From simulator's point of view those windows are treated as one entity. Hence, they will always display values of the same simulation time.

CAST.LISAS displays data values which are transmitted via communications links, or internal states of processors either permanently or temporarily. It is up to the user to specify values he is interested in, and to have them displayed. Beyond that, values of outputs and internal cell states can be changed interactively between clock ticks. Thus, the user can interactively arrange data flow patterns and manipulate results of cell operations. This, and the possibility to use symbolic cell operations (Figure 2.4.7) are further steps towards interactive design of systolic algorithms and architectures.

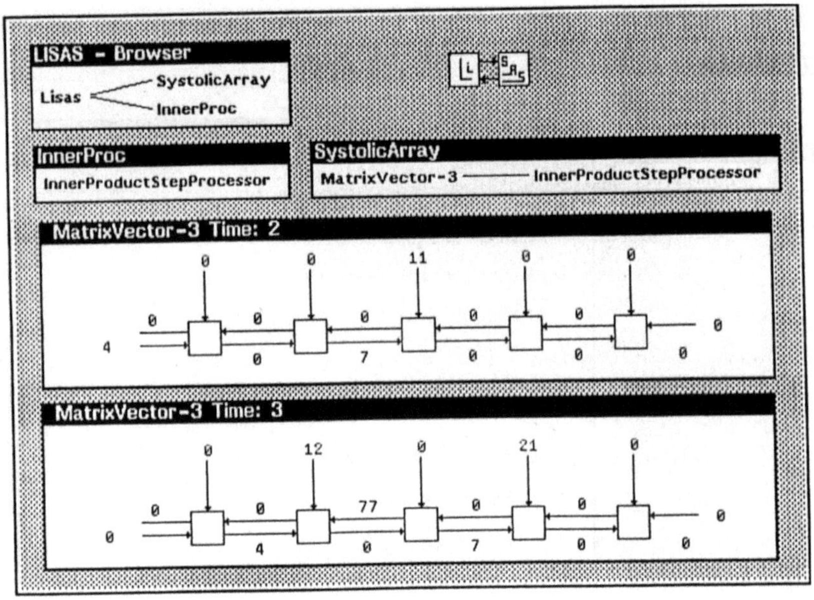

Figure 2.4.6 Two time-independent simulation windows.

Actual simulation runs may be done step by step, but simulation time may be increased by an arbitrary number of clock ticks without interruption, too. Stepwise simulation is suited for monitoring an array's internal operation, complete simulation runs to verify the I/O-behavior. In any case, values of output lines are gathered in lists for offline inspection or post-processing by means of user defined InterLISP-D functions.

2.4.4.3 Evaluation

If the main purpose of a simulation run is monitoring the flow of data using single step mode, the produced output data are less important. But if the simulation runs are intended to evaluate the array's response to certain input data, the LISP environment is very well suited to define and execute necessary post-processing.

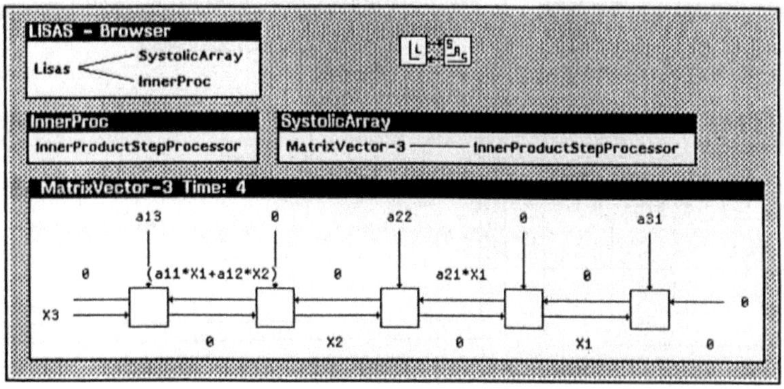

Figure 2.4.7 Symbolic simulation run for the matrix-vector multiplication problem.

2.4.5 Further Simulation Examples

2.4.5.1 LU-Decomposition

A graphical representation of a hex-connected two-dimensional array [Kung 80] is shown in Figure 2.4.8. Our array, however, is embellished in that the non-local output-lines have been eliminated, and distinct cell types are displayed differently. Cells on the left and upper boundaries are also inner-product-step processors, but their orientation is changed. Thus, in our model they differ from the standard inner-product-step processor.

2.4.5.2 Cellular Automata

Recently, CAST.LISAS was adapted to simulate cellular automata and to visualize state trajectories for one-dimensional and state snapshots for two-dimensional systems. Cellular automata and their applications are introduced and discussed in [Wolfram 86]. However, a very short and limited explanation of fundamentals will be necessary for our purpose.

Cellular automata contain a large number of simple identical components, arranged in one or two dimensions with local interconnections and a discrete state-variable at

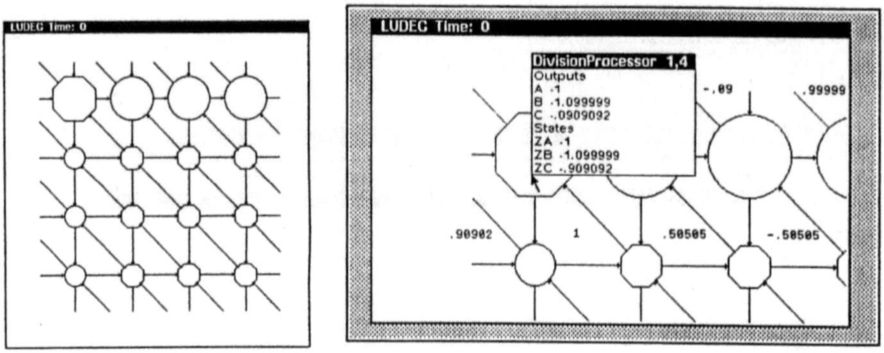

(a) (b)

Figure 2.4.8 *(a)* Hex-connected LU-decomposition array, *(b)* Values of cell (1,4).

each cell. A new cell state depends merely on the values of states in the neighborhood at the preceding time step. The main difference to systolic systems is the lack of data pipelining. There are no overall data streams pulsing through the system. In cellular automata, communication lines are only used to pass on a cell's current state to its neighbors.

The well known computer game "Life" is easily modelled as a two dimensional cellular automaton. However, one dimensional cellular automata with binary cell states are even primitive examples. A cell's next state is defined by an overall rule and is calculated using its own state and the states of both neighbors.

To avoid special border cells, CAST.LISAS facilities for torus-like networks are used: the cellular automaton is a closed ring of cells, lacking external communication lines. The purpose of simulation runs is monitoring the cell states and analyzing their trajectory in time. The visualization of the state trajectories, as it is done in [Wolfram 86], is even possible using CAST.LISAS now. Figure 2.4.9 shows the new types of simulation windows for trajectories and snapshots of binary cell states.

2.4.6 Experiences and Conclusion

LISAȘ was used to model and simulate most of the designs proposed in [Leiserson 81]. These designs can roughly be classified as follows: A first class includes Leiserson's semi-systolic designs and the corresponding purely systolic arrays which can

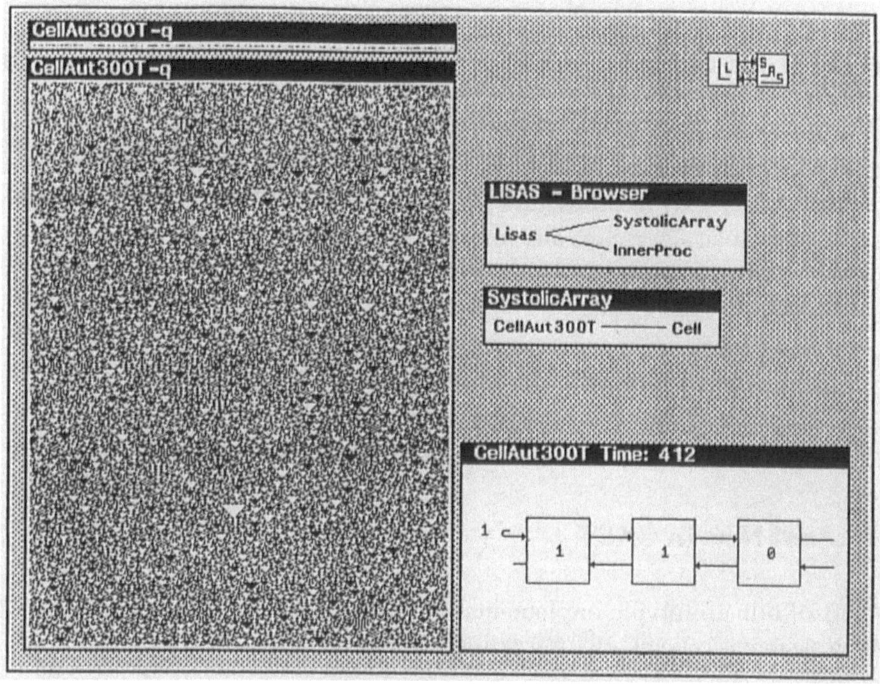

Figure 2.4.9 Linear system simulating a cellular automaton of 300 cells and evolving state trajectory (rule 150) after approximately 400 clock ticks.

be constructed by virtue of the "Systolic Conversion Lemma" [Leiserson 81]. Examples for designs of this class are systolic priority queues and counters. The common characteristics are simple topology (typically one cell type, simple interconnection schemes), but rather sophisticated processor functions. On the contrary, the second class of networks exhibits more complex data flow patterns and array topologies (more cell types); cells, however, are quite simple. This class comprises matrix computation and signal processing arrays.

When simulating architectures of the first category, the user is able to carefully examine cell operations; it is, then, easy to understand how the individual cells cooperate in order to realize the desired behavior of the entire system. In the second case, the key to understand a systolic algorithm is to follow the flow of data into and through the network.

In all cases, CAST.LISAS served as a valuable tool to gain a thorough understanding of the underlying algorithms. However, CAST.LISAS is even more useful in designing and testing new systolic algorithms than in studying existing ones. As an example, it was heavily employed in validating new systolic designs for the Generalized Discrete Fourier Transform [Hellwagner 89]. In summary, CAST.LISAS' modelling, simulation, and display facilities have proved useful for understanding the rhythmic dynamics of systolic algorithms and for investigating cellular automata.

2.5 CAST.PN – AN EDITOR AND SIMULATION TOOL FOR PETRI NETS

2.5.1 Petri Nets in CAST

One part of our prototypic implementation CAST.FSM deals with Petri nets. This software system is called CAST.PN [Mittelmann 88].

The Systems Theory knowledge is organized in CAST.FSM by means of the system type hierarchy, as explained in Section 2.2. The root is the most general kind of system, which specializes into five main system types, namely *black box*, *algorithm*, *generator* , *dynamic*, and *network* (see chapter 1 and section 2.1). These system types together with the transformations between them are implemented in CAST.FSM (as we have seen already) as the general framework. Each type and transformation has been specialized to more concrete kinds of systems several times and in different ways. The result is a tree-like graph with the general system as the root, and special types as the nodes, while the (directed) edges of the graph show the specialization relation.

One special system type is the Petri net [Brauer 80], especially suited for modelling of information flow, casual dependencies, and parallelism. Since its introduction by C.A. Petri, net theory greatly evolved. There are many abstractions and specializations of those original Petri nets. In the field of designing complex digital systems, two relevant net types are Condition/Event systems and Place/Transition nets [Reisig 86; Schwärtzel 81; Schwärtzel 85]. Both are implemented in the current version of CAST.PN. Petri nets are best modelled as systems of the general type generator, so the Petri net part of CAST makes up a subgraph of the generator branch of the system class hierarchy.

CAST.PN [Mittelmann 86] is a self-contained LOOPS program and as such, runs on any Dandelion (Siemens 5815) workstation with InterLISP-D/LOOPS version Harmony/Buttress or later [Bobrow, Stefik 83]. The CAST.PN program system uses the LOOPS class browser as the initial command center. It does not need any part of the CAST.FSM implementation. On the other hand CAST.PN is fully compatible with CAST.FSM and if both systems are loaded and used together, CAST.PN looks like an integral part of CAST.FSM.

2.5.2 Condition/Event Systems

An introduction to net theory may be found in [Reisig 86]. Here we will only declare those terms, which are essential for the following.

A *net* is a triple (S, T, F) consisting of the set S of *S-nodes*, the set T of *T-nodes*, and the relation F being a subset of $(S{\times}T) \cup (T{\times}S)$. S and T have to be disjoint non-empty sets. We alternatively could say, (S, T, F) is a bipartite graph with node sets S and T and the set of arcs F.

A *Condition/Event system* (S, T, F, C) is a net with the additional set C of cases. A *case* is a subset of S, denoting those S-nodes "carrying a token," using the metaphor that a S-node is some kind of place which is able to hold a thing, a token. The S-nodes of a Condition/Event system are called *conditions*, because they are well suited to model the state of Boolean condition: If the condition holds in a case, a token is placed on the node, otherwise if the condition fails at some point in time (a different case), the node carries no token. The T-nodes are called *events*.

Because each condition may have one or no token in any case, snapshots of the state of the net called *markings* are representable by subsets of S. The set of cases C is composed of some *initial* marking and all *reachable* markings of this initial marking under the used firing rule.

If a net N is in case c of C, an event e of T has *concession*, i.e. "is allowed to fire", if all preconditions are fulfilled and all postconditions are not fulfilled. By precondition of e we mean a condition with an outgoing arc leading directly to e. Similarly, a postcondition of e is a condition to which an arc leads from e. A condition is said to be fulfilled in c, if it carries a token, i.e. if it is an element of c.

If an event e has concession in the case c, it may *fire*. This means, it takes all tokens from its preconditions and places one token on each of its postconditions. The net is then in a new case c', which can be computed from c by removing the preconditions of e from c and then adding the postconditions of e to c. In this way the events model the local state change of the system.

A set of events with concession can make a *step*, i.e. can fire simultaneously, if they are pairwise independent, that is they have no preconditions or postconditions in common.

2.5.3 Place/Transition Nets

A *Place/Transition net* is a net together with an initial marking c, a capacity mapping *cap* and a weight mapping w. The S-nodes of this kind of net are called *places*, the T-nodes are called *transitions*. The capacity mapping *cap*: $S \rightarrow \mathbb{N} \cup \{\infty\}$ gives for each place p the maximal number of tokens it can hold, the *capacity* of p. The mapping w: $F \rightarrow \mathbb{N}$denotes for each arc how many tokens are to be transferred when firing. An arc with *weight* i may be seen as an abbreviation for i parallel arcs.

The term *concession* has to be defined according to these generalizations: An event e of a Place/Transition net has concession, if *(1)* for each preplace p of e, there are at least as many tokens on p as the weight of the arc (p,e) leading from p to e demands, and *(2)* for each postplace p of e, there are at least as many free capacity as the weight of the arc leading from e to p.

2.5.4 Implementing the Net Theory

Nets are composed objects. The incarnation and specialization dependencies of some general LOOPS objects as well as CAST.PN objects are visualized in the Figure 2.5.1. The predefined LOOPS mechanism for handling composed objects is not flexible enough for this application, so a new notion has been introduced. The general concept of metaclasses made this construction very simple. The incarnation of an object of a class C in LOOPS is done by sending the class C the message *New*. The class C itself is an incarnation of its metaclass, where the method *New* for this class is defined.

In our case, for creating a composed net object of the class *Condition/Event* system, we have to modify the inherited standard *New* method. We do this by specializing an appropriate metaclass, say, *PetriMetaClass* and specialize its inherited method *New* to create the main net object and the node objects, and set the related instance variables for proper interconnection.

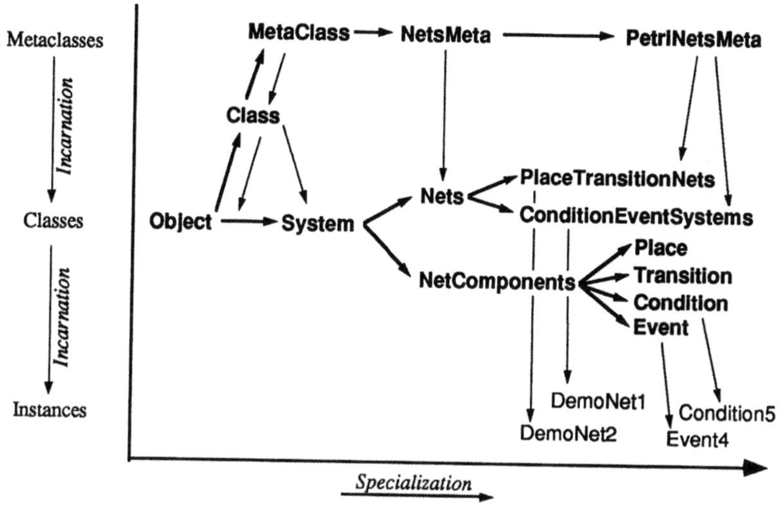

Figure 2.5.1 The LOOPS way of inheritance and incarnation.

The main classes are *Nets*, with specializations *ConditionEventSystems* and *Place-TransitionNets*, and *NetComponents*, with specializations *Condition, Place, Event* and *Transition*. Arcs of nets do not own any specific methods, so they are implemented as a list structure in a slot of the net object. If the arcs were to be selectable by the user interface, they should have been implemented as objects by a separate class.

Properties of the nets such as simplicity, purity, conservativeness, etc. and properties of their components such as having concession in the actual case translate directly into simple methods for the corresponding classes.

Simulation of a net means selecting events with concession to make up a step, and then firing the events of this step. The number of tokens of a S-node is reflected in one of its instance variables, and an instance variable of a T-node records wether the node is selected or not. Stepping the net is done simply by sending the message *Step* to the net, which in turn, after checking for concession and conflicts, fires each event in the step. As in any other part of CAST.FSM, a history list is maintained to make all state changes of the net undoable. Undoing steps is frequently used by a user checking out the effect of design changes, the main application of this CAST package.

Petri nets modeled in CAST.PN have an additional feature: By setting a certain flag (an instance variable of the net), the net may be set to be "active". Each T-node may have a prefire program and a postfire program stored in its corresponding instance variables. If the net is set active, and an event *e* is fired, the event *e* first calls its prefire program, if there it has one, then performs the firing by removing and adding the tokens, then calls its postfire program if this node has one. The fire programs may be arbitrary LISP (or LOOPS) programs called with the node as argument. Normally these programs will not be used to modify the net in any way, but to control some other system by side effects, e.g. by sending messages to another LOOPS object. In this way, CAST.PN Petri nets may be directly used to control other systems.

2.5.5 Implementing the User Interface

All parts of CAST.PN have two interfaces: a graphical user interface (GUI) and a "programmatic" interface. The latter allows any other InterLISP-D program, especially the other parts of CAST.FSM, to use the program modules to construct, analyze, and evaluate Petri nets.

In this section, we discuss the GUI.

The design goals were to make it

- easy to learn,
- easy to use,
- compatible with InterLISP-D, LOOPS, and CAST conventions,
- and to use object-oriented representations.

The primary user interface device in LOOPS (and hence in CAST) is the browser. A browser is an active graphical representation of objects and some dependencies between those objects. By clicking an object with the pointing device (mouse), the user gets a pop-up menu with the operations making sense for the selected object in the current situation.

This browser concept has been extended in CAST.PN. Nets are created by choosing a command from the usual LOOPS class inheritance browser. The result is a new window, were the nodes and arcs of the net are created and edited graphically by pointing and dragging with the mouse. The keyboard is used only for typing in the names of new objects. From the beginning, the net looks like the familiar Petri net textbook figures, so there is no need for the user to learn and type any new syntax for editing the nets. But this graph on the screen is not only the output medium. Like any browser, it is also the input device for all user input. The net as a whole, as well

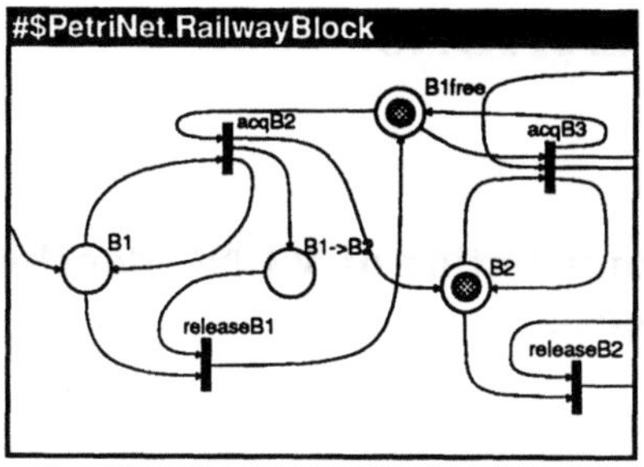

Figure 2.5.2 Petri net example screen window.

as the components (nodes) may be selected for command menus. By only using the mouse, one can

- change the graphical layout of the net;
- change the structure of the net by creating or deleting nodes or arc's;
- create copies of the net or transform it into a different type;
- activate the net, edit the firing functions;
- change the marking (number of tokens on S-nodes);
- let T-nodes with concession be highlighted;
- select T-nodes to make up a step;
- perform stepping and undo stepping (any depth);
- run the net with randomly created steps or other firing rules.

As stated above, the keyboard is only used for typing names for new objects. One may have multiple nets in several windows at the same time, but each net may have only one graphical representation open at any time. Shrinking a net window turns the window into a small icon on the screen and converts the graphical data to a suitable format for easy backup and restore.

2.6 CAST APPLICATIONS

2.6.1 Algebraic Decomposition of MCNC Benchmark FSMs for Logic Synthesis

2.6.1.1 Introduction

Logic synthesis systems which use a high-level hardware description language for design specification guarantee correct results, but limit the designers control of the final design. In a completely closed system the designer can either alter the high-level input specification or modify the synthesized design to achieve alternatives. The first approach requires deep understanding of the system's internal algorithms to control the final effects. Even worse, changing the final design without any support does not longer guarantee the correctness of the achieved design. The theory-based operations of a CAST method bank which are applicable interactively within the synthesis process help to preserve the design's correctness and help to work out design alternatives.

In the field of controller synthesis, decomposition efforts have considerable optimization potential. They are motivated by reducing the total area of hardware implementations and by more flexibility in floor-planning as the component machines are smaller than the original machines, even if the total area increases. Shortening the critical paths resulting in higher clock rates of the submachines and improving the testability by a reduced numbers of test vectors are further goals of decomposition efforts. Finally, the heuristics for state assignment and logic optimization work more efficiently on reduced problem sizes.

In CAST.FSM Algebraic Structure Theory [Hartmanis, Stearns 66] is used to derive possible decompositions. State assignment is not done for the original machine resulting in a decomposed structure, but the decomposition is done symbolically and state assignment is applied to the factor machines. Finally, modern heuristics for state assignment and logic minimization of the components target two-level or multi-level implementations.

Though Algebraic Structure Theory is referenced again in recent publications [Ashar et al. 89; Ashar et al. 90; Saucier, Poirot 89], no practical results of its applicability on industrial controllers or benchmark FSMs [Lisanke 88] have been presented. However, we demonstrate its applicability and its successful utilization for controller synthesis tasks.

Unfortunately, it was impossible to handle complex FSM objects by means of the first implementation of CAST.FSM because of the workstation's restricted memory size and slow performance. Thus, selected algorithms have been ported to Common LISP/CLOS/C on Apollo workstations, yielding faster calculations, but loosing the graphical and user-friendly interface. Instead, a basic Common LISP window toolkit based upon the Apollo window system was used (see Section 4.3) that provides enough functionality to display graphics and handle mouse activities for man-machine interaction. The flexibility in choosing and performing operations on FSMs is preserved by means of the LISP interpreter. A predefined set of operations is provided. However, its functionality is easily extended or modified: New operations, evaluations, or heuristic search algorithms can be specified, without the necessity of changing existing LISP functions.

2.6.1.2 Related Work on Controller Decomposition

In the past few years several approaches for FSM decomposition have been proposed. In [Devadas 88] a general form of decomposition is presented that is based on identifying subroutines of the original machine. Two algorithms, one optimum, one heuristic, for a general decomposition into dependent submachines are proposed in [Ashar et al. 89] minimizing the sum of the number of product terms in the one-hot coded and logic minimized submachines. Unified algorithms for decomposition and re-decomposition allowing a uniform treatment of different decomposition topologies are described in [Ashar et al. 90]. These approaches do not use the number of states as cost-function but consider the special target implementation, e.g., for two-level implementations the number of product terms is taken into account.

2.6.1.3 Preliminaries

Before the general procedure for our decomposition approach is discussed, some formal definitions are given. A finite state machine (FSM) is a quintupel $M = (I, O, S, \delta, \lambda)$, where I, O, and S are a finite set of inputs, outputs and states respectively, where δ is the state transition function, which maps $S \times I$ into S, and where λ is the output function, mapping $S \times I$ into O.

A *partition* π on S is a set of disjoint subsets S_i whose set union is S. The S_i are called the *blocks* of π. A partition π on the set of states S is said to have *substitution property* (SP) iff $s \equiv t(\pi) \Rightarrow \delta(s,i) \equiv \delta(t,i)(\pi)$ for all and $s, t \in S$. The terms SP-partition, closed partition, or congruence relation are used synonymously. A partition π on the set of states S is said to be *output-consistent* if, for every input and every block of π, all states of a block have the same/compatible outputs.

The set of all SP-partitions Q and the canonical addition and multiplication form a *lattice* $L(M)$. A *Hasse-diagram* illustrates the implied partial ordering relation on Q.

The greatest output-consistent partition of $L(M)$ is the *reduction partition* π_r.

The machine $M/_\pi = (I, S/_\pi, S/_\pi, \delta/_\pi, \lambda/_\pi)$ is called the *π-factor machine* of M, where $S/_\pi$ is the quotient set of S induced by π, where $\delta/_\pi$ maps $S/_\pi \times I$ into $S/_\pi$ such that $\delta/_\pi([s]_\pi, i) = [\delta(s,i)]_\pi$, and $\lambda/_\pi$ maps $S/_\pi \times I$ into $S/_\pi$ with $\lambda/_\pi([s]_\pi, i) = [s]_\pi$.

The serial and parallel decomposition used in our approach is based upon a composition of π-factor machines as shown in Figure 2.6.1. π_1, π_2, π, τ are partitions; π_1, π_2, and π must have substitution property where $\pi_1 \cdot \pi_2 \leq \pi_r$ and $\pi \cdot \tau \leq \pi_r$. For a thorough theoretical discussion we refer to [Hartmanis, Stearns 66].

2.6.1.4 FSM Decomposition

Decomposition structures with loop-free interconnections of the component machines can be classified in parallel, serial (or cascade) and cross (general) decompositions. Parallel and serial decompositions of FSMs are defined using the lattice of closed partitions on the set of states (see Section 2.2). The more general form of cross decomposition is handled by a partition pair algebra, but is not treated here. The proposed procedure for deriving FSM decompositions is outlined as follows:

1. *Computation of the lattice L(M) of the original machine M*

   In Figure 2.6.2a the Hasse diagram of the lattice of the MCNC benchmark FSM *tbk* is shown. One can derive from the graph whether the machine is state reduced and what factor machines exist for decomposition. The nodes represent SP-partitions, labeled with the number of blocks and the size of the largest block. The greatest partition contains just a single block, the least partition contains as many blocks as elements. The edges of the graph denote an ordering relation: Two nodes p, q are connected by an edge if p is smaller than q and

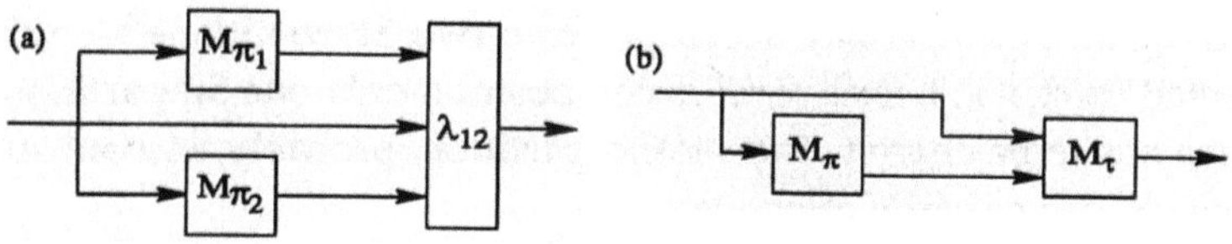

Figure 2.6.1 Parallel *(a)* and serial *(b)* decomposition.

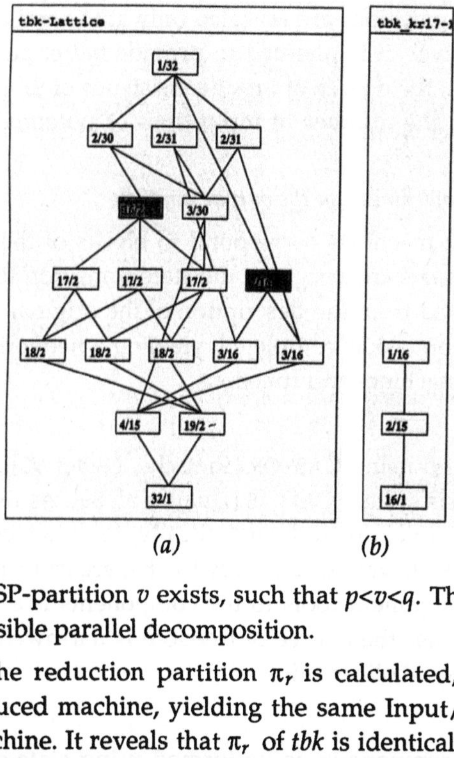

(a) *(b)*

Figure 2.6.2 Lattice of *tbk (a)* and state reduced *tbk (b)*.

no SP-partition v exists, such that $p<v<q$. The gray shaded nodes represent a possible parallel decomposition.

If the reduction partition π_r is calculated, it can be used to build a state reduced machine, yielding the same Input/Output behavior as the original machine. It reveals that π_r of *tbk* is identical with one SP-partition marked in Figure 2.6.2a for the parallel decomposition (16/2). Obviously, the sub-machine with 2 states is obsolete. It only separates redundant states of the original machine. Figure 2.6.2b shows the lattice of the reduced machine. Without further efforts (e.g., state splitting [Booth 67]) no parallel decomposition is possible, anymore. However, the existence of a non-trivial SP-partition allows a further serial decomposition of the reduced machine.

2. *Selection of partitions, defining the component machines*

Here we are concerned with the problem what the "best" partitions for a decomposition will be. The selection of partitions is supported interactively by

CAST.FSM. For a *parallel* decomposition two SP-partitions π_1, π_2 have to be selected ($\pi_1 \cdot \pi_2 \leq \pi_r$), for a *serial* decomposition only one SP-partition from the lattice has to be chosen. The second partition, probably without substitution property, can be easily calculated.

A decomposition with a minimal sum of states, i.e., a minimal sum of blocks tends to have a good realization with small area, but one should consider the number of bits necessary for binary encodings, too. Further evaluation criteria for possible decompositions are currently not available at this stage. The quality and suitability of selected partitions are revealed only at the end of the implementation process. However, it is planned to provide better support for human decision making. E.g., the degree of input-consistence of SP-partitions will be calculated, denoting the number of input-lines of potential submachines.

3. *Computation of the component machines and the output function*

   The states of the component machines correspond to blocks of the selected partitions. After symbolic names are assigned, the state transition tables and output functions are computed from the description of the original machine. At this time the decomposition task is completed, yielding a network of symbolically interconnected submachines and functions.

4. *State-Assignment*

   State assignment is performed using CASTOR [Rietsche, Neher 90], which is part of the high-level synthesis system CALLAS [Duzy et al. 89]. As the flow of data and component states are expressed symbolically, tools which exploit symbolic input, output, and state values [Lin, Newton 89] are fruitful. The sequence of applying state assignment tools to the components is a potential optimization parameter because the degree of freedom is narrowed sequentially by assigning values to symbolic entities.

5. *Logic optimization*

   Logic optimization for the components is performed using state-of-the-art tools ESPRESSO and MIS [Brayton et al. 87b, 87b].

2.6.1.5 Implementation Aspects

The essence of the development and evaluation of serial or parallel decompositions, as depicted in Section 2.6.1.4, is the knowledge of favorable SP-partitions. The general procedure to generate the lattice $L(M)$, given a sequential machine M, was taken from [Hartmanis, Stearns 66] and involves two major steps:

1. *Generation of all basic partitions.* A basic partition is the least closed partition, with a certain pair of states s and t being in a common block. To determine basic partitions of $L(M)$ an implication graph $G(M)$ is used. The nodes in this directed graph represent all possible sets of two states $\{s, t\}$, the arcs describe transitions from nodes $\{s, t\}$ to nodes $\{\delta(s,i), \delta(t,i)\}$ for all $i \in I$. To calculate a least SP-partition, starting from a node $\{s, t\}$, all reachable nodes are traversed and each pair of corresponding next-states is put into a common block of the resulting partition.
2. *Detection of all possible sums of SP-partitions.* Instead of calculating all possible sums of partitions, a level based algorithm is used. The set of basic partitions represents level 1. Each level i+1 consists of all sums of level i. The level based summation is iterated until no new SP-partition of $L(M)$ is found. The correctness of the algorithm was proven by Foster [Foster 88]. However, the computational complexity of the algorithm and the size of necessary internal data structures may limit the application for large FSMs.

The detection of all nontrivial SP-partitions of a FSM is often regarded as an exhaustive and tedious process. But the necessary computational effort depends on the cardinalities of S and I and in particular on the underlying state transition function δ. Very small machines may lead to an incomputable amount of partitions, whereas reasonably large controller machines have computationally tractable lattices. The worst case occurs when the state transition function maps all states into one constant next state, describing a machine with no structure at all. Thus, the arising number of closed partitions equals the number of possible partitions of S and can be expressed by means of the *Stirling Numbers of the Second Kind* $S_n^{(m)}$:

$$Size(n) = \sum_{m=1}^{n} S_n^{(m)} = \sum_{m=1}^{n} \frac{1}{m!} \sum_{k=0}^{m} (-1)^{m-k} \binom{m}{k} k^n$$

Some illustrative numbers are given in Table 2.6.1:

Table 2.6.1 Number of possible partitions of a set of n elements.

n	Size(n)
4	15
8	4140
12	4.213.597
50	1,857 * 10^{48}

The investigation of such machines is practically not relevant but illustrates the maximum (intractable) complexity of lattices of SP-partitions. If it is not possible to calculate the whole lattice we propose another strategy: As it is not necessary to know all possible decomposition we need not calculate the complete lattice. A initial step is the computation of all basic SP-partitions, followed by heuristically (interactively) guided additions to derive possible decompositions.

Not only the number of partitions, but also the size of manageable machines impacts software implementation decisions: For some FSMs (e.g., *scf*) of the MCNC benchmark set it is neither possible to allocate memory for a full state table nor is it possible to perform an exhaustive iteration over all inputs. To avoid huge state transition tables, the external KISS-style representation is used as internal FSM representation, too. Thereby, the amount of data is reduced, but different manipulation algorithms are required. The generation of the implication graphs works on pairs of states (s, t) which have to be manipulated simultaneously for all $i \in I$. As the state transition of s and t may be defined by different input-cubes denoting different set of actual input-values, products with current states s and t cannot be arbitrarily combined to calculate $\delta(s,i)$ and $\delta(t,i)$: Pairs of successor states would be generated for which no actual input i exists. However, the following algorithm (in Common LISP) solves that problem and iterates over all inputs:

```
(dolist (product-s (get-product-list fsm s))
  (dolist (product-t (get-product-list fsm t))
    (when (compatible-p
              (get-input-descriptor product-s)
              (get-input-descriptor product-t))
      ;actual calculations
  )))
```

The predicate `compatible-p` takes two input-cubes and checks within $O(n)$ time, if there is a non-empty intersection. If the state transition function is completely specified, it can be shown, that each input-value is treated at least once by the algorithm, as it is by the conventional iteration over all inputs [Müller-W. 91].

2.6.1.6 Experimental Results

The decomposition results compare favorably to related work in [Ashar et al. 89; Ashar et al. 90; Foster 88]. Table 2.6.2 gives an overview of FSM-sizes, CPU-runtimes for the lattice calculations using an Apollo DN3500 workstation, numbers of SP-partitions, and numbers of possible parallel/serial decompositions. Some examples of

Table 2.6.2 Possible decompositions of MCNC FSMs.

	lumped FSM							reduced FSM		
Name	i/o/s	lat (sec)		par.	parallel: s/s	ser.	serial: s/s	s	par	ser.
bbara	4/2/10	2,6	7	1	4/4	5	2/9 3/8 4/7	7	1	2
bbsse-e	7/7/16	15,9	35	110	2/14 3/14 5/12	33	2/13 3/10 4/10	14	1	4
dk16	2/3/27	1380,0	586	7984	17/26 18/24 20/23	584	2/19 3/18 4/16	-	-	-
dk17	2/3/8	1,8	2	0	–	0	–	-	-	-
dk512	1/3/15	13,6	46	124	2/14 3/14 12/12	44	2/9 2/11 3/8	-	-	-
ex2-e	2/2/19	5,2	5	3	10/10	3	2/10 10/10	10	0	0
opus	5/6/10	3,1	4	0	–	2	2/9	9	0	2
planet	7/19/48	473,1	19	10	43/46 43/47 44/46	17	2/41 2/42 2/43	-	-	-
s1	8/6/20	22,7	10	4	18/18 18/19 19/19	8	2/16 2/17 2/18	-	-	-
shiftreg	1/1/8	5,4	30	47	4/7 5/5 5/6	28	2/4 3/4 4/4	-	-	-
tbk	6/3/32	49,5	18	53	2/16 3/16 4/16	16	2/16 3/16 4/15	16	0	1

parallel/serial decompositions are given, too. If a FSM is reducible, the last three columns show the resulting size and the number of decompositions after state reduction.

A two-level (Table 2.6.3) and a multi-level (Table 2.6.4) realization were done for some FSMs. State assignment of component machines was performed using CASTOR [Rietsche, Neher 90]. For multi-level implementations JEDI 1.1 [Lin, Newton 89] was used. Logic optimization was achieved by means of ESPRESSO [Brayton et al. 87a] followed by MIS 2.1 [Brayton et al. 87b] for multi-level implementations.

As the presented approach for the calculation of lattices of SP-partitions is restricted to FSMs with completely specified state transition functions, we had to do a simple completion for some benchmark FSMs. If only few state-transitions were missing, we added $\delta(q,i) = q$, i.e., "next-state = present state" and extended the name of the FSM by "-e".

Most of the lattice calculations succeeded within some seconds, *planet* and *dk16* needed minutes, and only *scf* did not finish within a day. However, *scf* was successfully reduced to 97 states, by a rapid prototyped algorithm: First, only the basic SP-partitions were calculated. By a repeated random depth-first addition of output-consistent SP-partitions a result with 97 blocks was found within few minutes. Thus, a reduced machine with 97 states could be specified.

Table 2.6.3 Two-level implementations of MCNC FSMs.

	FSM		decomposed FSM			
Name	s	Area	Applied Operations	s/s	Comp. Area	Area
bbara	10	550	reduction	7		399
			reduction, parallel decomp.	4/4	126/112/108	346
			reduction, serial decomp.	4/4	112/280	392
bbsse-e	16	1056	reduction	14		990
			parallel decomposition	2/14	5/552/522	1079
ex2-e	19	798	reduction	10		324
			parallel decomposition	10/10	272/256/36	564
modulo-12	12	195	parallel decomposition	3/4	32/40/0	72
scf	121		reduction	97		
planet	48	4641	parallel decomposition	43/46	1920/2176/3249	7345
			serial decomposition	42/2	2080/3360	5440
tbk	32	1440	par. decomp., random assign.	2/16	27/960/391	1378

1) Area of M_π / M_τ for serial decompositions, $M_{\pi 1}/M_{\pi 2}/\lambda_{12}$ for parallel decompositions.

Table 2.6.4 Multi-level implementations of MCNC FSMs.

	lumped FSM			decomposed FSM				
Name	s	Lit.	Area	Applied Operations	s/s	Comp. Lit.[1]	Tot. Lit.	Tot. Area
bbara	10	71	59392	reduct., serial decomp	4/4	17/26	43	40368
bbsse-e	16	120	103936	parallel decomp.	5/12	16/63/55	134	122032
ex2-e	19	144	117856	parallel decomp.	10/10	58/58/6	122	102544
				parallel decomp.	10/10	55/61/5	121	105328
modulo-12	12	33	27840	parallel decomp.	3/4	9/8/0	17	15312
planet	48	552	471888	serial decomp.	41/2	235/57/315	607	511792
tbk	32	219	169824	reduct.,serial decomp.	2/15	10/175	185	164256

1) Literal count (factored form) of M_π / M_τ for serial decompositions, $M_{\pi 1}/M_{\pi 2}/\lambda_{12}$ for parallel decompositions.

For 4 of 7 examples the decomposition renders a smaller area than the lumped machine for two-level implementations (4%–63%). For all examples, the component machines are smaller than the original machine. For 4 of 6 examples the multi-level implementation of the decomposition is better than the original machine: 16%–48% fewer literals and 3%–45% smaller area (using the library *lib2* from [Lisanke 88] for library mapping). The results in terms of literal count are in the factored form. The critical path of the component machines was shorter than for the lumped machine.

2.6.1.7 Conclusion

A decomposition approach for controller synthesis based on algebraic structure theory, utilizing state-of-the-art tools for state assignment and logic optimization, has been presented. The theory based interaction of the method bank CAST.FSM achieves flexibility within a closed logic synthesis process, preserving the guaranteed correctness of designs. Experimental results have been presented for some MCNC benchmark FSMs. Future work will be the investigation of evaluation criteria and heuristics for choosing good SP-partitions, as well as adapting theoretical work for a computational treatment of incompletely specified machines within CAST.FSM.

2.6.2 CAST.FSM Applications in Cryptology

The advantage of using cellular machines which are realized by systolic arrays in the field of signal processing is advertised in technical literature with the following arguments: The hardware design of machines with cellular architecture is easy and fast; the hardware overhead to reach their testability is minimal. The algorithms and, consequently, the software which can be implemented on such machines allow a high degree of parallelism and therefore realize effective computational methods. Such properties have led to practical success in signal processing applications [Kung 82]. Of special interest is the realization of signal processing operation of matrix-type, e.g. the discrete Fourier transform [Hellwagner 88; Beth 84]. A further example for the use of cellular machines in signal processing is given by the 1-D binary cellular finite state machines proposed by [Wolfram 86]. The unfolding of the state-process of such machines (of relatively simple construction but highly non-linear state transition) may generate 2-D patterns of a complex kind.

In this section we want to deal with the design of a cellular machine *CM* for application in cryptography. The individual cells of *CM* are specified as finite state machines. The topology of the cellular network is that of a grid, such that a regular cell has four neighbors.

The cellular machine *CM* which is to be designed should realize a one-way function f of high cryptological quality for any initial state. It should transform a binary input-vector $\mathbf{x} = (x_1,x_2,..,x_n)$ of known structure into an output-vector $\mathbf{y} = f(\mathbf{x})$, $\mathbf{y} = (y_1,y_2,...,y_n)$ which has no structure, such that it is computationally very hard to compute $\mathbf{x}$ from $\mathbf{y}$.

For modelling of the cellular machine *CM* we will take the following approach: We use CAST.FSM to find proper finite state machines $M(i,j)$ for the cells of *CM*. For model-application we depend on simulation experiments performed with CAST.LISAS.

2.6.2.1 The Basic Cellular Architecture

The basic goal is the design of a one-way cipher operation f which realizes a Boolean function with 16 input-lines and 16 output-lines. The inner architecture of f should be realized by a 2-D cellular network of 16 x 8 cells which are specified as binary finite state machines. For each cell machine $C(i,j)$, $i = 1,2,...,16$; $j = 1,2,...,8$, we assume to have exactly 32 inner states. Therefore a 5-bit register is needed to realize the memory of each $C(i,j)$.

In Figure 2.6.3 we see the architecture of the cellular network *CN*. Figure 2.6.4 shows the internal structure of each cell C(i,j). The pair (a_1,a_2) consisting of the horizontal and vertical binary input a_1 and a_2, respectively, is the input for an EXOR gate. The result $a_1 \oplus a_2$ is the input of a finite state machine $M(i,j)$ with initial state $q^{(0)}(i,j)$. The binary output of $M(i,j)$ is delayed for a time-unit and then branches to define the horizontal and vertical outputs b_1 and b_2, respectively.

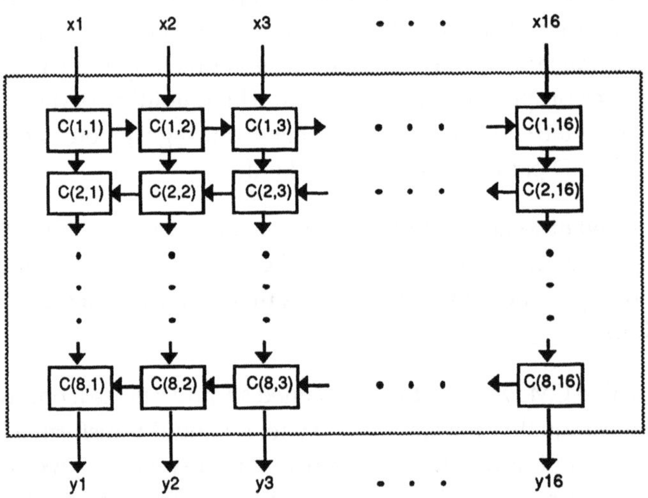

Figure 2.6.3 Basic Architecture of *CN*.

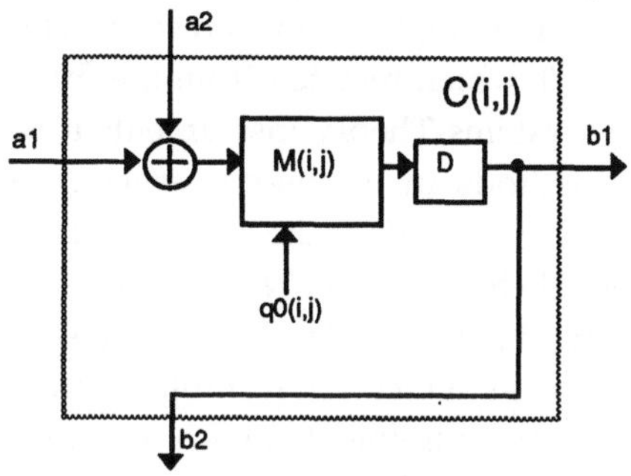

Figure 2.6.4 Internal structure of *C(i,j)*.

2.6.2.2 Functional Requirements

The cellular network *CN* should be designed in such a way, that it realizes (as it has been pointed out earlier) a cryptographically strong one-way function *f*. To achieve this, *f* has to fulfill the following necessary requirements:

1. The computation of $\mathbf{y} = f(\mathbf{x})$ from the value $\mathbf{x}$ should be "fast".
2. The determination of the value $\mathbf{x}$ from $\mathbf{y} = f(\mathbf{x})$ should be computationally "extremely difficult".
3. The property (2) of *f* is extended by the requirement that the determination of any known property $P(\mathbf{x})$ of $\mathbf{x}$ from the given value $\mathbf{y} = f(\mathbf{x})$ should also be computationally "extremely difficult".

In requirements (2) and (3), which the one-way function *f* has to fulfill, we assume that the construction of the cellular network *CN* is completely known and that this knowledge can be used to develop a computational method for the determination of $f(\mathbf{x})$ and $P(\mathbf{x})$, respectively. What is assumed to be not known, however, is the initial state $q^{(0)}$ of *CN*.

The design of *CN* can be divided into two main phases. The first phase deals with the design of the individual finite state machines *M(i,j)* which are used in each cell *C(i,j)*. The proper selection of the finite state machines *M(i,j)* such that the requirements (1) to (3) are fulfilled is crucial. With the help of FSM theory, a systematic search to find suitable FSMs to realize *CN* can be done. We will use CAST.FSM as a tool for this task.

The second phase in the design of *CN* is devoted to the construction of the cellular network using the cells *C(i,j)*, which resulted from the first phase. In order that this design gets a support by Systems Theory instruments it is necessary to be able to apply Systems Theory methods in model-construction and model-application. Depending on the kind of FSMs found for the construction of the cell machines *C(i,j)*, it will be more or less possible to evaluate *CN* (or substantial parts of it) with respect to requirements of the kind (1) to (3). Again, if we are able to associate corresponding problems to these requirements which are formulated in terms of FSM theory, we can use CAST.FSM to assure the fulfillment of such requirements. In addition, we are also able to use CAST.LISAS to simulate the I/O-behavior of *CN* and to evaluate the one-way function *f* by means of simulation experiments.

2.6.2.3 Design of *M(i,j)* by CAST.FSM

The following properties (P1)–(P6) of a FSM seem to be important for the evaluation of cell machines *M(i,j)* for cells *C(i,j)*:

(P1) degree of invertibility,

(P2) linear complexity,

(P3) existence of equivalent states,

(P4) finite memory property,

(P5) decomposability,

(P6) state graph complexity.

(P1) deals with the question, in what degree it is possible to construct to *M(i,j)* an inverse machine $(M(i,j))^{-1}$ which inverts the I/O operation. In order to certainly meet (P2) and (P3) it is desirable that *M(i,j)* (or parts of it) are not invertible. CAST.FSM allows the evaluation of a given FSM with respect to inversion.

(P2) is concerned with the question of the possibility of constructing a linear finite state machine *LM(i,j)* which is equivalent to *M(i,j)*. The dimension of the state space of *LM(i,j)* is called the "linear complexity"of *M(i,j)*. Certainly, to meet requirements of the kind (3) it is desirable that the linear complexity of *M(i,j)* is very large. Then, even when a property *P*(**x**) of **x** could be derived from *f*(**x**) by using the existing theory of linear FSM's, the large dimension of state space ensures that the actual computation of *P*(**x**) is not feasible.

By (P3) we evaluate the cell FSMs *M(i,j)* with respect of the existence of equivalent states, which are, from the cryptological point of view, not desirable. Efficient methods to check if a given FSM is reduced, that is, no equivalent states exist, are available.

The finite memory property (P4) of a FSM deals with the question whether one is able to uniquely determine the internal state of the FSM by using an I/O-experiment of certain length. Since the states of the cell machines $M(i,j)$ contain information on the used *key* (= initial state of *CN*) it is desirable that they have no finite memory.

Generally speaking, decomposition methods allow the construction of *divide-and-conquer* methods in problem solving. In case of the cell machines $M(i,j)$ it is desirable that they are not decomposable. FSM theory offers methods to check the degree of decomposability. Such methods allow the evaluation of $M(i,j)$ – and in consequence of $C(i,j)$ – with respect to property (P5).

The structure of a FSM is reflected by the graph-theoretical structure of the associated state graph. For example, linear machines have a very regular structure, where cycles of certain length are dominant. We consider the state graph of a FSM as being more complex the less it contains cycles and the more it looks tree-like. With (P6), the structure of the state graph of $M(i,j)$ (assuming constant input) is evaluated in that sense.

The CAST method bank system CAST.FSM contains model-transformations such that all properties (P1) to (P6) of a FSM can be checked easily. Furthermore, by its interactive use it is possible to develop modelling algorithm to find FSMs which fulfill the properties (P1) to (P6) to a degree, such that they qualify as a cell machine $M(i,j)$.

It should be quite clear to the reader that such a search needs computer assistance. But that is not enough it needs the instrumented assistance as it is provided by the CAST tool CAST.FSM.

2.6.2.4 Simulation experiments with CAST.LISAS

To ensure the fulfillment of requirements of the kind (1), (2), and (3) it is necessary to evaluate the final design of the cellular network *CN* by I/O simulation experiments. The goal in this kind of model-application is to ensure that requirements of the kind (1) to (3) which we formulated for the one-way function f are met.

As a CAST tool for the implementation of the simulator of *CN* and for "running" the simulation experiments, we can use CAST.LISAS. Since the input $\mathbf{x}$ and also the output $\mathbf{y}$ of f are elements of the set $\mathbb{B}^{16}$ ($\mathbb{B} = \{0,1\}$) and the set of initial states $q^{(0)}$ of *CN* is $\mathbb{B}^{640}$ ($640 = 5 \times 128$), it follows that any I/O-experiment of f can be constructed by relations R from $\mathbb{B}^{16}$ to $\mathbb{B}^{16}$ which are parameterized by binary numbers from $\mathbb{B}^{640}$. It is a matter of cryptological knowledge to select from the many possible relations of the kind R the proper one on which the simulation experiments should be based. The following metric distances between **x** and **y** are of special interest in this context:

(H1) the Hamming distance $d_{H1}(\mathbf{x},\mathbf{y})$ between **x** and **y**;

(H2) the Hamming distance $d_{H2}(\mathbf{y},\mathbf{y}')$, where $\mathbf{y} = f(\mathbf{x})$ and $\mathbf{y}' = f(\mathbf{x}')$ with $d_H(\mathbf{x},\mathbf{x}') = 1$;

(H3) the Hamming distance $d_{H3}(\mathbf{y},\mathbf{y}^*)$, where $\mathbf{y} = f(\mathbf{x})$ at time t and $\mathbf{y}^* = f(\mathbf{x})$ at time $t+1$.

In all cases (H1–H3) we assume that the initial state $q^{(0)}$ of CN is not changed during experimentation. By the size of vectors involved in experiments based on (H1–H3) it should be quite clear that the verification of the initial requirements can only be done by statistical means.

2.6.2.5 VLSI Design of CN

After discussing the design of *CN* (finishing the modelling phase of the task) we would also like to shortly discuss the phase of implementation. As an implementation tool the CAD system VENUS [Hörbst et al. 87] was used. Due to the complexity of the circuit, a design using standard cells of the VENUS library got preference. To further reduce the complexity of the design (to be more efficient in circuit-simulation and in layout generation) the circuit has been divided into 4 macro cells, each containing 32 cell machines $C(i,j)$. The design steps can efficiently be realized by the modular functionality and regular network topology of the circuit, in order to reach layout data suitable for production.

2.6.2.6 Modelling and Implementation Results

To evaluate the sensitivity of the chosen metric distances d_{H1}, d_{H2} and d_{H3} with respect to the type of cell machines $M(i,j)$ used in *CN*, the following simulation experiments were performed [Gassner 89]:

A. 16 cell machines with 32 states and 2 input lines were randomly generated and randomly placed in *CN*.

B. The cell machines were constructed according to the properties (P1–P6) of Section 2.6.2.3 by using CAST.FSM.

C. A specific finite state machine in the form of a clock-controlled shift register was used as the cell machine $M(i,j)$.

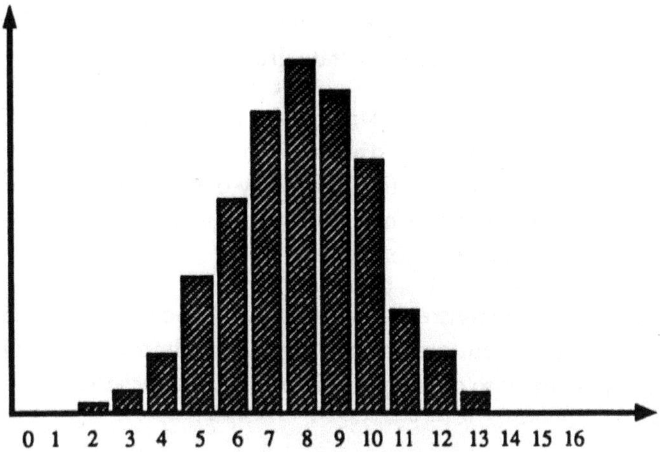

Figure 2.6.5 Typical histogram for metric distances d_{H1}, d_{H2}, and d_{H3} in *CN*.

The simulation results show that the statistical properties of the different metric distances do not depend strongly on the type of cell machine. Figure 2.6.5 shows a histogram which is typical for each simulation result.

The simulation experiment (B) to obtain cell machines $M(i,j)$ with properties (P1–P6) can be successfully instrumented by FSM theory. The search for suited machines $M(i,j)$ was successfully supported by CAST.FSM. As an example, we consider the finding of finite state machines M which have property (P6) – an accepted degree of state graph complexity $SGC(M)$. $SGC(M)$ is measured by the state graph of M. We consider the degree of state graph complexity $SGC(M)$ as acceptable for our purpose if for each individual input a of M the associated state graph $SG(M,a)$ of M has only 1 or 2 circles of maximal length. We represent in CAST.LISAS $SG(M,a)$ by the so-called *structure graph*.

In Figure 2.6.6, we have as an example a specific randomly chosen cell machine *M6.C* for which $SG(M6.C,0)$ has 2 circles of length 2 and 1, respectively, and $SG(M6.C,1)$ has 5 circles of length 8, 1, 4, 2, and 1, respectively. *M6.C*, therefore, does not fulfill our state graph complexity requirements.

In Figure 2.6.7, however, we have an example of $M(i,j) = M17.C$ which was selected for use in *CN*, with an accepted degree of state graph complexity.

Finally, we mention some details concerning the implementation of *CN* as a VLSI chip. The whole cellular network was divided into 4 macro cells as shown in Figure 2.6.8. In order to improve testability, the possibility of decomposition of the 4 macro cells by well-known multiplexing techniques was considered. In the final prototype of *CN*, however, this approach was not used.

In Table 2.6.5 we indicate the figures of the designed *CN*-chip. Figure 2.6.9 shows the final VENUS layout. The results demonstrate that an improved layout and more careful design is still needed to develop a chip with a smaller area.

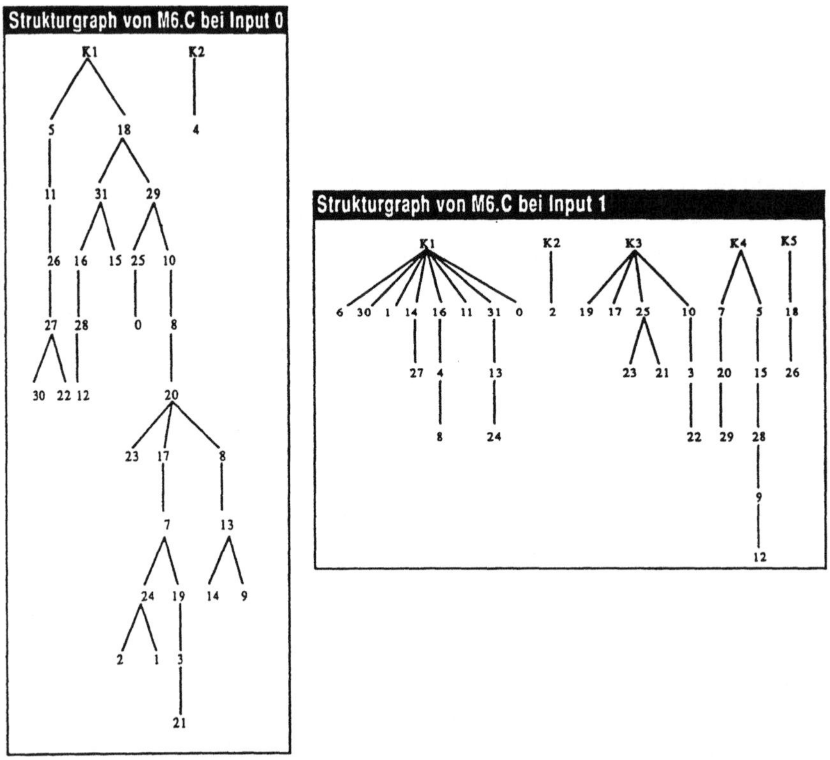

Figure 2.6.6 Structure graphs of *M6.C* with non-accepted degree of state graph complexity.

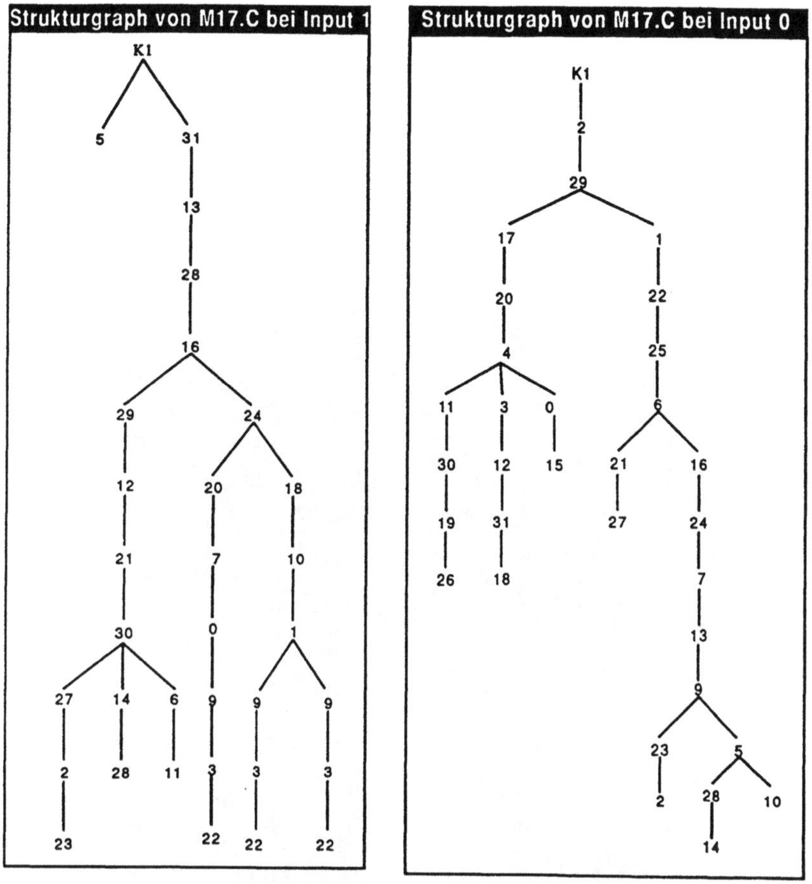

Figure 2.6.7 Structure graph of *M17.C* of accepted degree of state graph complexity.

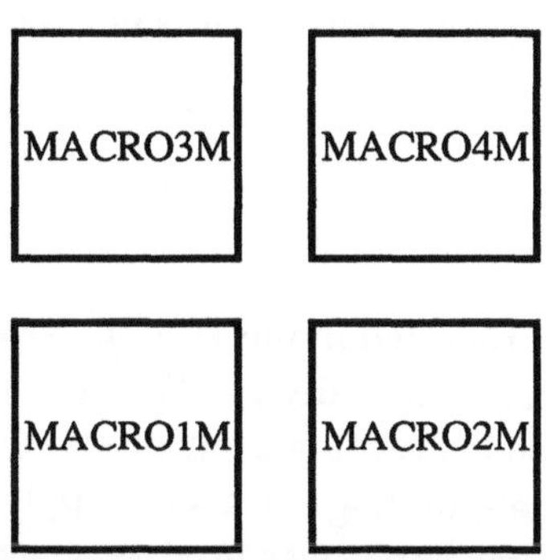

Figure 2.6.8 Decomposition of *CN* into macro cells.

Table 2.6.5 *CN*-chip figures.

number of pads	52
number of cells	13.894
gate equivalence	44.288
number of transistors	189.662

Figure 2.6.9 VENUS layout of the *CN*-chip.

2.6.2.7 Conclusions

The goal of the investigation reported in this section, was to give an example for the application of CAST tools in practical design. The task of designing a Boolean one-way function *f* for applications in cryptography offers enough chances for the application of CAST tools in model-building and model-application. The concept of a cellular network of finite state machines was used as a specific system type to realize the black box *f*. Such a network has theoretical and practical advantages. From the theoretical point of view it allows the application of FSM theory to design the specific cells *C(i,j)* of the cellular network *CN*. Furthermore, the regular grid architecture also allows theoretical insights beyond the cell-level and efficient simulation. CAST.FSM and CAST.LISAS were successfully applied to give instrumental support to that type of system-theoretical design tasks.

The practical advantage of cellular architecture is assured by its efficient implementation as a systolic array. The design of an ASIC chip using a commercial CAD tool can easily be made by that type of architecture. The VENUS CAD-software brought satisfying results in our case study and the feasibility of the design was proven. In summary, it can be claimed that the design of *CN* to realize a one-way function *f* of high cryptographic quality requires the application of CAST tools. CAST.FSM and CAST.LISAS have been proven to be useful tools for that purpose.

REFERENCES – CHAPTER 2

[Ahmed, Rao 75]
N. Ahmed, K. R. Rao, *Orthogonal Transforms for Digital Signal Processing*, Springer Verlag, Berlin, 1975.

[Ashar et al. 89]
P. Ashar, S. Devadas, A. R. Newton, "Optimum and Heuristic Algorithms for Finite State Machine Decomposition and Partitioning," *Proc. of the Intern. Conf. on Computer-Aided Design*, 1989.

[Ashar et al. 90]
P. Ashar, S. Devadas, A. R. Newton, "A Unified Approach to the Decomposition and Redecomposition of Sequential Machines," *Proc. of the 27th Design Automation Conference*, 1990.

[Bainbridge 75]
E. S. Bainbridge, "The Fundamental Duality of System Theory," in *Systems: Approaches, Theories, Applications*. Ed. W. E. Hartnett, D. Reidel Publishing Company, Dordrecht, Chapter 3, pp. 45–61, 1975.

[Beauchamp 84]
K. G. Beauchamp, *Applications of Walsh and Related Functions*, Academic Press, London, 1984.

[Beth 84]
T. Beth, *Verfahren der schnellen Fourier-Transformation*, Teubner, Stuttgart, 1984.

[Bobrow, Stefik 83]
D. G. Bobrow, M. Stefik: *The LOOPS Manual*, Xerox PARC, 1983.

[Bongiovanni et al. 76]
G. Bongiovanni, P. Corsini, G. Frosini, "One-dimensional and Two-dimensional Generalized Discrete Fourier Transforms," *IEEE Trans. Acoustics, Speech, and Signal Processing*, vol. ASSP-24, pp. 97–99, Feb. 1976.

[Booth 67]
T. L. Booth, *Sequential Machines and Automata Theory*. John Wiley and Sons, New York, 1967.

[Bracewell 84]
R. N. Bracewell, "The Fast Hartley Transform," *Proc. IEEE*, vol. 72, no. 8, pp. 1010–1018, 1984.

[Brauer 80]
W. Brauer (ed.), *Net Theory and Applications*, Lecture Notes in Computer Science, vol. 84, Berlin, Springer Verlag, 1980.

[Brayton et al. 87a]
R. Brayton, G. D. Hachtel, C. T. McMullen, A. Sangiovanni-Vincentelli, *Logic Minimization Algorithms for VLSI Synthesis*, Kluwer Academic Publishers, Dordrecht, 1984.

[Brayton et al. 87b]
R. Brayton, R. Rudell, A. Wang, A. Sangiovanni-Vincentelli, "MIS: A Multilevel Logic Optimization System," *IEEE Trans.on Computer Aided Design*, vol. 6, no. 6, pp. 1062–1081, 1987.

[Bretschneider et al. 90]
F. Bretschneider, H. Lagger, B. Schulz, "Infrastructure for Complex Systems CAD Frameworks", in *Computer Aided Systems Theory – EUROCAST '89*, F. Pichler, R. Moreno-Diaz (eds.), Lecture Notes in Computer Science, vol. 410, Springer Verlag, Berlin, pp. 125–133, 1990.

[Burger 88]
W. Burger, "Programmierung von Computer Vision Systemen durch graphische Instruktion," in *Statistik und Mustererkennung*, F. Pichler, A. Pinz (eds.), R. Oldenbourg, Wien, pp. 136–149, 1988.

[Burger 89]
W. Burger, "On Interactive Programming of Computer Vision Systems," in *Wissensbasierte Mustererkennung*, A.Pinz (ed.), R. Oldenbourg, Wien, pp. 23–35 1989.

[Burger 91]
Burger W., "A Versatile Development Tool for Knowledge-Based Computer Vision," *Cybernetics and Systems*, vol. 22, no. 3, pp. 313–329, 1991.

[Casti 77]
J. L. Casti, *Dynamical Systems and their Applications: Linear Theory*, Academic Press, New York, 1977.

[Devadas 88]
S. Devadas, "Decomposition and Factorization of Sequential Finite State Machines," *Proc. of the Int. Conf. on Computer-Aided Design*, pp. 148–151, 1988.

[Devadas et al. 89]
S. Devadas, H.-K. T. Ma, A.R. Newton, A. Sangiovanni-Vincentelli, "The Relationship between Logic Synthesis and Test," *Proc. of VLSI 89 Conf.*, Munich, pp. 175–186, 1989.

[Duzy et al. 89]
P. Duzy, H. Krämer, M. Pilsl, W. Rosenstiel, T. Wecker, "CALLAS - Conversion of Algorithms to Library Adaptable Structures," *Proc. of the VLSI 89 Conference*, Munich, pp. 197–218, 1989.

[Foster 88]
M. Foster, "Partitioning Real Finite Automata: A Progress Report," *Technical report*, AT&T Bell Laboratories, 1988.

[Gajski, Kuhn 83]
D. Gajski, H. Kuhn, "Guest Editors Introduction, New VLSI Tools". *Computer*, vol. 16, no. 12, pp. 11–14, 1983.

[Gassner 89]
F. Gassner, *Entwurf und Implementierung eines komplexen systolischen 16x8 Schaltwerkes mit dem VLSI-Entwurfssystem VENUS*, Master Thesis (in German), Johannes Kepler University, Dept. of Systems Science, 1989.

[Gonzalez, Wintz 87]
R. C. Gonzalez, P. Wintz, *Digital Image Processing*, Addison-Wesley, 2nd edition, 1987.

[Harmuth 69]
F. H. Harmuth, *Transmissions of Information by Orthogonal Functions*, Springer Verlag, New York, 1969.

[Hartmanis, Stearns 66]
J. Hartmanis, R. E. Stearns, *Algebraic Structure Theory of Sequential Machines*, Prentice Hall, Englewood Cliffs, 1966.

[Hellwagner 88]
H. Hellwagner, "A Systolic Array with Constant I/O Bandwith for the Generalized Fourier Transform," *Int. Conference on Systolic Arrays*, San Diego, CA, USA, pp. 207–216, 1988.

[Hellwagner 89]
H. Hellwagner, *Systolic Architectures for the Generalized Discrete Fourier Transform*, Ph.D. Thesis (in German), Johannes Kepler Univ., Linz, Verband der wissenschaftlichen Gesellschaften Österreichs (VWGÖ), Vienna, 1989.

[Hofestädt, Feiten 90]
H. Hofestädt, W. Feiten, "Embedding Test Pattern Generation into Design". in *Computer Aided Systems Theory – EUROCAST '89*, F. Pichler, R. Moreno-Diaz (eds.), Lecture Notes in Computer Science, vol. 410, Springer Verlag, Berlin, pp. 381–398, 1990.

[Hörbst et al. 87]
E. Hörbst, C. Müller-Schloer, H. Schwärtzel, *Design of VLSI Circuits, Based on Venus*, Springer Verlag, Berlin, 1987.

[Kalman 69]
R. E. Kalman, P. L. Falb, M. A. Arbib, *Topics in Mathematical Systems Theory*, McGraw-Hill, New York, 1969.

[Kanal, Kumar 88]
L. Kanal, V. Kumar (eds.): *Search in Artificial Intelligence*, Springer Verlag, New York 1988.

[Klir 85]
G. J. Klir, *The Architecture of Systems Problem Solving*, Plenum Publishing, New York, 1985.

[Klir 91]
G. J. Klir, *Facets of Systems Science*, Plenum Publishing, New York, 1991.

[Kohavi 70]
Z. Kohavi, *Switching and Finite Automata Theory*. Mc Graw Hill, New York, 1970.

[Korf 85]
Richard E. Korf: *Learning to Solve Problems by Searching for Macro-Operators*, Pitman Publishing, 1985.

[Kung 80]
H. T. Kung and C. E. Leiserson, "Algorithms for VLSI Processor Arrays," in C. Mead, L. Conway, *Introduction to VLSI Systems*, Addison-Wesley, 1980.

[Kung 82]
H. T. Kung, "Why Systolic Architectures," *IEEE Computer*, pp. 37–46, January 1982.

[Kunz 77]
H. Kunz, *Approximation optimaler, linearer Transformationen durch eine Klasse schneller, verallgemeinerter Fourier-Transformationen*, Dissertation, ETH Zürich, 1977.

[Leiserson 81]
C. E. Leiserson, *Area Efficient VLSI Computation*, Ph.D. Thesis, Carnegie-Mellon University, October 1981.

[Lin, Newton 89]
B. Lin, A. R. Newton, "Synthesis of Multiple-level Logic from Symbolic High-Level Description Languages," *Proc. of the VLSI 89 Conference*, Munich, pp. 187–196, 1989.

[Lisanke 88]
B. Lisanke, "Logic Synthesis and Optimization Benchmarks," *Technical Report*, MCNC, Research Triangle Park, North Carolina, December 1988.

[März et al. 89]
S. März, K. Buchenrieder, P. Duzy, R. Kumar, T. Wecker, "CALLAS – A System for Automatic Synthesis of Digital Circuits from Algorithmic Description". *Proc. of the EUROASIC Conf.*, Grenoble, 1989.

[Melhelm 85]
R. Melhem, "A Language for the Simulation of Systolic Architectures," in *Proc. 12 th Ann. Int. Symp. on Computer Architecture*, pp. 310–314, 1985.

[Mesarovic, Takahara 75]
M. D. Mesarovic, Y. Takahara, *General Systems Theory: Mathematical Foundations*, Academic Press, New York, 1975.

[Mesarovic, Takahara 89]
M. D. Mesarovic, Y. Takahara, *Abstract Systems Theory*, Lecture Notes in Control and Information Sciences, vol. 116, Springer Verlag, Berlin, 1989.

[Mittelmann 86]
R. Mittelmann, *Petrinetze in LOOPS / InterLISP-D*, Technical Report (in German), University of Linz, Austria, 1986.

[Mittelmann 87]
R. Mittelmann, *STIPS, CAST, Knowledge Based Systems*, Technical Report, University of Linz, Austria, 1987.

[Mittelmann 88]
R. Mittelmann, "Object-oriented Implementation of Petri Nets Concepts," *Cybernetics and Systems Research Conference '88* (R.Trappl ed.) Kluwer Academic Publishing, pp. 759–766, 1988.

[Mittelmann 89]
R. Mittelmann, *CAST.FSM User's Guide*, Technical Report 89-3, University of Linz, Austria, 1989.

[Mittelmann 90a]
R. Mittelmann, "Object-oriented Design of CAST Systems," in *Computer Aided Systems Theory – EUROCAST '89*, F. Pichler, R. Moreno-Diaz (eds.), Lecture Notes in Computer Science, vol. 410, Springer Verlag, Berlin, pp. 69–75, 1990.

[Mittelmann 90b]
R. Mittelmann, "Tools for Advanced CAST Environments," Proceedings of the Workshop on Modern Trends in Cybernetics and Systems, Temesvar, CSFR, 1990.

[Müller-W. 86]
T. Müller-Wipperfürth, *LISAS – A LOOPS-Implementation of a Systolic Array Simulator*, Master Thesis (in German), Univ. Linz, Dept. of Systems Science, 1986.

[Müller-W. 91]
T. Müller-Wipperfürth, *An Efficient Implementation of the FSM Lattice Computation and Decomposition Using Common LISP/Flavors/C*, Technical Report, University of Linz, Dept. of Systems Science, 1991.

[Newell, Simon 72]
A. Newell, H. A. Simon, *Human Problem Solving*, Prentice Hall, Englewood Cliffs, 1972.

[Nicholson 71]
P. J. Nicholson, "Algebraic Theory of Finite Fourier Transforms," *Journal of Computer and System Sciences*, vol. 5, pp. 524–547, 1971.

[Nilsson 71]
N. J. Nilsson: *Problem Solving Methods in Artificial Intelligence,* Tioga, Palo Alto, 1971.

[Nilsson 80]
N. J. Nilsson: *Principles of Artificial Intelligence,* Tioga, Palo Alto, 1980.

[Nussbaumer 82]
H. J. Nussbaumer, *Fast Fourier Transform and Convolution Algorithms,* Springer Verlag, Berlin, 1982.

[Ören 90]
T. Ören, "CAST nomenclature", personal communication, 1990.

[Pearl 84]
J. Pearl, *Heuristics: Intelligent Search Strategies for Computer Problem Solving,* Addison-Wesley, 1984.

[Pichler, Moreno-Diaz 90]
F. Pichler, R. Moreno-Diaz (eds), *Computer Aided Systems Theory – EUROCAST '89,* Lecture Notes in Computer Science, vol. 410, Springer Verlag, Berlin, 1990.

[Pichler, Praehofer 88]
F. Pichler, H. Praehofer, "CAST.FSM – Computer Aided Systems Theory: Finite State Machines," *Cybernetics and Systems Research Conference '88* (R.Trappl ed.), Kluwer Academic Publishing, pp. 737–742, 1988.

[Pichler, Schwärtzel 90]
F. Pichler, H. Schwärtzel, *CAST: Computerunterstützte Systemtheorie,* Springer Verlag, Berlin, 1990.

[Pichler 75]
F. Pichler, *Mathematische Systemtheorie,* De Gruyter, Berlin, 1975.

[Pichler 80]
F. Pichler, "Fast Linear Methods for Image Filtering," in *Applications of Information and Control Systems,* D.G. Lainiotis, N.S. Tzannes (eds.), Reidel, Den Haag, pp. 3–11, 1980.

[Pichler 81]
F. Pichler, "Schnelle Faltung von Bildern," in *Mustererkennung und Bildverarbeitung in Österreich,* F. Leberl, H. Ranzinger (eds.), OCG, Wien, pp. 13–30, 1981.

[Pichler 86]
F. Pichler, "Model Components for Symbolic Processing by Knowledge Based Systems: The STIPS Framework", in *Modelling and Simulation Methodology in the Artificial Intelligence Era,* Elzas, Ören, Zeigler (eds.), North Holland, Amsterdam, pp. 133–143, 1986.

[Pichler 88a]
F. Pichler, "CAST – Computer Aided Systems Theory: A framework for interactive method banks," *Cybernetics and Systems Research Conference '88* (R.Trappl, ed.), Kluwer Academic Publishing, pp. 731–736, 1988.

[Pichler 88b]
F. Pichler, "Konstruktion Korrelationsimmuner Schaltfunktionen und Schaltwerke mittels Walsh-Fourieranalyse," *Contributions to General Algebra*, vol. 6, Verlag Hölder-Pichler-Tempsky, Wien, pp. 213–222, 1988.

[Pichler 90]
F. Pichler, "CAST: Computer Aided Systems Theory, Perspectives for Research, Development and Education," *Syst. Anal. Model. Simul.*, vol. 7, pp. 251–266, 1990

[Pichler 91]
F. Pichler, "Modellbildung und Modellanwendung in der Informationstechnik – Systemtheoretische Erfordernisse," in *Angewandte Informatik und Software*, H. Schwärtzel (ed.), Springer-Verlag, Berlin, pp. 175–185, 1991.

[Pratt 78]
W. K. Pratt, *Digital Image Processing*, Wiley, New York, 1978.

[Reisig 86]
W. Reisig, *Petrinetze – Eine Einführung*, Springer Verlag, Berlin, 1986.

[Rietsche, Neher 90]
G. Rietsche, M. Neher, "CASTOR: State Assignment in a Finite State Machine Synthesis System," *Proc. of the IFIP Working Conference on Logic and Architecture Synthesis*, Paris, pp. 15–19, 1990.

[Saucier, Poirot 89]
G. Saucier, F. Poirot, "State Assignment Using a New Embedding Method Based on an Intersecting Cube Theory," *Proc. of the 26th Design Automation Conference*, pp. 321–326, 1989.

[Schwärtzel 81]
H. G. Schwärtzel: *Ein konzeptioneller Ansatz zur Verwendung Dynamischer Konstruktionen als Instrumente für die Modellierung von Systementwurfsprozessen*. Dissertation (in German), Johannes Kepler University, Verband der wissenschaftlichen Gesellschaften Österreichs (VWGÖ), Vienna 1981.

[Schwärtzel 85]
H. G. Schwärtzel, "Discrete Event-Dynamic Systems for Modelling Dynamic Valued Nets," *International Journal on General Systems*, vol. 10, pp. 153–162, 1985.

[Sommer, Savage 85]
M. E. Sommer and J. E. Savage, "SAS – A Systolic Array Simulator," Techn. Report *CS-85-02*, Brown University, 1985.

[Walker, Thomas 85]
R. A. Walker, D. E. Thomas, A Model of Design Representation and Synthesis. *Proc. of the 22nd Design Automation Conf.*, pp. 453–459, 1985.

[Wolfram 86]
S. Wolfram, *Theory and Applications of Cellular Automata*, World Scientific Publishing Co. Pte. Ltd., Singapore, 1986.

[Wymore 67]
A. W. Wymore, *A Mathematical Theory of Systems Engineering. The Elements*, Wiley, New York, 1967.

[Zadeh, Desoer 69]
L. A. Zadeh, C. A. Desoer, *Linear System Theory: The State Space Approach*, McGraw-Hill, New York, 1963.

[Zeigler 84]
B. P. Zeigler, *Multifacetted Modelling and Discrete Event Simulation*, Academic Press, London, 1984.

3 MODELLING AND SIMULATION

3.1 INTRODUCTION

3.1.1 The Enterprise of Modelling and Simulation

The general scientific problem solving process

Reviewing the *model-based approach* to scientific problem solving from a very general point of view, we observe the following pattern of actions (Figure 3.1.1): Starting with the *real world* situation, we recognize a problem we want to eliminate or improve. To tackle this problem using scientific methods we first build a *model*, and try to find a solution of the problem in the world of the model. The obtained solution in the model world has to be interpreted in the real world and finally we have to implement the solution in the real world to eliminate the problem.

Model building, the process of deriving a model of the real system given some information of the real system, is not a trivial task. The main reason for that is that through the complexity of the real system it is impossible to comprehend and represent a real system as a whole or even to get complete information of it [Zeigler 76]. Although model building is much easier for artificial systems or systems to be designed, this is even true here because uncertainties in the physical realization always exist. A real system therefore should be conceived as nothing else as an infinite source of observable data from which in finite time we are able to retrieve just a finite portion.

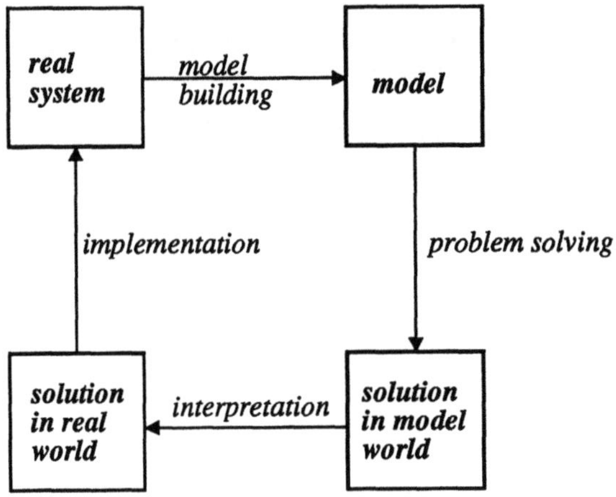

Figure 3.1.1 General problem solving process

Model building always means *idealization, abstraction* and *simplification*. The question if the model, the simplified, idealized representation of the real system, is a valid representation of the real system is crucial and always should be answered in connection with specifying the purpose the model is built for - the *objectives* of the problem solving enterprise [Zeigler 76, 84a; Pritsker, Pegden 79]. So, we say that a model validly represents a real system according to particular objectives. A model may be valid to answer a particular type of question of a real system while it may fail to answer another. Hence, the model building process always should be guided by the objectives at hand and should prescribe the level of abstraction and simplification needed to answer the questions. Always the most simple and most abstract level which is still appropriate should be taken to reduce model and problem solving complexity.

By model-application we summarize all activities to answer certain questions of the real system employing models (see Chapter 1). For our purpose we classify model-application methods in two main different types. These are

- analytical methods,
- simulation.

In analytical problem solving, the solution is deduced by employing inference rules which are known to be valid in the language the model is formalized in. The solutions are of general form in the sense that they allow generally valid statements about the problem. Individual solutions can be derived from them by replacing the symbolic coefficients of the general solution by individual values.

The applicability of analytical methods is restricted because, through the complexity of many real world problems, analytical inference rules are not known for the associated models or their application would require to simplify the model too much so that the obtained conclusions would be too coarse to answer the questions. In such situations, simulation as the other form of model-application offers another possibility to obtain insight into the behavior of the system.

Simulation

Simulation operates by building another real system which can be thought to be an implementation of the model and offers the possibility to undertake experiments with it. This second real system, which is behaviorally equivalent with the model and hence with the real system, is termed *real model* or *simulation model*. Experiments conducted with the simulation model provide insight into the dynamical behavior of the real system and, in that way, general statements about the real system are induced. Hence we define simulation in the following way:

> Simulation is the reproduction of the dynamical behavior of a real system using a (real) model to arrive at conclusions which are applicable to the real world.

Simulation models can be categorized into [Schmidt 85]

- physical simulation models,
- graphical simulation models,
- computer simulation models.

Physical simulation models are actual, physical re-implementations of the real world system, which is either to be built more easily or to be handled and experimented with more easily. A famous example of a physical simulation model is the wind tunnel. A graphical model is a pictorial representation of the real system. In computer simulation analog or digital computer programs are employed as simulation models. The advent and progress in computer technology provides a means to imitate the behavior of real system with a degree of precision not known so far. Although in the beginning of computer technology, the analog computer played an important role to simulate continuous systems, the digital computer superseded it. Sophisticated numerical integration methods allow to simulate continuous systems with a high degree of precision and fast execution speed on digital computers. For other types of simulation the digital computer is just the only means. We will exclusively deal with digital computer simulation here.

Application of computer simulation

The general nature of simulation offers a wide range of possible applications for different decision making tasks. According to the the degree of intervention the decision maker is able to exert on the real system, we define four levels in which simulation models can be employed [Pritsker, Pegden 79; Zeigler 84a]:

- explanatory,
- forecast,
- improvement,
- design.

In the first level, simulation is used to get more insight into the operation of the real system. In the forecast level, a simulation model of an existing system is used to get information as to how it will behave in the future. Weather forecast or world modelling are typical examples belonging to this level. In the third level, we use a model of an existing real system to analyze its operation and then to study different alternatives to improve its operation. Studying alternatives is done in the model world and only the best realizable alternative is implemented. In the design level, no real system exists and the models serve as design blueprints which are tested by simulation.

Computer simulation has acquired its position as an important tool in many scientific disciplines. Table 3.1.1 gives a short, non-comprehensive survey of them (see [Pritsker, Pegden 79] for a more detailed survey). However, with the progress in computer, software, and simulation technology, it can be expected that the degree of

Table 3.1.1 Survey of areas of application of computer simulation.

manufacturing system design	*control engineering*
transport system design	*chemical industries*
traffic control	*physics*
computer system design	*robotics*
communication system design	*environmental studies*
electronics hardware design	*economics and social sciences*

application will grow very much in the future. Especially in system design, the development and employment of simulation methods is just at its beginning [Rozenblit, Zeigler 88; Buerger 86].

Types of computer simulation models

Ören [Ören 89a] defines a comprehensive taxonomy of simulation models for which he uses the following classification criteria:

- the time set of the model,
- trajectory of descriptive variables,
- existence and range of variables,
- functional relations of variables,
- spatial distribution,
- organization of component models, and
- goals to be pursued.

Since dynamical behavior of simulation models is of central interest, we state that from these criteria the first two are the most important, viz. how is time modeled and how do the changes in the descriptive variables occur. Figure 3.1.2 shows a taxonomy according to these two criteria [Pritsker, Pegden 79; Schmidt 85; Ören 89a].

According to the criteria time, we speak of *continuous-time* models when time is represented by the real numbers and we speak of *discrete-time* models when time is represented by the integers or a set isomorphic to the integers. According to the second criteria, we speak of *continuous-change* models when state changes occur continuously over time and we speak of *discrete-change* models when they occur in discrete jumps. Pairwise combinations of one category of each criteria define well known modelling concepts. Continuous-time and continuous-change models are *differential equation specified models*, discrete-time and continuous-change models are *difference equation specified models*, discrete-time and discrete-change models are *finite state machine specified models* and discrete-change and continuous-time models represent the class of *discrete event* models.

In the literature the general term *discrete* is used to specify discrete-change models and the term *continuous* to specify continuous-change models. Furthermore, we call *combined discrete / continuous* or just *combined* models models resulting from a combination of discrete and continuous elements.

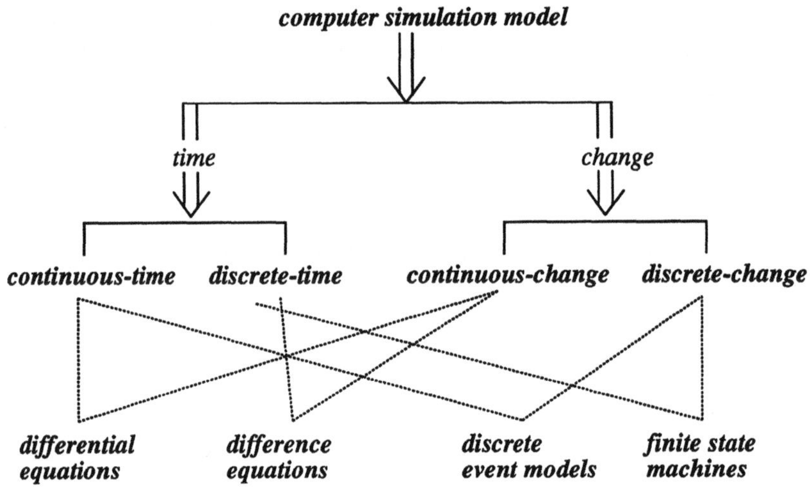

Figure 3.1.2 Simulation model taxonomy.

Elements involved in computer simulation

Zeigler depicts three main elements involved in computer simulation [Zeigler 76, 84a] (Figure 3.1.3):

- the real system,
- the model,
- the computer.

The real system is regarded to be nothing more than a source of data. The model is a set of instructions capable of generating such data and the computer is a means of carrying out these instructions to actually generate such data. A direct relation exists between the real system and the model and is summarized by the term *modelling*. A direct relation between the model and the computer exists and this relation can be denoted by *simulation*.

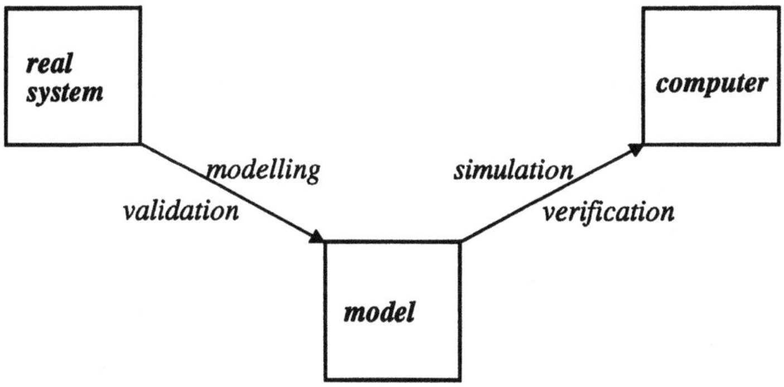

Figure 3.1.3 Digital computer simulation.

It is essential for the model-based methodology that between the real system and the simulation model we can identify a model. The reasons to separate the model from the computer implementation are manyfold and it can be compared to the advantages it has to separate an algorithm from the actual program which is implementing it. First of all, it supports the simulation modelling process. A model will be much easier to build and to comprehend than a computer program since all the program language details can be omitted. The actual implementation of a model, possibly represented already in a formal way, will be much easier or can be automated at all. Likewise the aim of the development of high-level computer languages was to free the programmer from computer details so that he was able to concentrate himself on the creative high level problem solving tasks, it should be the aim of the developments of simulation tools to bring the language for simulation model implementation as close as possible to the language the model is formalized in. Thereby, we do not promote a natural language interface but we state that the model should be described in a formal way, viz. in some sort of formalism. This formal description can serve as an unambiguous documentation in man-to-man communication as well as serving as a basis for a computer implementation. We promote that tools should be provided to translate such a formal description readily into a computer-readable form. Systems Theory provides a comprehensive, application independent language to formally describe models.

A further advantage results from the above observations. The process of *validation* is supported by such a scheme. By once *verifying* that the modelling language is correctly interpreted by the computer, the validation process can be restricted to showing that the model is a valid representation of the real system (and not that the program is a valid representation of the real system). Thus showing that the simulation program correctly simulates the real system is a two step process where the validation of the model comes first and then verification of the computer program (see Figure 3.1.3).

Using formal languages for model representation provides further possibilities. Theories coming with the formal language can be employed for symbolic model manipulation. Also, after model description there is still the possibility to decide if analytical methods are applicable and shall be used or if we have to recourse to simulation methods.

3.1.2 The Systems Approach to Modelling and Simulation

General system concepts

Systems Theory uses the fact that real systems can obey the same "system" laws and show similar patterns of behavior although they are physically very different [Bossel 87]. Hence, it is possible to use equivalent representations for different real systems. An example is the representation of oscillation of mechanical, electrical, chemical, or biological systems. In that sense, Systems Theory is an interdisciplinary science which tries to provide general concepts for model building and problem solving. It fulfills its task, either by extracting the general principles from problem solving methods developed in specialized scientific disciplines, and in that way, making them usable for other disciplines, or by setting up new constructions and recommending them as general principles [Pichler 75].

A generally agreed definition of what a *system* is does not exist [Laux 77]. Basically, we distinguish two differing interpretations which we adapt for our needs and summarize by the following definitions:

- A system (real system) is a combination of one or more structural related elements whose states are influenced by states of itself or other elements and which influence states of other elements [Bossel 87; Laux 77; Lerner 67].
- A system (dynamical system) is every formal construct which provides general modelling concepts for various kinds of disciplines [Pichler 75; Zeigler 76, 90a; Wymore 76].

According to the first definition, we use the term system mainly to denote the part of the real world we are interested in as well as its representation, viz. its model. Central to our point of view is that the system consists of elements and that there is some sort of interrelation between these elements. As the interrelations among elements crucially contribute to the system behavior, the system as a whole defines more than just the sum of its elements [Simon 62]. Another important issue is that a system does not mean the real world as a whole but we always only look at a part of reality. The *system boundary* therefore encloses the part of reality which we are interested in and excludes everything else.

According to the second definition, we use the term system to denote every formal construct which is used to model some part of reality and has been developed according to system theoretical principles, i.e., which is interdisciplinary in its nature. So a "system" in this sense is a system theoretical means to model a "system" in the sense of the first definition. When the meaning is not obvious, we will distinguish these two interpretations by denoting the first *real system* and the second *dynamical system*, as in simulation we exclusively deal with a system showing some sort of dynamical behavior.

Figure 3.1.4 presents a system and introduces several terms used in the connection with systems. As we have already mentioned, the system boundary encloses everything which we want to be part of our system and excludes everything else. Everything else is called the *system environment*. There cannot be a impervious demarcation between the system and its environment but we have to allow to model how the environment affects the system and the system can take some influence on the environment. The *system inputs* allow one to represent the influences the environment takes on the system and through the *system outputs* we have a means to represent the influence the system takes on its environment.

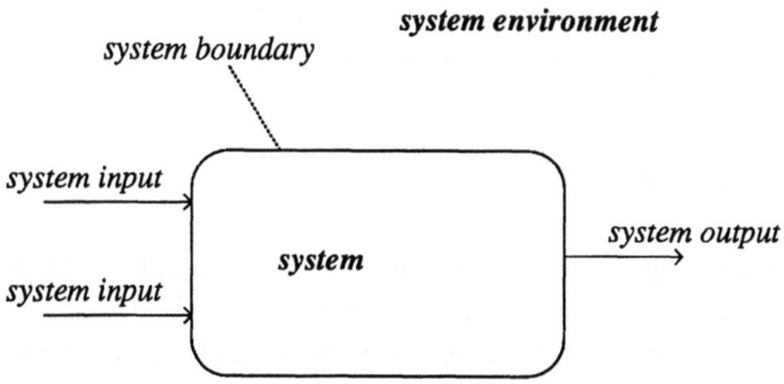

Figure 3.1.4 Systems concepts.

The system itself can be represented with a varying degree of concreteness and precision. Below we will outline a hierarchy of levels of system descriptions which go from a very abstract to a very concrete level of description and which is related to the definition of the general model types introduced in Chapter 1. According to the concreteness, the system manifests in different ways and we use different terms to speak about it. By *system behavior* we denote the way the system manifests on its boundary, i.e., how it reacts to inputs by outputs. The interior of a system is described by the *system state* and the *system dynamic*. The system state represents the condition the system is in at a particular time and the system dynamic means the way this state changes over time. When the state is represented by one or more variables we speak of *state variables*. The dynamic of the system can be determined by some sort of *state transition function* which depends on the input and the state itself. How the state and current input is matched to the output is determined by the *output function*. When we identify several elements corresponding to parts of the real world we speak of *system elements* and the state of each of the elements is represented by the *system element state*. The interdependencies of these elements contributing to the system dynamic is called the *system structure*. When we go further and represent parts of the system by systems themselves and their interdependencies by coupling these parts, we speak of *system components* and *system couplings*. Such a system is called *multi-component system* or *coupled system*.

The state plays a central role in the description of a system. It represents a memory for the past of a system. A state has to be able to summarize the past in such a way that it represents a full description of the interior of a system at any point in time and it is possible to compute the future of the system just knowing the current state and the future inputs [Pichler 75; Bossel 87; Zeigler 76]. Hence, the state variables represents just that part which has to be initialized at the beginning of each experiment. When a experiment is interrupted, the state variables are sufficient to continue the experiment and gain the same results as if no interruption had occurred [Zeigler 76].

Levels of system descriptions

Several systems theoretic frameworks exist [Pichler, Schwärtzel 90; Klir 85; Mesarovic et al. 70]. Zeigler in [Zeigler 76] defines a *hierarchy of levels of system descriptions* which exposes different means to describe a system and stratifies them according to their level of concreteness - from a very abstract level to a very concrete level. Each level introduces more concreteness into the description of the internal structure of a system. In the following we review this hierarchy starting from the most abstract level and proceed to the most concrete level of description.

Level 0: Observation Frame $O = <T, X, Y>$ Using an observation frame O we just define the system boundary. The set X is the input interface in form of a *set of inputs* and Y is the output interface in form of a *set of outputs*. The set T is called the *time base*. It is used to order events and represents the observation times, i.e., points in time when we are able to define system behavior and system dynamic. Usual time bases are the real numbers - the *continuous time base* - or the integers - a *discrete time base*.

Level 1: Relation Observation $IORO = <T, X, \Omega, Y, R>$. The relation observation $IORO$ defines the behavior of the system using a relation R of possible input values at particular times and their possible output values. Hence there is not an unequivocal mapping of input values to outputs but for a particular input there can be several different outputs.

Level 2: Function Observation $IOFO = <T, X, \Omega, Y, F>$. This level is equal to level 1 with the only difference being that the relation R is partitioned into a set F of functions f and each function defines a unique output response for a particular input segment. Hence, when we have knowledge of the function f, we have knowledge of the "initial state" to get a unique output response.

Level 3: Input / Output System (or dynamical system) $IOS = <T, X, \Omega, Q, Y, \Delta, \Lambda>$. Whereas in level 0 to 2 it only was possible to represent the system boundary and the system behavior, in this level we are able now to model the interior of the system. For that we use the set of states Q. The *global state transition function* Δ realizes the dynamic of the system. The output function Λ is used to model how the current state manifests at the output interface. Level 3 incorporates our most important modelling concept for simulation modelling and therefore we also denote an input / output system *dynamical system*.

Level 4: Structured Input / Output System $SIOS= <T, X, \Omega, Q, Y, \Delta, \Lambda>$. A structured input / output system is an input / output system where the set of inputs, outputs, and states are multi-variable sets, i.e., that we are able to identify several input, output and state variables.

Level 5: Coupled System $CS = <T, X, Y, Components, Coupling>$. In a coupled system the interior of the system is represented by identifying several component systems and the couplings between them. The coupling defines how the coupled system inputs and the component system outputs are mapped into component system inputs and coupled system outputs. The state of the coupled system is built up by the states of all component systems. The system dynamic is built up by the dynamic of the components and the coupling scheme. As a coupled system also has an input / output interface and is a system itself, it can appear as a component in a bigger coupled system.

The levels of system descriptions can be related to the general model types introduced in Chapter 1 in the following way: Observation Frame O, Relation Observation IORO, Function Observation IOFO are black-box model types, Input / Output System IOS and Structured Input / Output System SIOS are dynamics model types, and the Coupled System CS corresponds to the network type.

Modular, hierarchical model construction

The most powerful way of constructing models is to build them up by coupling together simpler component models using coupled system specification [Zeigler 89; Bossel 87; Ören 85]. As every system defines its system boundary in form of an input / output interface, a system represents a *modular* building block. As component models themselves can be coupled systems, the construction of *hierarchical models* is possible. Zeigler defines a hierarchical model in the following way [Zeigler 90a]:

- an *atomic model*, i.e., every model which is not a coupled model, is a hierarchical model;
- a *coupled model* whose components are hierarchical models is a hierarchical model;
- nothing else is a hierarchical model.

The *static structure* - the structure representing the system without its dynamical behavior - of a hierarchical model can be represented by a finite tree - the so-called *composition tree* [Zeigler 84a, 90a] (Figure 3.1.5). The root node is the "biggest model". The interior nodes are coupled systems and their children represent their components. To make sense, the leaves have to be atomic models. An input / output interface is associated with every node. With the interior nodes we associate the definition of the couplings, while with the leaves we associate the state descriptions.

Modular, hierarchical model construction offers several advantages not known in conventional modelling approaches [Zeigler 90a]:

- component systems can be constructed and tested independently before their integration into bigger systems;
- a model once developed for a particular purpose can be reused in a bigger model just by using it as a component;
- modular, hierarchical models support the development of model bases.

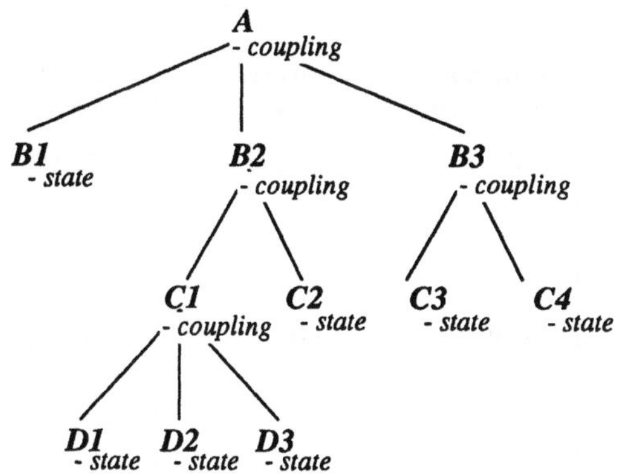

Figure 3.1.5 Composition tree representation of the static structure of a hierarchical model.

The last point is especially important when the problem is not the development of one single model but when the problem necessitates the development and the maintenance of a whole family of related models [Zeigler 84a, 90a] (see Section 3.5).

System formalisms

When we define constraints on the parts a dynamical system description, we derive different *system formalism*. A system formalism is a short-hand mean to specify a system. Only those parts which distinguish this particular system from others have to be specified [Zeigler 84a, 84b]. A system formalism specifies a modelling language that can be used to express a class of systems [Ören 89b]. Basic system formalisms are *differential equation specified systems, discrete time specified systems,* or *discrete event specified systems.* Starting from these system formalisms we are able to derive more specialized formalisms by further specialization. The level of system descriptions and system formalisms build an orthogonal relation (Figure 3.1.6) where every combination of system formalism and system level represents a possible modelling concept. Figure 3.1.7 introduces the constraints imposed by the system formalisms at the input / output system level of description. The formalisms impose appropriate constraints on the time base, input, output, and state sets, and input, output, and state trajectories.

levels \ *formal*	***differential equation***	***discrete time***	***discrete event***
coupled system	*DESN*	*DTSN*	*DEVN*
structured system			
input output system	*DESS*	*DTSS*	*DEVS*
function observation			
relation observation			
observation frame			

Figure 3.1.6 [Zeigler 84a] Orthogonality of levels of system description and system formalisms.

	differential equation	***discrete time***	***discrete event***
time base T	*continuous reals*	*discrete integers*	*continuous reals*
basic sets X, Y, Q	*real vector space*	*arbitrary*	*arbitrary*
input segments	*piecewise continuous*	*sequences*	*discrete event segment*
state trajectory	*continuous*	*sequences*	*piecewise constant*
output trajectory	*continuous*	*sequences*	*discrete event segment*

Figure 3.1.7 [Zeigler 84a] Constraints imposed by system formalisms at the input / output system level.

	differential equation	*discrete time*	*discrete event*
time base T	*continuous reals*	*discrete integers*	*continuous reals*
basic sets X, Y	*real vector space*	*arbitrary*	*arbitrary*
input segments	*piecewise continuous*	*sequences*	*discrete event segment*
components	*differential equ systems*	*discrete time systems*	*discrete event systems*
couplings	*no algebraic cycles*	*no algebraic cycles*	*no direct feedbacks*

Figure 3.1.8 Constraints imposed by system formalisms at the coupled system level.

Figure 3.1.8 introduces the constraints imposed by the system formalisms at the coupled system level of description. Now the constraint on the state set is replaced by constraints on the components as well as on the couplings.

For simulation modelling the higher levels of descriptions are employed. Especially, simulation models are built using the structured input / output system level and the coupled system level. When only these higher levels of descriptions are important, it is convenient to take a different point of view where the general dynamical system - the system description at the input / output system level - represents a basis from which different system formalisms can be derived by constraining the parts of the dynamical system. We also call this process *specialization*. In this way we obtain a complex hierarchy of system formalisms where the dynamical system is the root and the three basic system formalisms are the first descendants. Figure 3.1.9 shows a section of this hierarchy for different discrete time system formalisms. The *dynamical system* is the root and the *discrete time formalism* its direct descendant. From that formalism, we derive the more specialized formalisms *coupled discrete time system*, *sequential machine*, and *difference equation*. A further prominent specialization of coupled discrete time system is *cellular automata* and one of difference equation is *linear difference equation*.

It has to be mentioned that this hierarchy is in no way complete as well as it is not the only possible hierarchy. Several different possibilities exist. The hierarchy only has to obey the rules that descendants of a formalism always are built by specializing it and that more specialized formalisms always also define a system of the more general formalisms. So, e.g., a cellular automaton defines a system of the coupled discrete time formalism, as well a system of the discrete time formalism and of course it defines a dynamical system.

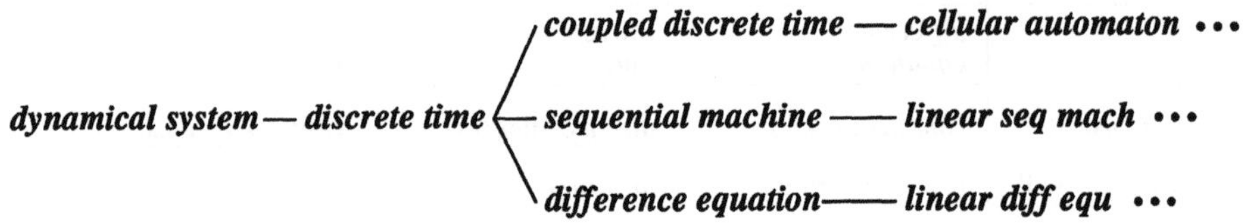

Figure 3.1.9 Hierarchy of discrete time system formalisms.

Simulation from the Systems Theory point of view

Looking at Figure 3.1.3 in Section 3.1.1, we see three elements involved in computer simulation. How can we interpret these elements in the context of our Systems Theory framework outlined above (Figure 3.1.10)? We conceive the real system as a source of observable data. Hence the real system depicts itself at the relation observation level. The model is built by using the modular, hierarchical modelling approach and system specification formalisms.

The computer now has to interpret the model. For that purpose Zeigler defined the *abstract simulator concept* [Zeigler 84a]. The abstract simulator for discrete event simulation takes a modular, hierarchical discrete event model and runs a simulation experiment. So the modular, hierarchical model can be readily transferred into an executable simulation program. The abstract simulator consists of objects, reflects the hierarchical structure of the model, and works by message passing. In the course of developing concepts for multi-formalism modelling and simulation, abstract simulator concepts for modular, hierarchical discrete time and differential equation specified systems have been developed. These abstract simulators then have been integrated to facilitate simulation of multi-formalism models [Praehofer, Zeigler 90; Praehofer 91].

In our framework only one part is still missing. As we always deal with open systems and models, i.e., they receive inputs from and send outputs to the environment, a model is not a self-contained unit and in a simulation experiment somebody has to provide the inputs and receive the outputs. And in a simulation experiment, something has to control the experiment, i.e., in which condition to start it, when to stop it and which kind of data to deliver to the user.

All these tasks are fulfilled by the so-called *experimental frame* [Zeigler 76, 84a] which represents all the objectives of the experiment. In our approach, the experimental frame itself is implemented by a modular, hierarchical model consisting of special types of models. *Generators* generate the input segments. *Acceptors* and *transducers* receive the outputs. Acceptors use the output trajectories to control the experiments. Transducers use the outputs to compute some sort of interesting data, in particular to compute statistics.

Before the simulation experiment, the implementation of the experimental frame in form of a modular, hierarchical model is coupled to the system model (Figure 3.1.11). After setting initial states appropriately, the resultant is ready to run simulation experiments. The modularity of the model and experimental frame coupling has several advantages. By coupling different models to the same experimental frame, different models can be evaluated under the same experimental conditions. By coupling different experimental frames to the same model, one model can be evaluated under varying experimental conditions.

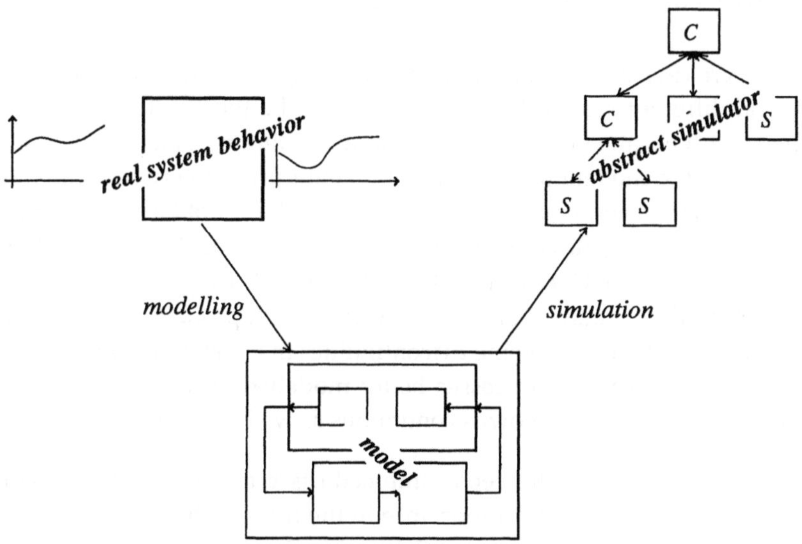

Figure 3.1.10 Real system, model and computer from the Systems Theory point of view.

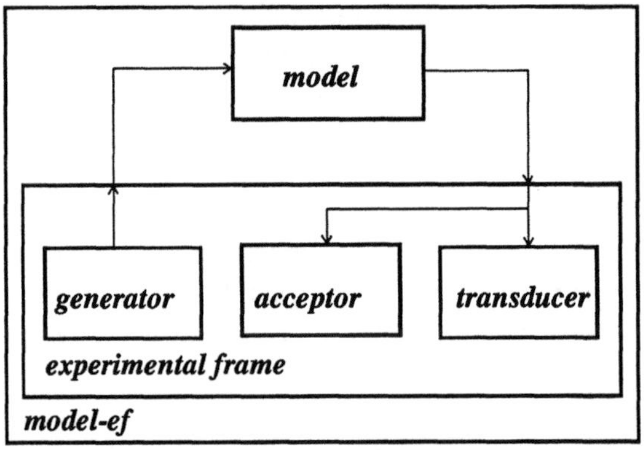

Figure 3.1.11 Model / experimental frame coupling.

3.1.3 Multi-Formalism System Modelling

Recall that there exist several different types of computer simulation models and we have several different system formalisms to describe them. Often it is possible to model one and the same system by different modelling methodologies [Cellier 86]. The objective of the modelling and simulation enterprise determines the methodology to be used. However, there exist problems which cannot be conveniently represented by either a purely discrete or purely continuous model [Fahrland 70; Golden, Schoeffler 73; Pritsker 74; Garzia et al. 86; Cellier 79, 86]. They are best modelled by a combination of discrete and continuous elements. Such problems range from piecewise continuous systems with some discontinuities in the state variables, to queuing models containing some subprocesses better modelled in detail at the continuous level, to coupled models which owns continuous as well as discrete components.

One area of application which needs combined discrete-continuous modelling and simulation will be of increasing importance in the future. This is the design and the analysis of *intelligent control systems* [Cellier 86; Garzia et al. 86; Zeigler et al. 88; Antsaklis et al. 89; Zeigler 90b]. Such control systems take over functions requiring intelligence from the human operator. While the system to be controlled and the lower level control schemes are best modelled by a continuous approach, the higher level decision making tasks will require discrete, especially discrete event modelling techniques.

To facilitate combined discrete-continuous modelling and simulation, we develop concepts for *multi-formalism system modelling and simulation*. Systems Theory so far has dealt only with homogeneous models, i.e., models which are built up by only one kind of system formalism. It was not possible to deal with systems built up by several different formalisms. In the following sections we will introduce a system theoretical approach to deal with multi-formalism systems. We define a multi-formalism system to be a system defined employing different system formalisms and hence showing features of different system formalism. A coupled system whose components are specified in different system formalisms is the most prominent representative of a multi-formalism system and is called *coupled multi-formalism system*. Beside them we also will get to know *atomic multi-formalism systems*.

3.2 SYSTEM SPECIFICATION FORMALISMS

3.2.1 A Brief Review of System Specification Formalisms

In this section we review the traditional system specification formalism for simulation modelling - the *differential equation specified system* (DESS) and the *discrete time specified system* (DTSS) formalism - as well as the system specification formalism for event oriented simulation modelling introduced by Zeigler - the *discrete event system specification* (DEVS) formalism [Zeigler 76, 84a]. We will do this review from two points of view - first from a Systems Theory point of view and second from a simulation point of view. In particular, we will take a perspective as taken by Zeigler [Zeigler 76, 84a] and Ören [Ören 85, 89a]. Most important is that each of these formalisms is closed under coupling, i.e., one can couple systems together so that the resultant again specifies a system [Zeigler 84a]. Thus, coupled systems can in turn be coupled together and hence the construction of modular hierarchical models is possible.

Differential Equation Specified System Formalism

An *atomic differential equation specified system (DESS)* is a structure [Zeigler 76; Ören 85]

$$DESS = (X, Y, Q, f, \lambda)$$

where

X	the *set of inputs* is a real vector space
Y	the *set of outputs* is a real vector space
Q	the *set of states* is a real vector space
$f: Q \times X \to Q$	is the *rate of change function*
$\lambda: Q \to Y$	is the *output function.*

The time base of DESS are the real numbers. The input segments for a DESS have to be piecewise continuous segments. The output segments produced by DESS as response are continuous segments over Y and the state segments are continuous segments over Q.

The output value y_t produced by the DESS at any time t where s_t is the state at time t is given by the function $y_t := \lambda(s_t)$. A state trajectory *straj* over $(Q, \Re)$ of a DESS for a given input segment $\omega \in (X, \Re)$ has to satisfy at any point in time t of the domain of ω the condition imposed by the differential equation $d\ straj(t) \,/\, dt := f\,(straj(t), \omega(t))$. Unique state trajectories exist if the derivative function f satisfies the Lipschitz condition.

Discrete Time Specified System Formalism:

An *atomic discrete time system (DTSS)* is a structure [Zeigler 76; Ören 85]

$$DTSS = (X, Y, Q, \delta, \lambda)$$

where

X	is the *set of inputs*
Y	is the *set of outputs*
Q	is the *set of states.*
$\delta: Q \times X \to Q$	is the *state transition function*
λ	is the *output function*

and λ is a function either

$$\lambda: Q \to Y$$

in the case of *Moore-type* components or

$$\lambda : Q \times X \rightarrow Y$$

in the case of *Mealy-type* components.

With atomic DTSS we associate the following dynamical behavior: As time base of DTSS we take the integers $\mathfrak{I}$ (= $1 * \mathfrak{I}$) or a set $ta * \mathfrak{I}$ isomorphic to the integers where *ta* is a constant time advance. When a DTSS is in state s_t at time t and the input at time t is x_t, then it gives out the value $y_t := \lambda(s_t)$ or $\lambda(s_t, x_t)$ in the case of Moore- and Mealy-type components respectively and transits to state $s_{t+ta} := \delta(s_t, x_t)$. s_{t+ta} is the state at time $t + ta$.

Discrete Event System Formalism:

An *atomic discrete event systems (DEVS)* is a structure [Zeigler 76, 84a]

$$DEVS = (X, Y, S, \delta_{ext}, \delta_{int}, \lambda, ta)$$

where

X	is the *set of external events*
Y	is the *set of outputs*
S	is the *set of sequential states*
$\delta_{ext} : Q \times X \rightarrow S$	is the *external transition function*
$\delta_{int} : S \rightarrow S$	is the *internal transition function*
$\lambda : S \rightarrow Y$	is the *output function*
$ta : S \rightarrow \Re_0^+ \cup \infty$	is the *time advance function*

and

$Q := \{ (s, e) \mid s \in S, 0 <= e <= ta(s) \}$ is the *set of total states.*

Discrete event systems have a continuous time base but inputs and state transitions occur at discrete time instances. So input segments are constraint to so-called *DEVS segments*, i.e., for a finite number of points in time there are inputs and between these times there is no input (or the so-called *non-event*). State and output trajectories are piecewise constant segments. The set of inputs, outputs and states are arbitrary sets. The state transition consists of two parts - the external transition function is executed whenever an input arrives and the times of execution of the internal transition function are scheduled by the time advance function.

If at time t the DEVS has been in state s for $e \leq ta\ (s)$ time units, then the DEVS is scheduled to undertake its internal transition at time $tn = t + (ta(s) - e)$. When there will arrive no external event until time tn, then the DEVS will stay in state s until that time so that it has been in s exactly $ta(s)$ time units and then will transit from s to $s' := \delta_{int}(s)$ at time tn. Right before undertaking the internal transition, the DEVS will put out the value $\lambda(s)$. After the internal transition, the DEVS is scheduled for the next internal transition at time $tn + ta(s')$.

Input values can arrive at arbitrary time instances at the input ports. If at time t the DEVS has been in state s for time $e \leq ta\ (s)$ and there arrives an external input, then the DEVS has to execute its external transition function and the DEVS will transit from s to $s'' := \delta_{ext}(s, e, x)$ where it will remain either to the next external event or $ta(s'')$ time units.

Well-defined DEVS must have the property of *legitimacy*. This property prevents the DEVS from getting into an infinite sequence of states in which the clock would not advance beyond a certain point [Zeigler 84a]. Formally we can define legitimacy in the following way:

> In every finite time interval $<t_{in}, t_f>$, and for every initial state $q \in Q$ and input segment ω:$<t_{in}, t_f> \rightarrow X$ there are only a finite number of state transitions.

Modelling Example: Single Server with FIFO-Queue

A model of a single-server station with constant processing time *10* and with a FIFO-queue shall serve as an introductory example to DEVS modelling. As usual we denote by $\{x, y, ..\}^*$ the set of all words over $\{x, y, ..\}$ and together with access functions *FIFOFirst*, *FIFORest*, *FIFOEmpty* and *FIFOInsert* implement a data type queue. We will present the state transition and output functions of our example in a pseudo-code description similar to a computer program implementation. In particular, we use several pseudo-code elements which should be self-descriptive and we regard the argument s of the state transition functions as an input and output parameter. Hence setting of a state variable in the transition functions means definition of a new state value.

```
FIFOStation = (X, Y, S, δext, δint, λ, ta)

X = {job | job ∈ {a, b, c, ... }}
Y = {job | job ∈ {a, b, c, ... }}
S = {(sigma, phase, queue) |
         sigma ∈ ℜ+0,∞, phase ∈ {idle, busy}, queue ∈ {a, b, c, ... }*}
δext (((sigma, phase, queue), e, job)
   queue := FIFOInsert (job, queue)
   model-in-phase
      idle:    phase := busy
               sigma := 10
      busy:    phase := busy
               sigma := sigma - e    -- continue with the current job

λ ((sigma, phase, queue)) :=
      FIFOFirst (queue)         -- output of the model is the first in the queue

δint ((sigma, phase, queue))
   queue := FIFORest (queue)
   if FIFOEmpty (queue)
      then phase := idle
           sigma := ∞     -- no jobs to process, hence passivate in idle
      else sigma := 10    -- begin to service next job

ta ((sigma, phase, queue)) :=
      sigma    -- the time advance is specified by the state variable "sigma"
```

Figure 3.2.1 shows the trajectories of an example simulation experiment. The first job comes in at time *5* and is immediately serviced because the server is idle; the phase becomes *busy* and sigma is set to the constant service time *10* scheduling an internal event at time *15*. As there is no further input event until time *15*, the internal event is executed resulting in the output of *a*, the removal of it from the queue, and, as there are no further jobs to service, the passivation of the server. Note that now, although the elapsed time is monotonically increasing, it never will reach infinity and hence no internal event will be executed. At time *20* the next job *b* arrives and awakes the server. The job is serviced but before completion there arrives a further job *c*. This job *c* has to be queued and wait until the server is free. However, this interrupt necessitates the update of *sigma* (*sigma := sigma - e*) because the elapsed time is set to *0* through the external event but only *3* time units are left until the completion of job *b* and this fact has to be reflected in the time advance function. At time *30* the internal event puts out and removes from the queue job *b* and starts the service of job *c*.

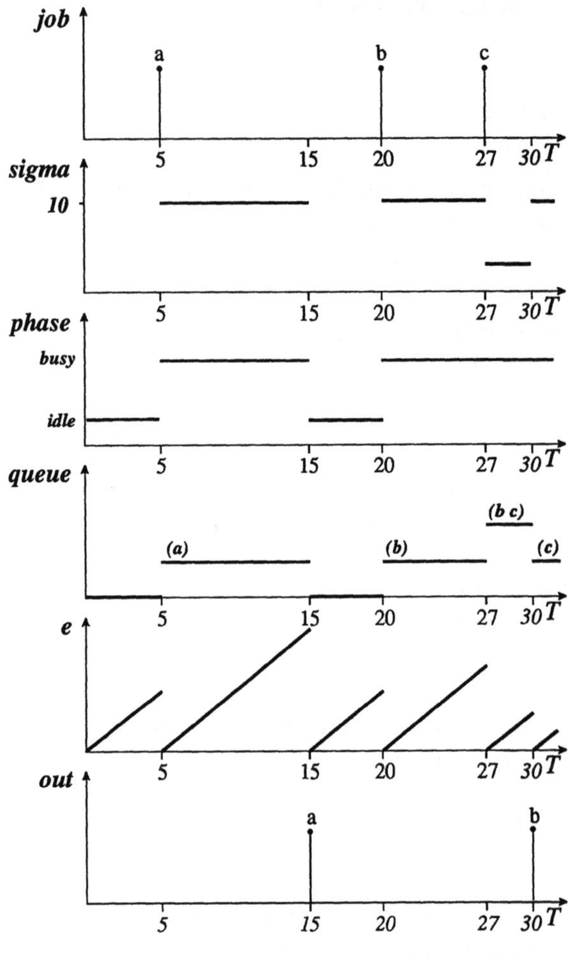

Figure 3.2.1 Single-server station trajectories.

Memoryless Models - Functions

A *memoryless model* or *function specified system (FNSS)* is a structure [Zeigler 76; Ören 85]

$FNSS = (X, Y, \lambda)$

where

X	is the *set of inputs*
Y	is the *set of outputs*
$\lambda : X \to Y$	is the *output function*

Memoryless Models can be defined in connection with DTSS and DESS networks. When the memoryless model receives an input x_t at time t it simple responds instantaneously with the output value $y_t := \lambda(x_t)$ at the same time t.

Coupled Systems or Networks

When coupling systems of our above defined system types together, we derive coupled systems or networks. The coupling of the components is exclusively done by connecting input and output ports. The network also has its own input ports and output ports. So we have to distinguish between three types of couplings:

- *external input coupling* - coupling of input ports of the network to input ports of the components;
- *external output coupling* - coupling of output ports of some components to output ports of the network;
- *internal coupling* - coupling of output ports of components to input ports of components.

To specify coupled systems, we use the following formalism [Zeigler 84a, 90a; Ören 85]:

$Netw = (X, Y, D, \{M_d\}, EIC, EOC, IC)$

where

X	is the *set of inputs* of the network
Y	is the *set of outputs* of the network
D	is the *set of the component names*
$\{M_d \mid d \in D\}$	is the *set of the component systems*
EIC	is the *external input coupling*
EOC	is the *external output coupling*
IC	is the *internal coupling*.

As we distinguish three system types for atomic system specification, we also derive three coupled system types - *DTSS networks, DESS networks,* and *DEVS networks* - when coupling together components of the same type.

DTSS network

Beside the specification of the components and the couplings, the definition of a DTSS network requires the specification of the time base. For a well defined DTSS networks, the following constraints must hold:

- the components are DTSS of type Mealy or Moore, FNSS or DTSS networks;
- every input port of a component must be influenced by exactly one port;
- every external network output port must be influenced by exactly one port;
- the time base of the network and all its components have to be equal;
- the network must not contain algebraic cycles, i.e., in a feedback loop there has to be at least one component whose output can be computed without knowledge of its input .

The last constraint is especially important. It is an extension of the well known rule [Zeigler 76] that in every feedback loop there must be at least one Moore-type component. This extension is necessary in hierarchical specified networks.

DESS networks

For a well defined DESS network the following constraints must hold:

- the components have to be of type DESS, FNSS or DESS network;
- every input port of a component must be influenced by exactly one port;
- every external network output port must be influenced by exactly one port;
- the network must not contain algebraic cycles, i.e., in a feedback loop there has to be at least one component the output of which can be computed without knowledge of its input.

DEVS networks

In DEVS the time base is implicitly given by the real numbers. DEVS networks require the addition of a tie breaking rule to the network definition. This tie breaking rule *Select* selects one component to undertake its internal transition function when more than one of the components are imminent to do so. So Select is a function [Zeigler 84a]

$$Select : subsets\ of\ D \rightarrow D\ .$$

For well-defined DEVS networks the following constraints must hold:

- The components have to be of type DEVS or DEVS network.
- No direct feedback loops are allowed, i.e., no output port of a component may be connected to an input port of itself.

3.2.2 Atomic Formalisms for Combined Discrete-Continuous Modelling

Systems Theory so far only dealt with one-formalism systems. Simulation modelling, however, requires capabilities to model systems using discrete and continuous elements together [Cellier 86; Pritsker, Pegden 79; Garzia et al. 86; Antsaklis et al. 89]. To support multi-formalism system modelling, we will proceed by introducing new system specification formalisms.

First we define a new atomic system formalism which is a combination of discrete event and differential equation specified system formalism and name it *combined discrete event and differential equation specified system (DEV&DESS)*. This system type is the most basic formalism for combined discrete-continuous modelling. In Section 3.3 we will present a simulation algorithm for that type of system.

Then, we will present several model prototypes which are based on this most general multi-formalism system which have proved to be most useful for simulation modelling. They should support modelling of a large variety of combined systems. These special formalisms are:

- *continuous-input-DEVS* - atomic discrete event systems with some continuous inputs;

- *DEVSwithDESS* - combination of atomic discrete event and differential equation specified systems which are first of all discrete event systems and also employ some continuous state variables;
- *DESSwithDEVS* - combination of atomic differential equation specified systems and discrete event systems which first of all are differential equation specified systems with state and time events causing discontinuities in the state and output trajectories;
- *DEVSwithAnalyticalDESS* - combination of atomic discrete event and differential equation specified systems with analytical solution which first of all are discrete event systems and also employ some continuous state variables for which analytical solutions are known.

Combined Discrete Event and Differential Equation Specified Systems (DEV&DESS)

A *combined atomic discrete event and differential equation specified system* is a structure as depicted in Figure 3.2.2 and is defined as follows:

$$DEV\&DESS = (X_M, Y_M, S, \delta_{ext}, \delta_{state}, \delta_{int}, \lambda_d, ta, f, \lambda_c)$$

where

$X_M = X_c \times X_d$	is the *set of inputs;* the cartesian product of *continuous inputs* X_c and *discrete inputs* X_d ;
$Y_M = Y_c \times Y_d$	is the *set of outputs;* the cartesian product of *continuous outputs* Y_c and *discrete outputs* Y_d ;
$S = S_c \times S_d$	is the *set of states;* the cartesian product of *continuous states* S_c and *discrete states* S_d;
$\delta_{ext} : Q \times X_M \rightarrow S$	is the *external transition function;*
$\delta_{state} : Q \times X_c \rightarrow S$	is the *state event state transition function;*
$\delta_{int} : S \times X_c \rightarrow S$	is the *internal state transition function;*
$\lambda_d : S \times X_c \rightarrow Y_d$	is the *discrete output function;*
$ta : S_d \rightarrow \Re_0^+ \cup \infty$	is the *time advance function;*
$f : S \times X_c \rightarrow S_c$	is the *rate of change function;*
$\lambda_c : S \rightarrow Y_c$	is the *continuous output function;*

with $Q_d := \{ (s_d, e) \mid s_d \in S_d , 0 <= e <= ta(s) \}$ is the *set of total discrete states;*

and $Q := S_c \times Q_d$ is the *set of total states.*

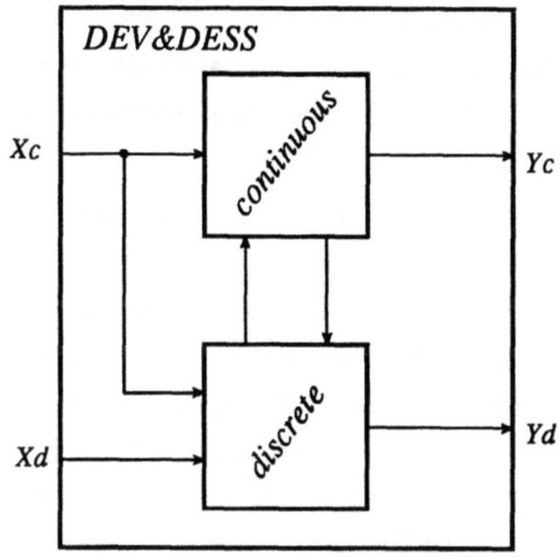

Figure 3.2.2 Combined DEV&DESS system.

Beginning at time t_i the system is in state q and there has not been a time scheduled internal event, external, or state event for e time units. The variable e defines the *elapsed time* since the last event. As it is in pure DEVS systems, the internal events, the executions of the internal transition function, are time scheduled by the time advance function *ta*. Internal events are also called *time events* [Cellier 79, 86] here. The external transition function of DEV&DESS is is employed to model the state changes caused by event inputs of the discrete inputs X_d. The state event transition function is used to model *state events* [Cellier 79, 86] caused by the continuous changes of the continuous inputs X_c and the continuous changes of the continuous states S_c. These state events are modelled by conditions specified in the state event transition function. Whenever the continuous changes cause a condition to become true in the state event transition function, this function has to be executed to update the state appropriately.

The event inputs arriving at the continuous input ports are saved in an additional state components *sx* with the affect that *sx* represents the piecewise constant representation of the DEVS input segment. The derivative function f, the continuous output function λ_c, but also the external and state event transition functions use these piecewise constant trajectories to retrieve the values. Between the events - external, state, or internal - the continuous state trajectories of the continuous states have to obey the constraints imposed by the differential equation f.

With that in mind, we can define the dynamical behavior of DEV&DESS in the following way: Given an initial state $q = (s_c, s_d, e)$ at initial time t_i and an input segment ω defined over interval $<t_i, t_f>$ then the final state is derived by the following scheme:

(1) If there is no event input in discrete and continuous inputs in the whole interval $<t_i, t_f>$, and there is no state event, i.e., there is no time t in $<t_i, t_f>$ where the application of the state event transition function would result in a change of the state, and the total elapsed time $e + t_f - t_i$ at time t_f is smaller than the value of the time advance function $ta(s)$, then there is no event at all and only the elapsed time e and the continuous states have to be updated. The elapsed time e is enhanced by the total elapsed time $t_f - t_i$ during the interval. The continuous states at the end of the interval are computed from the state at the beginning plus the integral of the derivative function f along the interval. As there are no event inputs the state component sx stays unchanged.

(2) If there is no event input and no state event until the execution of the first internal transition at time $t_i+ta(s)-e$, an internal, time scheduled event occurs and the internal transition function is executed. The continuous states at the time of the internal transition are computed from the state at the beginning plus the integral of the derivative function f along the interval until the time of the internal event. A discrete output unequal the non-event occurs only at internal events. As there are no event inputs, the state component sx stays unchanged.

(3) If a state event occurs in the interval before the first internal event and there is no event input prior, then the state event transition function has to be executed. The continuous states at the time of the transition are computed from the state at the beginning plus the integral of the derivative function f along the interval until the time of the event.

(4) If there is an event input at a discrete input port before the first internal event and the first state event, then the external transition function has to be executed. The continuous states at the time of the transition are computed from the state at the beginning plus the integral of the derivative function f along the interval until the time of the event. As there may be also an event input at a continuous port at the same time, the respective variable of the state component sx is updated by the value of the event input.

(5) If there is an event input at a continuous input port in the interval before the first internal event, then the respective variable of the state component sx is updated by the value of the event input.

After the internal, external, as well as state events, the recursive application of the procedure to the new state and the remaining segment defines the sequel.

To summarize the dynamical behavior of DEV&DESS specified above, we state the following:

- *time* or *internal events*: are time scheduled by the time advance function and modelled in the internal transition function;
- *external input events*: are caused by the event inputs of the discrete inputs X_d and are modelled in the external transition function and
- the values of the event inputs at the continuous input ports are stored in the state component *sx*;
- *state events*: caused by the continuous changes of the continuous inputs X_c and continuous states S_c are modelled in the state event transition function;
- like in DESS systems, between the events the trajectories of the continuous states have to be updated according to the rate of change function *f*.

To determine unique continuous state trajectories, the derivative function has to satisfy the Lipschitz condition as stated for pure DESS. The property of legitimacy can be specified analogously for DEV&DESS systems and must hold so that the DEV&DESS specifies a dynamical system. However both conditions are not sufficient for DEV&DESS systems because it is not guaranteed that the number of state events is finite in a finite interval. Therefore, we define the property of *state-event-legitimacy* in the way legitimacy is defined for pure DEVS systems.

> Let us define $\Sigma'(q, \omega)$ to be the number of state events, i.e., the events of execution of the state event transition function, in interval $<t_i, t_f]$ given the input segment $\omega: <t_i, t_f] \rightarrow X$ and the initial state q at time t_i. Then we define that a DEVS is *state-event-legitimate*, if and only if
>
> $$\forall\, \omega: <t_i, t_f> \rightarrow X \in \Omega, q \in Q : \Sigma'(q, \omega) < \infty,$$
>
> i.e., when the number of state events is finite.

A DEV&DESS specifies a dynamical system if and only if the derivative function *f* has unique solutions, if the system is legitimate, and if it is state-event-legitimate [Praehofer 91].

Modelling example

As an introductory example of a combined DEV&DESS model we present a model of a barrel generator (Figure 3.2.3). The model has one continuous input port *inflow* and one discrete input port *act/passivate*. It has one discrete output port *barrel* and one

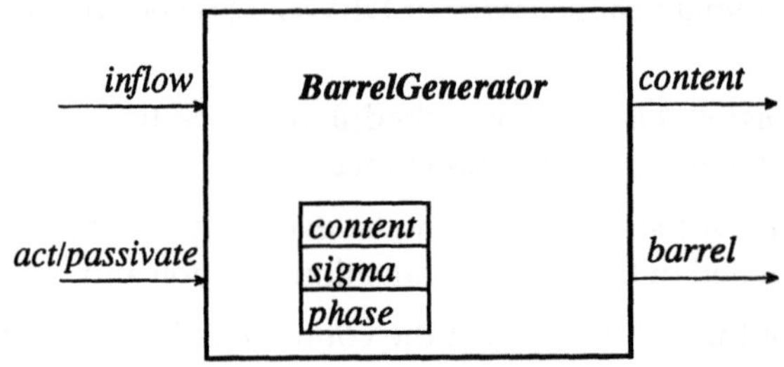

Figure 3.2.3 Example of a combined DEV&DESS model: Barrel generator.

continuous output port *content*. There is also a continuous state variable *content* and discrete state variables *sigma* and *phase*. The derivative of the continuous state variable *content* is given by the input value of *inflow* when the system is in the *active* phase, and it is zero when the system is in the *passive* phase. The system switches between these two phases according to the discrete input at the discrete input port *act/passivate*. Barrels are put out at the discrete output port *barrel* whenever the variable *content* reaches the value 10. After that, the value of *content* is initialized by 0. At the continuous output port *content* we put out the current value of the state variable *content*.

$\text{BarrelGenerator} = \langle (X_c \times X_d), (Y_c \times Y_d), (S_c \times S_d), \delta_{ext}, \delta_{state}, \lambda_d, \delta_{int}, ta, f, \lambda_c \rangle$

$X_c = \{\text{inflow} \mid \text{inflow} \in \Re\}$

$X_d = \{\text{act/passivate} \mid \text{act/passivate} \in \{\text{activate, passivate}\}\}$

$Y_c = \{\text{content} \mid \text{content} \in \Re\}$

$Y_d = \{\text{barrel} \mid \text{barrel} \in \{\text{10-liter-barrel}\}\}$

$S_d = \{(\text{sigma, phase}) \mid \text{sigma} \in \Re^{+}_{0\infty}, \text{phase} \in \{\text{active, passive}\}\}$

$S_c = \{\text{content} \mid \text{content} \in \Re^{+}_{0}\}$

δ_{ext} ((content, sigma, phase), e, (inflow, act/passivate))

```
   if act/passivate = activate  then
         phase := active
   elseif act/passivate = passivate   then
         phase := passive
   endif
```

δ_{state} ((content, sigma, phase), e, inflow)

```
   when-event  content   ≥ 10  then
      sigma :=  0   -- activate the model to put out the full barrel
```

```
λd ((content, sigma, phase), inflow) :=
   10-liter-barrel  -- send out 10-liter-barrel at port barrel

δint ((content, sigma, phase), inflow)
   content := 0
   sigma := ∞     -- passivate

ta ((sigma, phase)) := sigma

f ((content, sigma, phase), inflow)
   model-in-phase
      passive:  d content / dt := 0
      active: d content / dt := inflow

λc ((content, sigma, phase), inflow) :=
   content      -- the output at the discrete output port content
                -- is the value of the state content
```

Figure 3.2.4 shows the trajectories of a sample experiment of the barrel generator system. Initially the *phase* is *active,* the value of *content* is 0, *sigma* is infinity and the continuous input *inflow* is constant *1*. Thus the content increases linearly until it reaches 10 and a state event occurs, *sigma* is set to *0* causing an internal event. The first barrel is put out and *content* is initialized to *0*. The filling of the second barrel is interrupted by an external event at the discrete input port *act/passivate*. The model transits to the passive phase and no filling is done any more. After some time there is a second external event which activates the model again and the filling of the second barrel can be finished and the barrel can be put out at time *24*. During the filling of the third barrel, there is a discontinuous jump of the inflow from *1* to *2* resulting in increased filling rate. The filling of the fourth barrel is again interrupted by an external event and the model transits to the passive phase. The filling cannot be finished.

Note, that during the whole experiment, the value of state variable *content* is provided by the output at output port *content*.

In the following we will learn to know several combined discrete-continuous model prototypes which are based on the atomic DEV&DESS and which have proved to be most useful for simulation modelling.

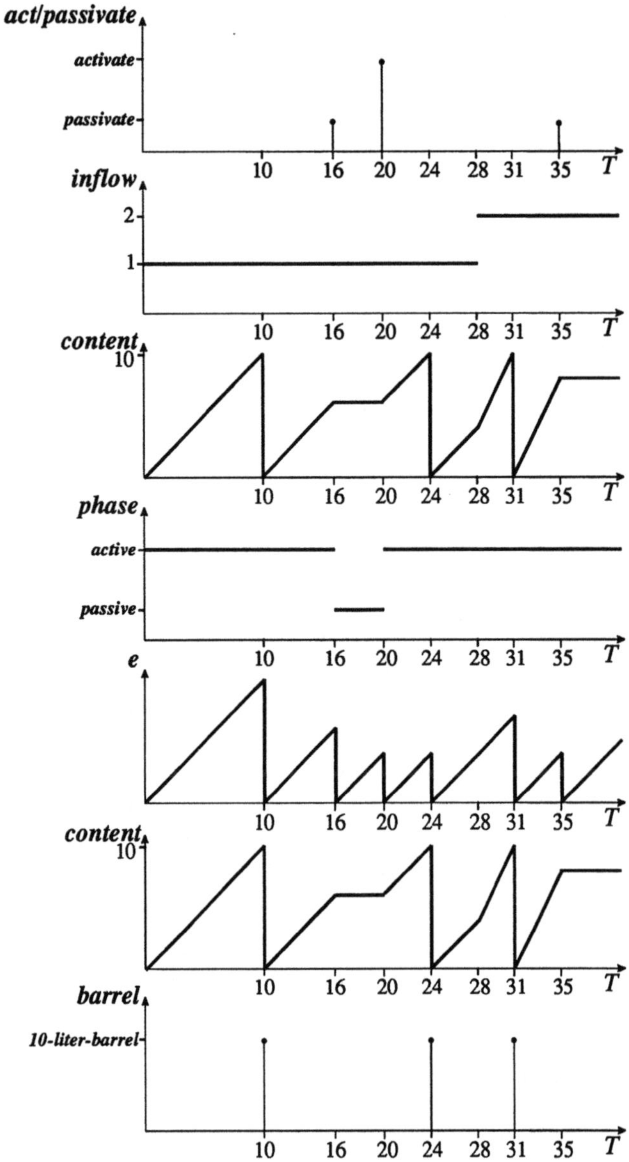

Figure 3.2.4 Barrel generator trajectories.

Continuous Input DEVS systems

Input segment of DEVS systems are constrained to DEVS segments. This guarantees that only a finite number of input events will occur and hence only a finite number of external transitions have to be executed. As we will see in Section 3.2.3, in a multi-formalism coupled system, when we couple an output of a DESS system to a DEVS system, the input to this DEVS is a continuous segment which represents an infinite number of external events and hence leads to an invalid DEVS. To overcome this problem, we use a special type of combined DEV&DESS system which allows continuous inputs and call it *continuous-input-DEVS.*

A continuous-input-DEVS is a structure as depicted in Figure 3.2.5 and is a combined DEV&DESS

$$DEV\&DESS = (X_M, Y_M, S, \delta_{ext}, \delta_{state}, \delta_{int}, \lambda_d, ta, f, \lambda_c)$$

with the following constraints:

- $Y_c = \{\}$;
- $S_c = \{\}$;
- $f : S \times X_c \rightarrow \{\}$, the rate of change function can be omitted;
- $\lambda_c : S \rightarrow \{\}$, the continuous output function can be omitted.

Hence, a continuous-input-DEVS is a combined DEV&DESS system which only has

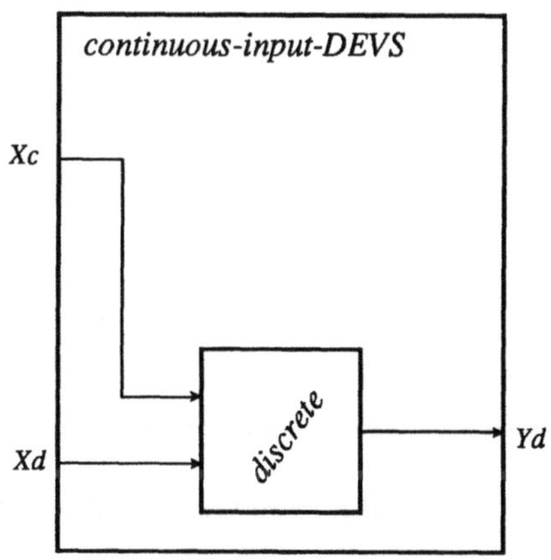

Figure 3.2.5 Continuous Input DEVS.

continuous inputs but no continuous states and no continuous outputs. It is possible to model state events which are caused by continuous changes of continuous inputs.

Modelling example: Model of an intelligent threshold sensor

The following continuous-input-DEVS is a model of a smart threshold sensor which is able to detect if the incoming input segment is stable around *0*. Such a sensor can be used to supervise an analog control process (see modelling example in Section 3.2.3). The continuous input to the sensor has to be the deviation of the actual control value from a nominal value. If the absolute value of the deviation is lower than a specified value *epsilon* and stays lower than this value *epsilon* for a specified time interval, then the signal is regarded to be stable.

```
StableSensor = <X, Y, S, δstate, δint, λd, ta>

X = {deviation | deviation ∈ ℜ}
Y = {sensorOut | sensorOut ∈ {off, on} }
S = { (sigma, sensor) | sigma ∈ ℜ+0∞, sensor ∈ {off, on}}
δstate ((sigma, sensor), e, deviation)
   when-event  sensor = off & |deviation| < epsilon  then
      sensor := on              -- wait timeInterval time units
      sigma := timeInterval     -- to put out the sensor value is "on"
   when-event  sensor = on & |deviation| >= epsilon  then
      sensor := off
      sigma := 0 -- put out the sensor value is "off" immediately

λd ((sigma, sensor), deviation) :=
   sensor         -- output of the model is the state variable "sensor"

δint ((sigma, sensor), deviation)
   sigma := ∞

ta ((sigma, sensor)) :=
   sigma    -- the time advance is specified by the state variable "sigma"
```

Figure 3.2.6 shows an example simulation run with a test deviation signal (a damped sinusodial signal) as input. Whenever the signal gets into the epsilon band, which is from -0.2 to 0.2 in our example, the state variable *sensor* is set to *on* and *sigma*, defining the value of the time advance function, is set to *5*. But as soon as the signal leaves the epsilon band, the sensor state value is set to *off* again. At time 13 the deviation

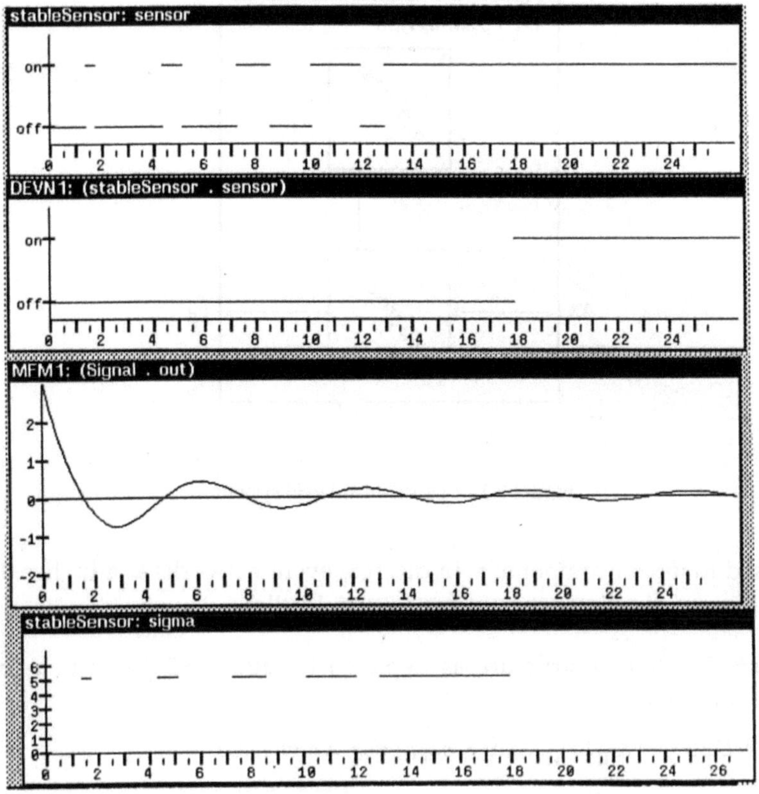

Figure 3.2.6 Input segment and state and output trajectories of stable sensor model.

enters the epsilon band and now stays within the band for more than 5 time units. Only now the model reacts with an output at time 18.

DEVSwithDESS systems

Atomic DEVSwithDESS systems are combined DEV&DESS systems but this type of systems is primarily a DEVS system which also employs some continuous state variables. That means, the input and output behavior is like that of DEVS systems but some of the interior states behave like that of DESS systems. This type of system is

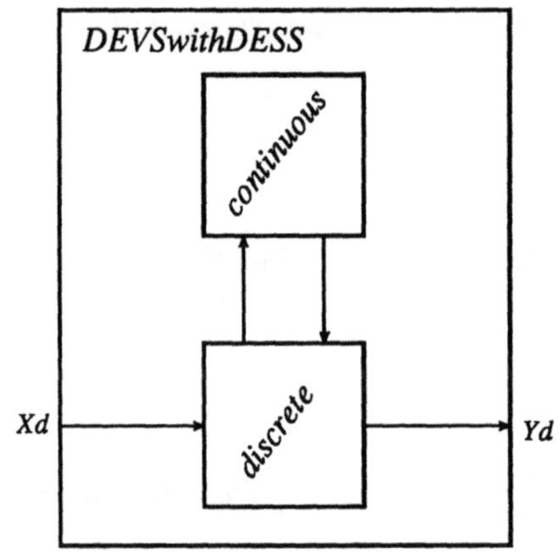

Figure 3.2.7 DEVS with DESS system.

mainly used to model components in discrete event networks at a higher degree of precision than a pure discrete event system would allow.

A *DEVSwithDESS* is a structure as depicted in Figure 3.2.7 and is a combined DEV&DESS system

$$DEV\&DESS = (X_M, Y_M, S, \delta_{ext}, \delta_{state}, \delta_{int}, \lambda_d, ta, f, \lambda_c)$$

with the following constraints:

- $X_c = \{\}$;
- $Y_c = \{\}$;
- $\lambda_c : S \rightarrow \{\}$, the continuous output function can be omitted;

Hence, a DEVSwithDESS is a combined DEV&DESS system which only has continuous states but no continuous inputs and outputs. It is possible to model state-events which are caused by continuous changes of continuous states.

Modelling example: Model of a continuous filling process

In queueing network simulation models it is sometimes required to model a process more detailed in order to get better information of the performance indices of the system under study. This may necessitate modelling a component at the continuous level and using it at the discrete level. An example of such a component is the model of a continuous filling process which is used in a network modelling a brewery [Schmidt 87].

The inputs to this model are the arrivals of empty barrels at the filling station. If the filling station is empty, then the the filling process starts, otherwise the barrel is queued before the station. Depending on the current content of the barrel, the filling rate is increased from an initial filling rate of 1 to a full filling rate of 6. Then the filling rate stays constant until the barrel is full. The current barrel is putted out immediately and a signal is sent out indicating that the next barrel can be serviced.

The external transition function is used to handle the arrivals of the barrels. There are two state events - one to specify the event when the filling process has to switch from the phase where the filling rate is accelerated to a full speed filling rate of 6 and the second to model the barrel-full event. In the derivative function *f*, the phase variable of the model is used to distinguish between the different filling rates. Depending on the value of the phase variable, different derivatives are defined for the continuous state variable *content*. Hence, this model also shows how the phase variable and state events can be employed to implement multi-models as defined by Ören [Ören 85, 89b].

FillingProcess = $< X, Y, S, \delta_{ext}, \delta_{state}, \lambda_d, \delta_{int}, ta, f>$

X = {in | in ∈ {20, 40, 60}}

Y = {(out, next) | out ∈ {20, 40, 60}, next ∈ {true}}

S_c = {(content) | content ∈ $\Re^+_0$}}

S_d = {(sigma, phase, barrel) | sigma ∈ $\Re^+_{0\infty}$,
    phase ∈ {waitForBarrel, acceleratedFilling, fullSpeedFilling, done},
    barrel ∈ {20, 40, 60}}

δ_{ext} ((content, sigma, phase, barrel), e, in)
  when-input-event -on-port
    in: barrel := in
        phase := acceleratedFilling
        sigma := ∞

```
δstate ((content, sigma, phase, barrel), e, in)
   when-event  [(0.2 * content) + 1] >= 6  then
        phase := fullSpeedFilling    -- full speed filling now
        sigma := ∞
   when-event content >= barrel then
        phase := done
        sigma := 0 -- activate the model to put out the full barrel

λd ((content, sigma, phase, barrel)) :=
   (barrel, true)    -- send out barrel at port out and true at port next

δint ((sigma, phase, barrel, content))
   barrel := 0
   content := 0
   phase := waitForBarrel
   sigma := ∞    -- passivate in waitForBarrel

ta ((sigma, phase, barrel)) := sigma

f ((content, sigma, phase, barrel))
   model-in-phase
      waitForBarrel, done:   d content / dt := 0
      acceleratedFilling:    d content / dt := (0.2 * content) + 1
      fullSpeedFilling:   d content / dt : = 6
```

Figure 3.2.8 shows the piecewise constant trajectory of the state variable *phase* and the piecewise continuous trajectory of the state variable *content* of an example simulation run. Note the dependencies of these two variables. A filling process always starts with the phase *waitForBarrel* and a filling rate of 0. In the *acceleratedFilling* phase the content increases with an ever increasing rate (see the curvature at the beginning of the filling processes). After about 9 time units the system transits into the *fullSpeedFilling* phase and we observe a constant increase of the *content* variable. Finally, in the *done* phase *content* makes a discontinuous jump to 0.

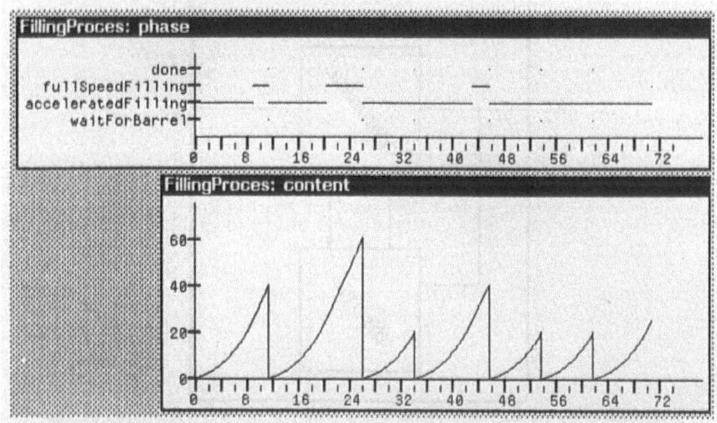

Figure 3.2.8 Trajectories of phase and content variables of filling process.

DESSwithDEVS systems

Atomic DESSwithDEVS systems are combined DEV&DESS systems which are primarily DESS systems. They employ a derivative function to specify the rate of change of the continuous state variables. The state event transition function and the internal transition function can be used to model events causing *discontinuities* in the continuous variables. Input segments as well as output segments are continuous or piecewise continuous and hence we regard the system to be continuous.

A *DESSwithDEVS* is a structure as depicted in Figure 3.2.9 and is a combined DEV&DESS

$$DEV\&DESS = (X_M, Y_M, S, \delta_{ext}, \delta_{state}, \delta_{int}, \lambda_d, ta, f, \lambda_c)$$

with the following constraints:

- $X_d = \{\}$;
- $Y_d = \{\}$;
- $\delta_{ext} : S \times X_M \rightarrow S$, the external transition function can be omitted;
- $\lambda_d : S \rightarrow \{\}$, the discrete output function can be omitted.

Hence, a DESSwithDEVS is a combined DEV&DESS system which only has discrete states but no discrete inputs and outputs.

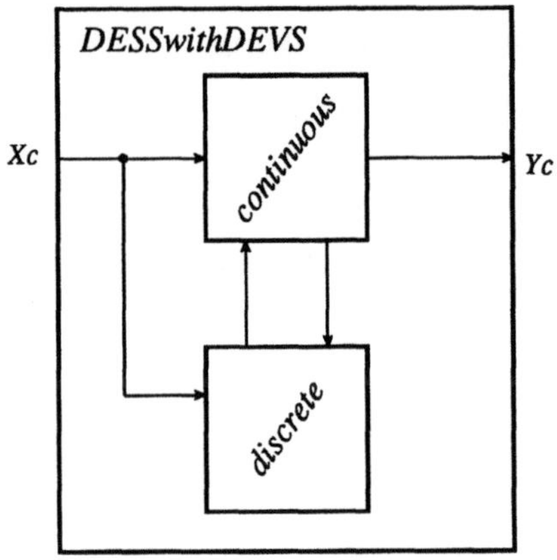

Figure 3.2.9 DESS with DEVS system.

Modelling example: Bouncing ball

A table tennis ball dropped from a specified height is a typical example of a situation best modelled by this type of system. The free fall is modeled by the DESS part while the impact of the ball on the ground is modelled by a state event in the state event transition function. Figure 3.2.10 shows an example simulation run.

```
BouncingBall = <X, Y, S, f, δstate, δint, ta, λc >
X = {}
Y = {pos | pos ∈ ℜ}
S = {(pos, vel) | pos, vel ∈ ℜ}
f ((pos, vel))
   d vel / dt := - g - resistance * vel    -- gravitation - resistance
   d pos / dt := vel
λc ((pos, vel)) :=
   pos    -- output of the model is the position of the ball
δstate ((pos, vel), e)
   when-event pos <= 0 then
         vel := - vel     -- invert the velocity
ta (()) := ∞     --time advance = ∞, so no time events
```

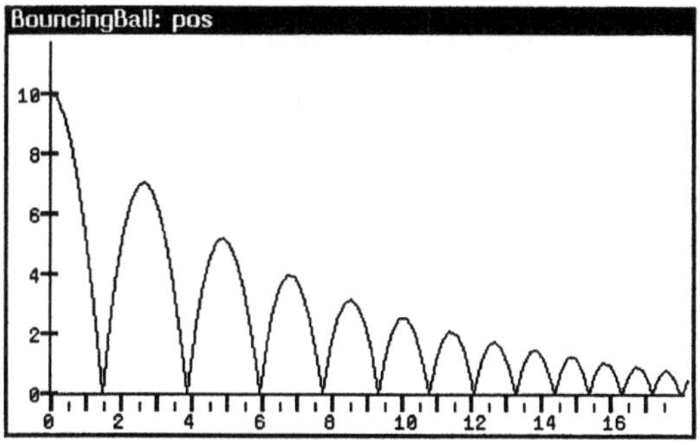

Figure 3.2.10 Trajectory of position variable of bouncing ball.

DEVSwithAnalyticalDESS systems

This special atomic system type for combined discrete-continuous modelling is also a special type of combined DEV&DESS system and is similar to the DEVSwithDESS system formalism. It is mainly a DEVS system only employing discrete inputs and discrete outputs but it owns some continuous state variables. In difference to the DEVSwithDESS type, the analytical solutions of the continuous state variables are known and are given in form of a function $F = \int f$. As the system only shows a discrete event behavior at the input / output interface, it is not necessary to compute the continuous state trajectories of the continuous state variables. It is sufficient to jump from one event to the next and just compute the values of the continuous states at the event times. This signifies a step to pure discrete event modelling. However, continuous states may cause state events. The solution F can be employed for the computation of the times of the state events and of the state at the event times. In Section 3.3.4 we will outline a simulation algorithm which uses these advantages.

A *DEVSwithAnalyticalDESS* is a structure as depicted in Figure 3.2.11 and is defined by

$$DEVSwithAnalyticalDESS = (X_d, Y_d, S, \delta_{ext}, \delta_{state}, \delta_{int}, \lambda_d, ta, F)$$

with $F: S \rightarrow S_c$

specifying the combined DEV&DESS system

$$DEV\&DESS = (X_M, Y_M, S, \delta_{ext}, \delta_{state}, \delta_{int}, \lambda_d, ta, f, \lambda_c)$$

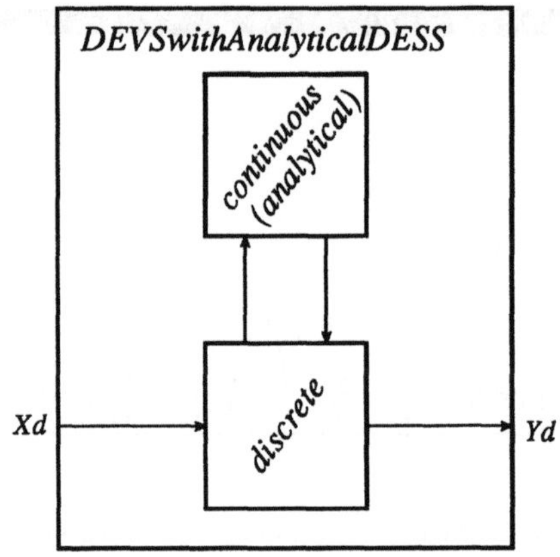

Figure 3.2.11 DEVS with Analytical DEVS system.

with the following constraints:

- X_d, Y_d, S, δ_{int}, δ_{ext}, δ_{state}, λ_d, ta have the same interpretation as in DEV&DESS systems;
- $X_c = \{\}$;
- $Y_c = \{\}$;
- $\lambda_c : S \rightarrow \{\}$, the continuous output function can be omitted;
- $f = dF / dt$, the derivative function f of the continuous state is given by the derivative of the analytical solution F;

Hence, a DEVSwithAnalyticalDESS is a combined DEV&DESS system which only has continuous states but no continuous inputs and outputs. It is possible to model state events which are caused by continuous changes of continuous states. As the input and output interface is like that of pure DEVS systems and values of the continuous states are only computed at event times, they can be used as components in DEVS networks.

With a DEVSwithAnalyticalDESS system we associate the following dynamical behavior: When there are no external events or state events, then the system transits from one internal transition to the next. As in DEVS, before every internal transition there is an output. An event input is handled like in DEVS, it activates the external transition function. At each event, the continuous states before the event are computed employing the analytical solution F. The analytical solution F also can be employed to compute the time of the states event. Knowing the value of the continuous state variable when the state event occurs, this value and the analytical solution can be set equal and the time of the event can be computed.

Modelling example: Tank system

A simple model of a tank with a sensor system taken from [Zeigler 89] will serve as an introductory example for DEVSwithAnalyticalDESS modelling. The system has three input variables *inflow, outflow* and *leakage* by which the water inflow, the outtake of water and some losses of water by leakage are provided from outside the system. The analytical solution for the continuous state variable *level* is defined by the old value of *level* plus the *inflow* minus *outflow* and *leakage* multiplied by the time difference Δt. Furthermore, there is an output variable *sensor* which models the sensor system. When the water level gets below a constant value of 3, the value of the sensor should become *empty*. When the level gets beyond 10, the value of the sensor should become *full*. Thus, there are two state events, viz., one when the level is 3 and one when the level is 10. The model looks the following:

$$Tank = < X, Y, S, \delta_{ext}, \delta_{state}, \lambda, \delta_{int}, ta, F>$$

$$X = \{(inflow, outflow, leakage) \mid inflow, outflow, leakage \in \Re^+_0\}$$

$$Y = \{sensor \mid sensor \in \{full, empty\}\}$$

$$S_C = \{(level) \mid level \in \Re^+_0\}\}$$

$$S = \{(sigma, phase, sinflow, soutflow, sleakage) \mid sigma \in \Re^+_{0\infty}, phase \in \{empty, filling, full, emptying\}, sinflow, soutflow, sleakage \in \Re^+_0\}\}$$

```
δext ((level, sigma, phase, sinflow, soutflow, sleakage), e,
     (inflow, outflow, leakage))
  when-x-on-port
        inflow:     sinflow := inflow
        outflow:    soutflow := outflow
        leakage:    sleakage := leakage

δstate ((level, sigma, phase, sinflow, soutflow, sleakage), e)
  when-event  level >= 10 & phase = filling  then
        phase := full
        sigma := 0     -- activate to put out sensor = full
  when-event level <= 3 & phase = emptying then
```

```
      phase := empty
      sigma := 0     -- activate to put out sensor = empty

λ ((level, sigma, phase, sinflow, soutflow, sleakage)) :=
  model-in-phase
    empty:   sensor := empty  -- output sensor = empty
    full:    sensor := full   -- output sensor = full

δint ((level, sigma, phase, sinflow, soutflow, sleakage))
  model-in-phase
    empty:  phase := filling
    full:   phase := emptying
  sigma := ∞    -- passivate
ta ((sigma, phase, sinflow, soutflow, sleakage)) := sigma
F ((level, sigma, phase, sinflow, soutflow, sleakage), Δt)
  level := level + Δt (sinflow - soutflow - sleakage)
```

Figure 3.2.12 shows sample trajectories of the tank system model together with the trajectories of an event-based controller of the tank and consumers and a leakage of the tank [Praehofer 91]. The first trajectory shows the water outtake from the tank at any particular time of the experiment which is the sum of the outtake of each of some consumers. The second trajectory shows the the outflow of water through leakage at any point in time. It can be seen that the leak in the tank gets bigger and bigger. The third trajectory now shows the control output of an event-based controller which is the water inflow to the tank. The water inflow is turned on when the controller receives the *empty* signal from the tank and it is turned off again when it receives the *full* signal. The sensor signals, which are provided by the output from the tank, are shown in the next trajectory. The trajectory just below shows the values of the continuous state variable *level* of the tank. Note that the values of the continuous state variable *level* are computed at the event times only. Therefore, a DEVS trajectory is used to put out the values of the continuous state trajectory at the event times. The last trajectory shows the phases the event based controller transits through.

Under normal circumstances it cycles through phases *empty, fullToEarly, filling, full, emptyToEarly, emptying* and *empty* again. Through the increasing leakage in the tank, water losses finally get so high that the tank becomes empty to early. An error in the emptying process is encountered by the event-based controller which causes the controller to transit to phase *errorInEmptying* and to put out an error message as show at the bottom of the figure.

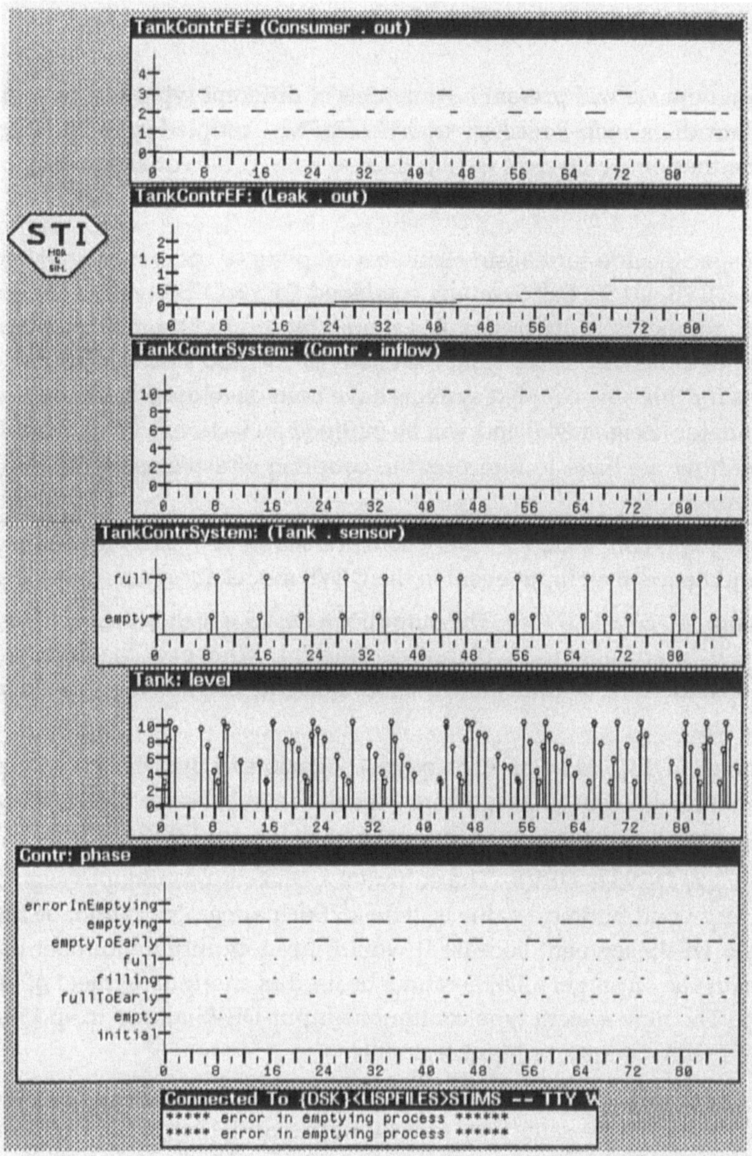

Figure 3.2.12 Simulation of the event-based tank control system.

3.2.3 Coupled Multi-Formalism System Formalism

In this section now we will present how models of different types can be coupled together and introduce *multi-formalism networks (MFN)* – coupled systems whose components are either of the discrete event, discrete time, differential equation or again of the multi-formalism type.

This system specification formalism facilitates coupling of components which can be either of the DEVS, DTSS, DESS, or any combined DEV&DESS type. It is possible to couple the components of different types simply by connecting their output and input ports in the same way as we build the one-type coupled systems. Special simulation concepts for this new coupled systems have been developed based on the DEVS abstract simulator [Zeigler 84a] and will be outlined in Section 3.3.5. In the following we illustrate how we have to interpret the coupling of systems of different types (Figure 3.2.13):

- *Coupling of DTSS to DEVS* Every state transition of a DTSS results in an output and hence in an input event in the DEVS model at regular time intervals.
- *Coupling of DEVS to DTSS* The output of a DEVS which is the input to a DTSS model is interpreted to be piecewise constant. Whenever the DTSS is scheduled to undertake its transition, it takes the value originated from the last output event.
- *Coupling of DEVS to DESS* The output segments of the DEVS are interpreted to be piecewise constant segments which are a subclass of piecewise continuous segments and hence are valid input segments for DESS models.
- *Coupling of DESS to DEVS* The output of a DESS is a continuous segment, i.e., for every point in time a value is defined. Such a segment cannot be interpreted as a DEVS segment because it would mean an infinite number of events and thus the output of a DESS cannot be used as an input segment of ordinary DEVS. The new system type continuous-input-DEVS allows to specify DEVS models with continuous input segments.
- *Coupling of DTSS to DESS* In such a coupling structure the output of a DTSS is interpreted in the same way as we interpreted the output of DEVS in DEVS to DESS couplings, namely as piecewise constant. The output at one transition of the DTSS specifies a constant input value for the DESS until the next transition.

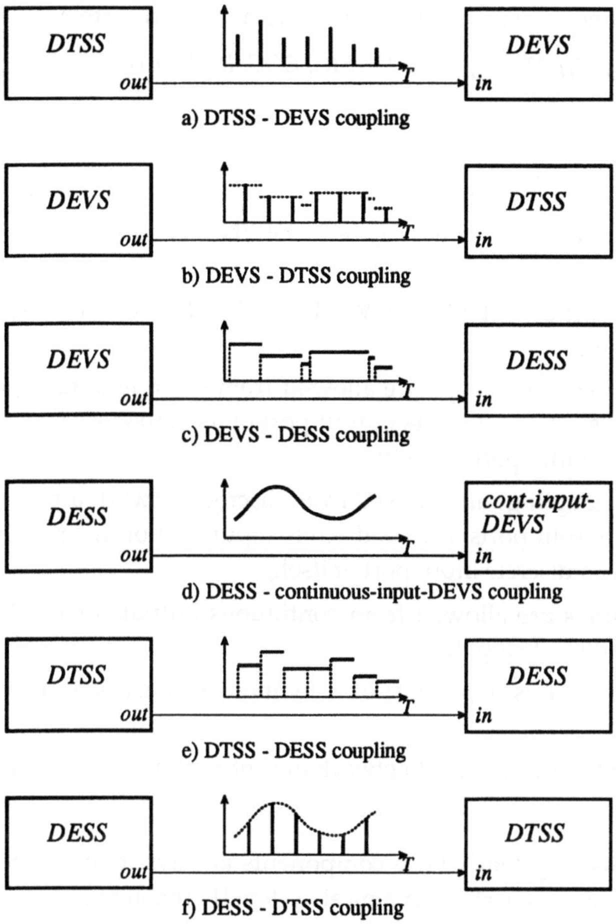

Figure 3.2.13 Multi-formalism system couplings interpretations.

- *Coupling of DESS to DTSS* The input of a DTSS coming from a DESS at one transition at time t is defined by the current value of the continuous output segment at time t.

Formally we define a *multi-formalism network* as a coupled system

$$MFN = (X_N, Y_N, D, \{M_d \mid d \in D\}, EIC, EOC, IC, Select)$$

where *Select* is a function

Select: subsets of $D \rightarrow D$

used for *tie-breaking* in the same way as in DEVN and subject to the following constraints:

- the components are DEVS, DEVN, DTSS, DTSN, DESS, DTSN, DEV&DESS, or MFN systems;
- no direct feedback loops are allowed from discrete output ports to discrete input ports, i.e., no discrete output port of a component may be connected to an discrete input port of itself;
- no direct couplings are allowed from discrete network input ports to discrete network output ports, i.e., no discrete input port of the network may be connected to an discrete input port of itself;
- no couplings are allowed from continuous outputs or continuous network inputs to discrete inputs;
- every continuous input port of a component must be influenced by exactly one port;
- every continuous external network output port must be influenced by exactly one port.

A multi-formalism system whose components are dynamical systems and which obeys the constraints is itself a dynamical system [Praehofer 91].

The resulting system can be regarded as a cross-product of the component systems. The resulting state is defined by the cross-product of the states of the dynamical system description of the components (which also includes a components sx_a to store the event inputs arriving at the continuous input ports of the component a) plus the set Y_c (state component sy) to store the event outputs arriving at the continuous network output ports. Whenever an event input arrives at a continuous port, this value is stored to define the constant value until the next input event at the continuous port. The piecewise constant trajectories then are used by transition and output functions. Together with the continuous input values defined by continuous trajectories, they define the current inputs. sy is used to define the continuous external outputs. The internal, external, and state event transition functions are defined analogously as in pure discrete event networks. We only have to care for the state events in the component systems.

Modelling example:
Combined analog and event-based motor speed control system

A simple model of a motor whose speed is first controlled by a conventional analog controller which is then controlled and supervised by an intelligent event-based control system [Zeigler 90a] will serve as an example of a combined discrete-continuous multi-formalism coupled model. A motor is controlled by a conventional PI-controller which in turn is controlled by an event-based controller.

A motor which drives a machine has to be brought to different speeds and different work loads are exerted to fulfill a predefined sequence of tasks. An event-based controller which knows this sequence of tasks in pursuit of a given goal is used. It issues the required nominal speeds as well as the work loads and supervises the process runs. To drive the motor to the nominal speeds, a conventional analog controller is used.

Figure 3.2.14 shows the structure of this control system. The analog part, which is modelled by a DESS coupled system, consists of the component modelling the motor as well as the model of the analog controller. The motor is modelled at the behavioral level by a simple atomic DESS and should be considered as a motor driven by a force input together with a tachometer which provides the current speed. The input *load* provides the current work load which functions like a resistance. The conventional controller is a combination of an integral and proportional controller whose outputs are added. The adder also amplifies the control signal to provide an appropriate force to drive the motor. A second output from the analog control part supplies the deviation of the current speed from the nominal speed.

The event-based control system is modelled by a DEV&DESS network and consists of three parts. There are two threshold sensors and the event-based controller itself. The threshold sensors receive the deviation from the analog controller in the form of a continuous signal. They are modelled by continuous-input-DEVS. For the model *stableSensor*, the model specified in the threshold sensor example of Section 3.2.2 is used. Using the deviation from the nominal speed and the actual speed, it is able to detect when the motor speed is stable around the nominal speed. The model *validSensor* is very similar, the only difference being that the sensor immediately reacts whenever the deviation is recognized to be without the band of normal operation.

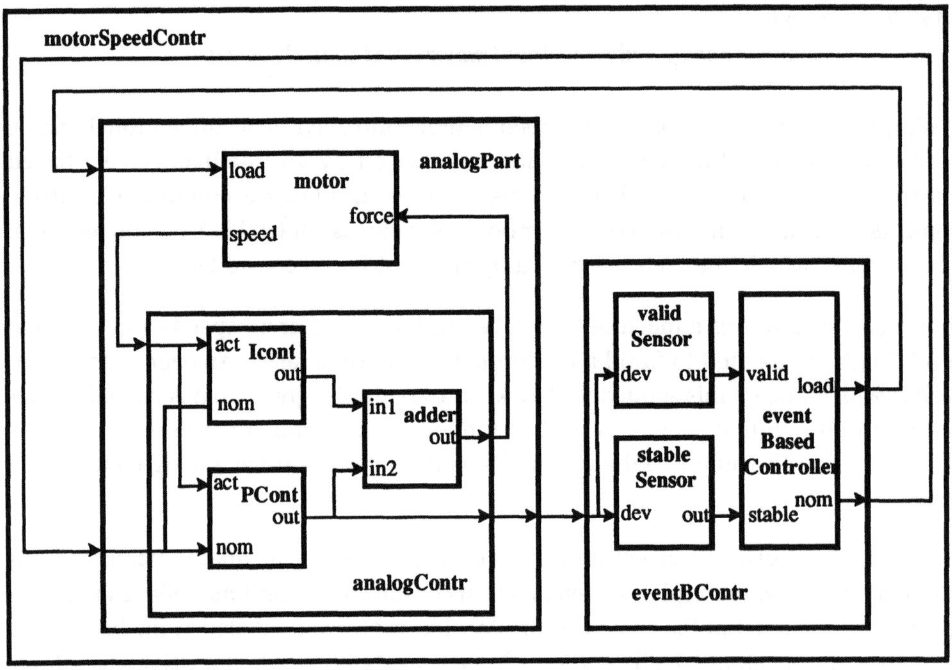

Figure 3.2.14 Model of an analog and event-based motor speed control system.

The event-based controller is modelled by a pure DEVS. It knows of a sequence of tasks and each task consists of a nominal speed, a work load and a working time. First it exerts the nominal speed. This makes the analog controller react and drive the motor to the new speed. In the event-based controller, a time window is used to supervise the analog control process. In the case that it takes too long to drive the motor to the new speed, a failure in the analog controller is detected and the process is stopped.

When the threshold sensor *stableSensor* reacts, the speed is at the right value and is stable, the real working process can begin, and the work load is exerted. The working load now causes a loss of speed which, if the work load is not too high, can be compensated by the analog controller. But if the work load is too high, the controller is not able to compensate the loss fast enough, the threshold sensor *validSensor* reacts and the load is withdrawn to allow a restabilization of the speed. This procedure is continued until the absolute stabilization time exceeds the working time. After that, the event-based controller prints an error message and continues with the next task.

Figure 3.2.15 shows a screen of the simulation of this control system. The four trajectories show nominal speeds and work loads issued by the event-based controller, the control output of the analog controller and the speed of the motor. The first task with nominal speed 4 and work load 2 is processed correctly. After some time the speed is stable and the work load is exerted. The loss in speed through the load can be compensated fast enough and the work is finished without any interruption. In the second task which is characterized by nominal speed 7 and work load 5, the work load is too high, resulting in a big and fast loss of speed, a reaction of the valid sensor, and hence in a withdrawal of the load. After some other trials the event-based controller prints an error message and continues with the next task.

3.3 ABSTRACT SIMULATOR CONCEPTS

Zeigler [Zeigler 84a] introduced simulation concepts for modular, hierarchical discrete event systems. The *abstract simulator for DEVS* has a hierarchical structure reflecting the structure of the hierarchical DEVS. It handles all the simulation needs. Thus a DEVS model can be directly transformed into an executable simulation program using the abstract simulator. Similar simulation concepts have been developed for discrete time and differential equation specified systems. We pursue an approach very similar to the DEVS abstract simulator. The *abstract simulator for DTSS* and *DESS* will also reflect the hierarchical structure of the model. The same methodology and terminology is used. The reason why we do this is mainly to be compatible to the DEVS simulator and to make preparations for the usage of them for the simulation of multi-formalism models.

Each of the different types of abstract simulators consist of two types of object - the *simulators* and the *coordinators* (Figure 3.3.1). With every atomic model we associate a simulator, with every coupled model we associate a coordinator of the appropriate type. A *root coordinator* is attached to the overall network coordinator which has the task of initiating the simulation cycles. Simulation proceeds by messages passed among the simulators and coordinators. It is important to note that simulators and coordinators receive the same kind of messages. Consequently, a coordinator does not have to distinguish if the subordinate is a simulator or coordinator. The abstract simulators consisting of objects and working with message passing is readily implementable in an object-oriented programming language.

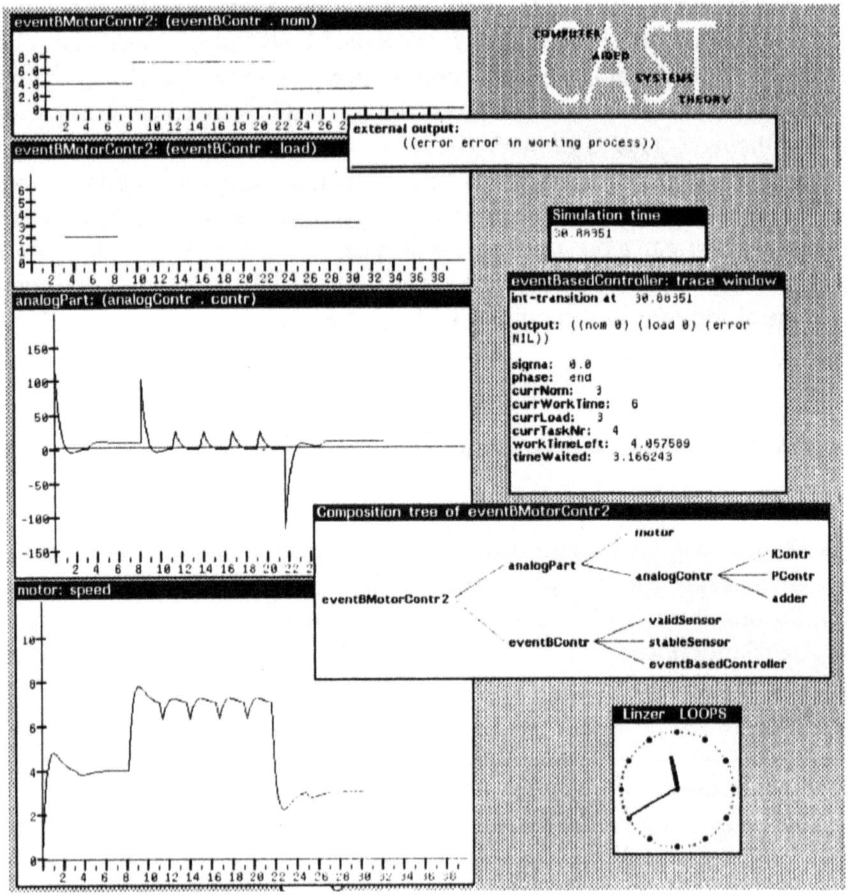

Figure 3.2.15 Simulation of the event-based motor speed control system.

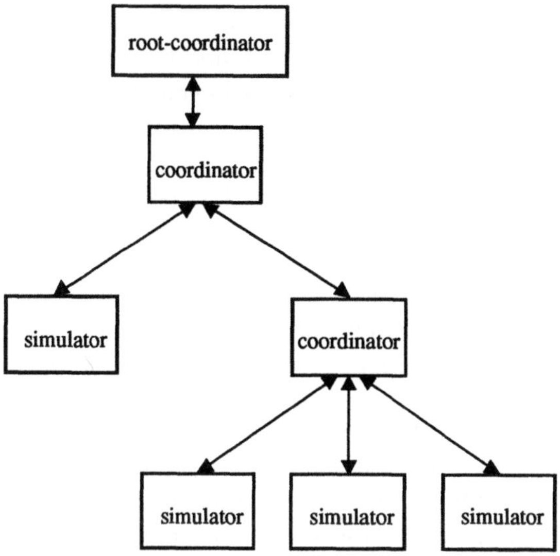

Figure 3.3.1 Abstract simulator attached to a hierarchical model.

In the following we will outline the abstract simulators for the different model types presented in Section 3.2. We will present the variables the different objects use, how they react to messages they receive and which messages they send.

3.3.1 Abstract Simulator of Discrete Event Systems

The *abstract simulator for modular, hierarchical DEVS* has been developed by Zeigler [Zeigler 84a] and used in the implementations of DEVS-Scheme [Zeigler 87, 90a] and STIDEVS [Praehofer, Spalt 88]. This concept has been slightly modified here by introducing methods for initialization purposes.

Simulation proceeds by several messages passed among the *devs-simulators* and *devs-coordinators* which carry information about internal and external events, about outputs of associated DEVS, as well as information for initialization (Figure 3.3.2). We have *i-messages* for initialization purposes and *-, x- and *y-messages* for normal simulation execution. All of these messages carry the current simulation time t.

An i-message causes the initialization of the abstract simulator. Initialization requires the computation of the time of last event *tl* and time of next event *tn* of all components. A *-message indicates that an internal event shall be executed. An x-message indicates the arrival of an external event and leads to the execution of the external transition function. Y-messages report the outputs of selected components. It is important to see that devs-simulators and coordinators are able to handle the same type of messages. This enables the simulation of hierarchical specified DEVS.

When a devs-simulator receives an i-message, it computes the time of last event by subtracting the elapsed time *e* from the time of the i-message. The time of next event *tn* is computed by adding the value of the time advance to the time of last event.

When a *-message is received by a devs-simulator, it computes the output and carries out the internal transition function of the associated DEVS. The output is sent back to the parent coordinator in a y-message. Finally the time of the last event *tl* is set to the current time and the time of the next event is given by the current time plus the value of the time advance function *ta*.

When an x-message is received by a simulator, it directly executes the external transition of the atomic DEVS. As by internal events, the time of the last event *tl* is set to the current time and the time of the next event is set to the current time plus the value of the time advance function *ta*.

```
Devs-simulator

variables:
   parent       -- parent coordinator
   tl           -- time of last event
   tn           -- time of next event
   DEVS         -- associated model
         with total state (s, e)
   y            -- current output value of the associated model

when receive i-message (i, t) at time t
   tl := t - e
   tn := tl + ta(s)
```

```
when receive *-message (*, t) at time t
   if t ≠ tn then
      error: bad synchronization
   y := λ(s)
   send value y in y-message (y, t) to parent coordinator
   s := δint (s)
   tl := t
   tn := tl + ta(s)

when receive x-message (x, t) at time t with input value x
   if not (tl ≤ t ≤ tn) then
      error: bad synchronization
   s := δext (s, t - tl, x)
   tl := t
   tn := tl + ta(s)

end Devs-Simulator
```

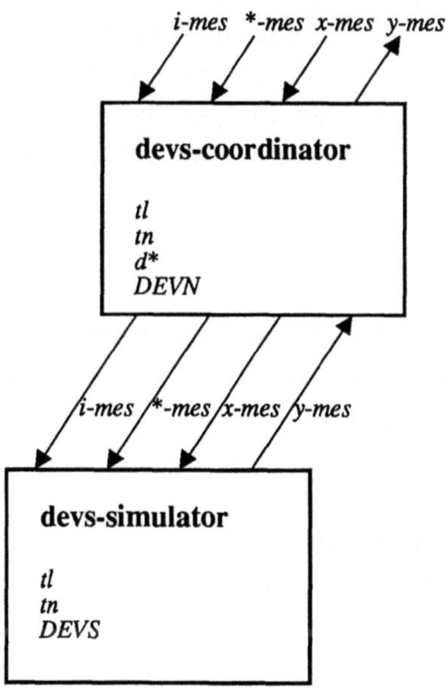

Figure 3.3.2 DEVS abstract simulator.

When a devs-coordinator receives an i-message, it forwards it to each of its children. When all the children have handled their i-messages, it sets its time of last event *tl* to the maximum of the times of last event of the children and its time of next event *tn* to the minimum of the times of next event of its children.

When a devs-coordinator receives a *-message, the message is transmitted to the subordinate representing the imminent component DEVS. If there is more than one imminent component, the tie breaking order *Select* is used to select one of them. After the completion of the internal event and all the external events caused by the output of the imminent component, the time of the next event is computed by the minimum of the next event times of its children.

When a coordinator receives a x-message, it consults the external input coupling of the DEVS network to generate the appropriate x-message for the subordinate influenced by the external event. X-messages are sent to all children influenced by the external input. After that, the time of the next event is computed.

When a coordinator receives a y-message which carries the output information of its imminent child, it consults the external output coupling of the DEVS network to see if it should be transmitted to its parent coordinator and the internal couplings to obtain the children and their respective input ports to which the message should be sent. In the first case, a y-message carrying the external network output is sent to the parent coordinator. In the latter case, the y-message is converted into a x-message indicating the arrival of an external event at the influenced component.

```
Devs-coordinator

variables:
   DEVN = (X, Y, D, {Md | d ∈ D}, EIC, EOC, IC, Select)
               -- the network associated
   parent      -- parent coordinator
   tl          -- time of last event
   tn          -- time of next event
   imminent    -- list of imminent children
   d*          -- selected imminent child
```

```
when receive i-message (i, t)  at time t
   for-each child d in D do
      send i-message to child d
   end for
   tl := max {tl_d | d ∈ D}
   tn := min {tn_d | d ∈ D}

when receive *-message (*, t) at time t
   if t ≠ tn then
      error: bad synchronization
   imminent := {d | d ∈ D ∧ tn_d = min {tn_e | e ∈ D}}
   d* := Select (imminent)
   send *-message (*, t) to d*
   tl := t
   tn := min {tn_d | d ∈ D}

when receive x-message (x, t) at time t with external input x
   if not (tl ≤ t ≤ tn) then
      error: bad synchronization
   -- consult EIC to get children influenced by the external input
   receivers := components r ∈ children influenced by external input x
   for-each r in receivers
      send x-messages (x_r , t) to r with input value x_r got from x through EIC
   tl := t
   tn := min {tn_d | d ∈ D}

when receive y-message (y, t) with output y from d*
   if the output from d* influences the external network output ports then
      send y-message (y_N, t) to parent with value y_N got from y through EOC
   -- check IC to get children influenced by the output y of d*
   receivers := components r ∈ children influenced by d*
   for-each r in receivers
      send x-messages (y_r , t ) to r with input value y_r got from y through IC

end Devs-coordinator
```

3.3.2 Abstract Simulator for Discrete Time Systems

The abstract simulator for hierarchical discrete time system has to implement the fixed step algorithm to compute the state and output values given the input value at

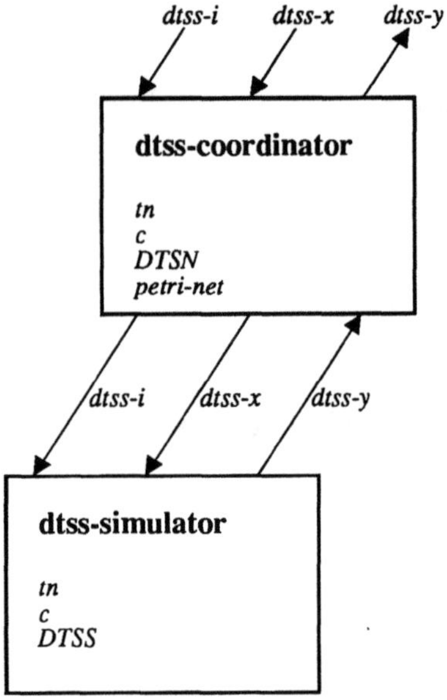

Figure 3.3.3 DTSS abstract simulator.

each time of the discrete time base. It consists of *dtss-simulators* of several types and *dtss-corrdinators* and uses *dtss-i-*, *dtss-x-*, and *dtss-y-messages* (Figure 3.3.3). The dtss-x-message causes the execution of the state-transition and the computation of the output value and carries the input value. The output value is sent back to the parent coordinator in a dtss-y-message. There is a main difference between the simulation of Moore-type systems and Mealy-type systems. As the output of Moore-type components only depends on the state value, the output for the next simulation cycle can be computed right after the state transition. This value is sent back to the parent coordinator where it is stored for the next simulation cycle. For the computation of the output of Mealy-type components, besides the state, also the input value is required. Thus, it is not possible to compute the output in advance. The simulation of function specified systems (FNSS) works in the same way as the simulation of Mealy components but without the state transition.

This difference in the simulation of Moore types on the one hand and Mealy and memoryless components on the other hand requires a special initialization of the hierarchical abstract simulator. For this purpose we use the dtss-i-messages. When a simulator of a Mealy or FNSS component receives such a message, it can ignore it. When a simulator of a Moore-type component receives a dtss-i-message it computes the output from the initial state and send it back in a dtss-y-message to the parent coordinator.

```
Dtss-moore-simulator
variables:
   DTSS = (X, Y, Q, δ, λ, c)    -- associated model
         with state q
   y               -- the current output value of the associated model
   tn              -- time of next state transition

when receive dtss-i-message (dtss-i, t) at time t
   tn := t
   y := λ (q)
   send y in dtss-y-message (y, tn) to parent coordinator

when receive dtss-x-message (x, t) with input value x
   q := δ (q, x)
   tn := tn + c
   y := λ (q)     -- output for the next simulation cycle
   send dtss-y-message (y, tn) to parent coordinator

end Dtss-moore-simulator
```

```
Dtss-mealy-simulator
variables:
   DTSS = (X, Y, Q, δ, λ, c)    -- associated model of type Mealy
         with state q
   y               -- the current output value of the associated model
   c               -- constant time step of the associated model
   tn              -- time of next state transition

when receive dtss-i-message (dtss-i, t) with time t
   tn := t

when receive dtss-x-message (x, t) with input value x
   y := λ (q, x)
   send y in dtss-y-message (y, t) to parent coordinator
   q := δ (q, x)
   tn := tn + c

end Dtss-mealy-simulator
```

The main task of the dtss-coordinator is to schedule the components' state transitions in the right sequence. The scheme is data-driven - only when the component's inputs at all the ports are available, it is ready to go.

A coordinator for a DTSS network may receive and send dtss-x-, dtss-y, and dtss-i-messages. When it receives a dtss-x-message, it first stores the input values and then schedules the computation of the next state and the output value of each of its components. The determination of a working sequence of the activation of the components is data driven. When all inputs for one component are available, the component can go. This can be visualized most appropriately by a Petri net [Reisig 86] where the places contain output values for times (y, t) instead of tokens.

For a network $N = (X_N, Y_N, D, \{M_d \mid d \in D\}, EIC, EOC, IC, c)$, this Petri net is constructed in the following way:

$$Petri\text{-}net_N = (P, T; F)$$

where

$$P = P_{EIC} \cup P_{EOC} \cup P_{IC}$$

$$P_{EIC} = \{p_{Nd} \mid d \in D \text{ and } d \text{ is influenced by the network input ports}\}$$
are the input places,

$$P_{EOC} = \{p_{dN} \mid d \text{ influences the external network output ports}\},$$
are the output places,

$$P_{IC} = \{p_{de} \mid d, e \in D \text{ and } d \text{ influences } e\},$$

$$T = D,$$

$$F = P \times T \cup T \times P = \{(p_{xd}, d) \mid p_{xd} \in P, d \in T\} \cup \{(d, p_{dx}) \mid p_{dx} \in P, d \in T\}.$$

and we define the function *has-concession (d, t)* by

$$has\text{-}concession(d,t) := \begin{cases} true & \text{if for all } p_{ed} \in preset(d) \text{ there is a value } (y_e, t) \\ false & \text{otherwise,} \end{cases}$$

meaning that on all places of the preset of d the values for time t are available, which in the data-driven scheduling scheme means that d can go for time t.

With every coupling from component e to component d we associate a place and with every component we associate a transition. We call the places associated with the external input and external output couplings input places and output places respectively. There is an arc from place p to transition d, iff the coupling represented by

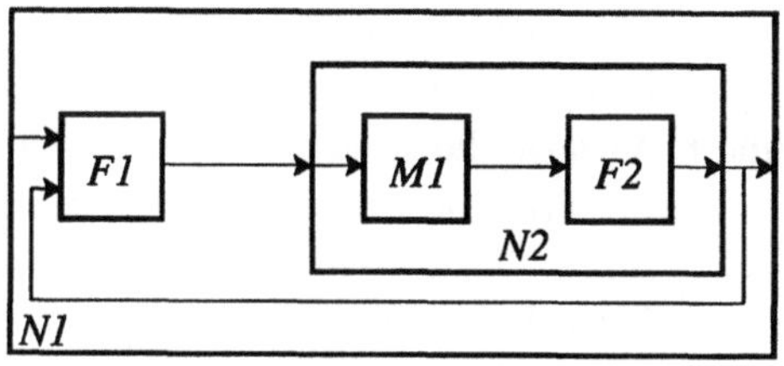

(a) network N1 with subnetwork N2

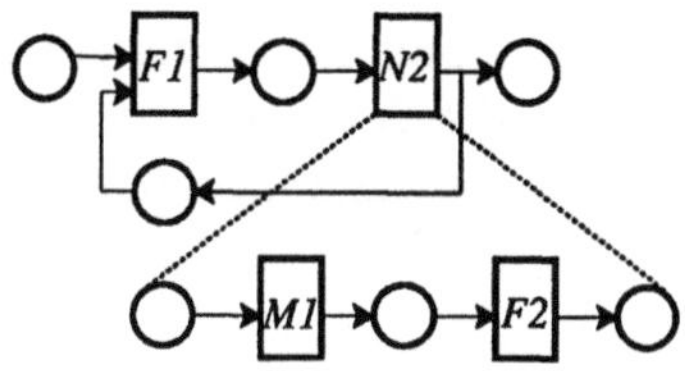

(b) petri net for network N1 and petri net for subnetwork N2

Figure 3.3.4 Networks and their Petri nets.

p ends at component associated with d. There is an arc from transition d to place p, iff the coupling represented by p starts at the component associated with d. Figure 3.3.4 shows the Petri nets constructed for network *N1* and its component network *N2*.

When a dtss-coordinator now receives a dtss-x-message, it first stores the input value on the input places associated with the external input couplings. Now each transition has to be fired exactly once. A transition has concession if all the places of its preset have got values, meaning that all the outputs of the influencers are available. Firing a transition d means taking all the values y_e of the preset and sending value $x_d = (.., y_e, ...)$ in a dtss-x-message to the simulator (or coordinator) of the component d. After receiving the output from this child in a dtss-y-message, the output is stored on the places of the postset of d.

The dtss-i-messages are used for initialization purposes. When a dtss-coordinator receives a dtss-i-message, it first send dtss-i-messages to each of its children and receives the output values from the Moore-type systems. After that, it tries to fire each of the mealy-components from which it has not received an output yet. But, of course, only those can be fired whose inputs do not depend on the network input. Finally, if all output places have values, this network is of type Moore [Praehofer 91]. It takes the values at the output places and sends these values back to the parent coordinator.

The dtss-x-message finds the situation that outputs of Moore-type components and components fired already in the initialization phase reside on the places for the components not fired for time *t*. As now the external input is available, also the Mealy components not fired in the initialization phase can be fired in a first loop. Now all outputs for time *t* are available and, if the component is of type Mealy, the output of the network can be sent back to the parent coordinator. In the next loop all the Moore-type components are fired and they make their state transition and send back the output for the next simulation time. Based on these outputs, all components which have been fired in the initialization phase are fired again for the next simulation time. If the network is of type Moore, the network output can be sent back to the parent. The dtss-x-message leaves the Petri net in the same situation for the next simulation cycle as it was before the message execution.

```
Dtss-coordinator

variables:
   N = (XN, YN, D, {Md | d ∈ D}}, EIC, EOC, IC, c)
               -- the network associated with the coordinator
   moore-comps := {d | d ∈ D, is of type Moore}
               -- components of type Moore
   mealy-comps := {d | d ∈ D, is of type Mealy or FNSS}
               -- components of type Mealy
   init-mealy
      -- mealy-type components already fired in the initialization phase
   not-init-mealy   -- mealy-type components not fired in the initialization phase
   petri-net        -- containing the places, transitions and arcs
                    -- constructed according to the coupling
   tn               -- time of next state transition
   type             -- type the network is moore or mealy

when receive dtss-i-message (dtss-i, t) at time t
   init-mealy := {}
   not-init-mealy := mealy-comps
   tn := t
   for d in D do
      send dtss-i-message (dtss-i, t) with time t to d
   endfor
```

```
    while there exists an element d in not-init-mealy which has-concession(d, t) do
       d : = select one element in not-init-mealy which has concession
       init-mealy := init-mealy ∪ {d}
       not-init-mealy := not-init-mealy - {d}
       take values (ye , tn) of places ped of preset (d) remove them from there
       and send dtss-x-message (xd, tn) with xd = (..., ye, ...) to d
    endwhile
    if all output places peN have values (ye , tn) then
       type := moore     -- Moore-type network
       take the values (ye , tn) of output places peN remove them from there and
       send dtss-y-message (yN, tn) with yN = (..., ye, ...) to parent coordinator
    else
       type := mealy     -- Mealy-type network

 when receive dtss-x-message (xN, t) with input value xN
    store input value (xN, t) on respective input places pNd
    -- fire all not-init-mealy at time t
    trans-to-go := not-init-mealy
    while trans-to-go ≠ {} do
       d : = select one element in trans-to-go which has-concession(d, t)
       trans-to-go := trans-to-go - {d}
       take values (ye , t) of places ped ∈ preset (d) remove them from there and
       send dtss-x-message (xd, t) with xd = (..., ye, ...) to d
    endwhile
    if type = mealy then          -- Mealy-type network
       take the values (ye, t) of place peN of output places remove them and
       send dtss-y-message (yN, t) with yN = (..., ye, ...) to parent coordinator
    -- fire all moore-comps at time t
    trans-to-go := moore-comps
    while trans-to-go ≠ {} do
       d : = select one element in trans-to-go which has-concession(d, t)
       trans-to-go := trans-to-go - {d}
       take values (ye, t) of places ped ∈ preset (d) remove them from there and
       send dtss-x-message (xd, t) with xd = (..., ye, ...) to d.
    endwhile
    tn := tn + c
    -- fire all init-mealy at time tn
    trans-to-go := init-mealy
```

```
    while trans-to-go ≠ {} do
       d : = select one element in trans-to-go which has-concession(d, t)
       trans-to-go := trans-to-go - {d}
       take values (ye , tn) of places ped ∈ preset (d) remove them from there
       and send dtss-x-message (xd, tn) with xd = (..., ye, ...) to d.
    endwhile
    if type = moore then   -- Moore-type network
       take the first values (ye , tn) of place peN of output places remove them
       send dtss-y-message (yN, tn) with yN = (..., ye, ...) to parent coordinator

 when receive dtss-y-message (yd, t) with output value yd from component d
    put output value (yd , t) on places pde ∈ postset (d)

 end Dtss-coordinator
```

The result of trying to compute as many output values as possible already in the initialization phase is that the components may operate at different times during one simulation cycle and it may be necessary to keep more than one output value on the places of the Petri net. But it should be emphasized that through this scheme it is possible to process all components by only once passing dtss-x-messages down. Otherwise, we would have to compute all the outputs of the Moore components first and then make the state transitions and compute the outputs of the others.

Figure 3.3.5 visualizes the processing of the initialization message by the Petri nets for network N1 and network N2. Sending the dtss-i-message to all components results in computation of the output of all Moore-type components (a). Then in network N2, all Mealy-type components have to be fired if possible (b). As the output of N2 is available then, network N2 is of type Moore and hence the output is sent back to N1 (c). Now also the output for N1 is available, network N1 is also of type Moore and the output is sent back to the parent (d). The inputs for all components which have not been fired are stored for the first simulation cycle.

Figure 3.3.6 visualizes the execution of one simulation step by the Petri nets. First the incoming input values are stored at the input places (e). Then, all components of type Mealy which have not been fired in the initialization phase are fired (f). Then all Moore components are fired (g). As N2 is a network, the input is sent to the N2. In N2 the Moore-type components are fired first and compute the output for the next simulation cycle (h). Then, the Mealy components fired already in the initialization phase are fired (i). The output of N2 is sent back (k). Finally the output of N1 is sent back to the parent (l) and we see that the Petri net is left in the same condition as after the initialization phase.

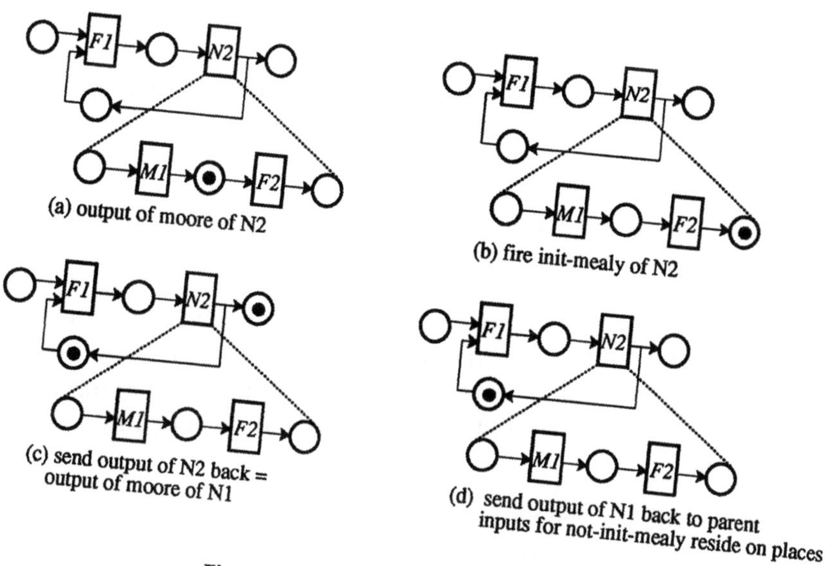

(a) output of moore of N2

(b) fire init-mealy of N2

(c) send output of N2 back = output of moore of N1

(d) send output of N1 back to parent inputs for not-init-mealy reside on places

Figure 3.3.5 Processing a dtss-i-message.

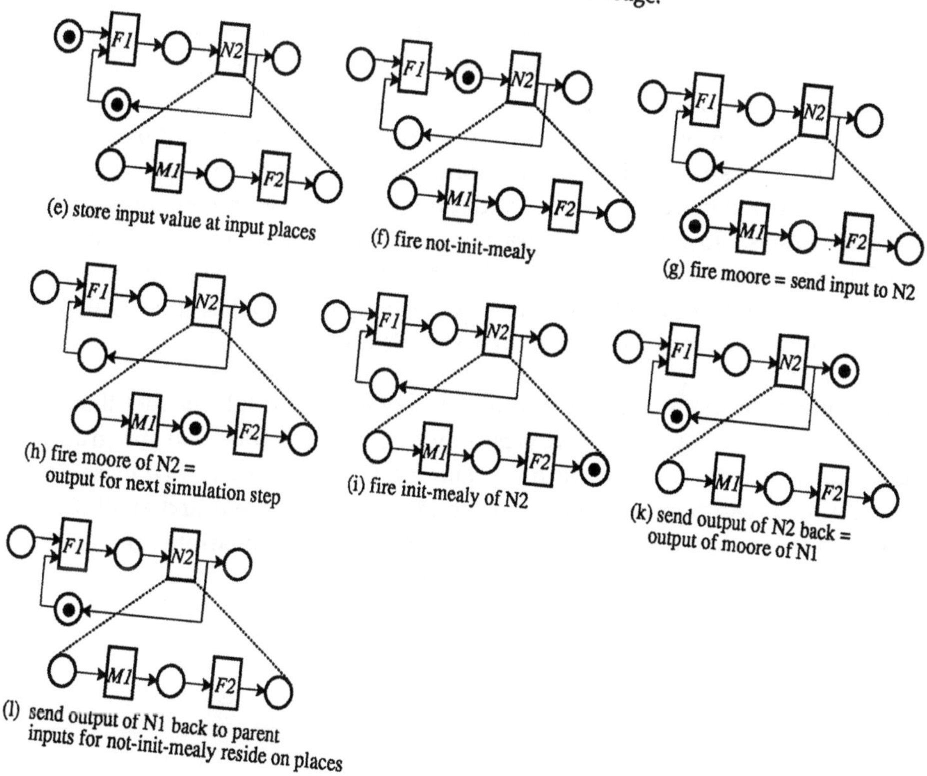

(e) store input value at input places

(f) fire not-init-mealy

(g) fire moore = send input to N2

(h) fire moore of N2 = output for next simulation step

(i) fire init-mealy of N2

(k) send output of N2 back = output of moore of N1

(l) send output of N1 back to parent inputs for not-init-mealy reside on places

Figure 3.3.6 Processing a dtss-x-message.

3.3.3 Abstract Simulator for Differential Equation Specified Systems

The abstract simulator of modular, hierarchical differential equation specified systems is a stepwise numerical simulation algorithm very similar in structure and operation to the DTSS abstract simulator. It consists of *dess-coordinators* and *dess-moore-*, *dess-mealy-*, and *fnss-simulators*. It employs *dess-i-*, *dess-x-*, and *dess-y-messages* as well as the same data-driven scheduling scheme. However, several changes have been made to meet the special needs of continuous simulation.

The first difference is that the dess-simulator uses a *numerical integration method* [Burden, Faires 89] to compute the next states of the atomic components. We assume that a method which is capable of appropriately computing the state trajectories of the atomic components is known [Burden, Faires 89] and do not address this question here.

The second difference is that the time step of the DESS abstract simulator is not fixed but can be varied by the integration method and that additional simulation cycles can be inserted when discontinuities in input segments have to be handled.

To be in the position to manage the above simulation scheme, a main extension has been made in the DESS abstract simulator. Instead of working with discrete state, input and output values, we use the trajectories between two consecutive simulation steps. In the computer implementation the trajectories can be represented by straight lines, by a sequence of straight lines or by a spline depending on the accuracy required. Those now are transmitted as output and input values between components and stored as state values. However, such a trajectory only represents a predicted development of the trajectory until the next simulation cycle which can be changed when an additional step is introduced before the scheduled next simulation time. Only the parts of the segments prior to the time of the last cycle are already valid.

Accuracy problems in the numerical simulation may arise through the fact that the atomic components exchange their values only once every numerical integration step. However, we justify our approach by claiming that accuracy can be increased by decreasing the integration interval. And the problem can be avoided at all if the coupled model is mapped to an atomic model which is always possible [Zeigler 76; Praehofer 91].

3.3.4 Simulators for Atomic Multi-Formalism Models

For the simulation of atomic combined models we introduce the *devs-dess-simulator* and the *devs-analytical-dess-simulator*. The devs-dess-simulator is employed to simulate general DEV&DESS models and the devs-analytical-dess-simulator is a special simulation algorithm which utilizes the advantages provided by the DEVSwith-AnalyticalDESS model type. Both employ the same messages as a devs-simulator. A *devs-dess-coordinator* which is very similar to the devs-coordinator will be employed to simulate couplings of multi-formalism models. As DEVSwithAnalyticalDESS can be used as components in DEVS networks, the devs-analytical-dess-simulator is constructed in such a way that it can be used as a direct subordinate of the devs-coordinator.

A challenge in the simulation of atomic combined models is the detection of state events. At the end of Section 3.3.4 we will discuss this problem.

Devs-dess-simulator

The devs-dess-simulator is used to simulate atomic DEV&DESS. As DEV&DESS systems are closed under coupling, they can be used as components of multi-formalisms networks. To simulate such networks, we will introduce a devs-dess-coordinator which will be very similar in structure and operation to the devs-coordinator in Section 3.3.5. In particular, it sends and receives the same kind of messages, viz. i-, x-, *- and y-messages. The devs-dess-simulator presented here also uses the same kind of messages.

When it receives an i-message at initial time *t* it computes the time of the last event by *tl* = *t* - *e* and schedules a first integration step at the same time *tn* = *t*. The time *t-next-integr* gives the scheduled time of the next integration step. The time *t-time-event* gives the time of the next time event determined by the time advance function. The time *t-state-event* gives the time of the next state event, i.e., an event caused by changes of continuous states and continuous inputs, not later as the time of the next integration step. The time *tn* determining the time of the next *-message is set to the minimum of these times.

Whenever a *-message is received at time *tn*, it may mean a state-event, a time-event but it will always mean an integration step. When a state-event occurs, i.e., *t-state-event* = *t*, the state event transition is applied to the current state at time *t*. When a time event occurs, i.e., *t-time-event* = *t*, the internal transition is applied to the current state at time *t* and the discrete output y_d is computed. In both cases, the time *tl* is set to *t* causing *e* to become *0*. After new time *t-next-integr* is determined and the continuous state trajectory until *t-next-integr* is computed, we compute the continuous output and new times of state and time events. The pair of continuous output and discrete output is sent back to the parent coordinator in a y-message.

Whenever a x-message is received at time *t*, with $tl \le t \le tn$, first the continuous inputs segments are stored in special state component *sx*, which has to be added to the state space to store the continuous inputs. In this variable *sx* continuous input segments are stored as they are and event-inputs at continuous input ports are stored as piecewise constant. However, these segments stored are not valid after time *t* - they only represent anticipated segments which can be changed by later inputs. In the following, the operations of the x-message are analogeous to the operations of the *-message. When there are event inputs in the discrete input set, the external transition function is executed. Afterwards, we compute the continuous state trajectories, we compute the continuous output and new times as in *-messages. The pair of continuous output and no-event for the discrete output is sent back to the parent coordinator in a y-message.

```
Devs-dess-Simulator

variables:
   parent          -- parent coordinator
   tl              -- time of last state transition
   tn              -- time of next state transition
   t-state-event   -- time of the next state event
   t-time-event    -- time of the next time event
   t-next-integr   -- maximum time of next integration step
   DEV&DESS = (X, Y, S, δext, δstate, δint, λd, ta, f, λc)     -- associated model
         with total state (sc, sd, e, sx)
   yc              -- current continuous output value
   yd              -- current discrete output value
   integr-step     -- integration interval defined by the numerical method

when receive i-message (i, t) at time t
   tl := t - e
   t-next-integr := t       -- an integration step at initial time t
   t-time-event := tl + ta(sd)
   t-state-event := minimum({t| t ∈ [t, t-next-integr],
                                          st ≠ δstate (st, t - tl, sxt) } ∪ {∞} )
   tn := min {t-next-integr, t-state-event, t-time-event }
```

```
when receive *-message (*, t) at time t
   if t ≠ tn then
      error: bad synchronization
   yd := Ø

   if t-state-event = t then                    -- execute a state event
      s_t := δ_state (s_t, t - tl, sx_t)
      tl := t            -- elapsed time is 0
   elseif t-time-event = t then                 -- execute a time event
      yd := λ_d (s_t, sx_t)
      s_t := δ_int (s_t, sx_t)
      tl := t            -- elapsed time is 0
   endif
   t-next-integr := minimum{t+integr-step, t2},
            where t2 is upper bound of interval of input sx_[t1, t2)
   sc_(t, t-next-integr ] := NumIntegration(f(s_t, sx_(t, t-next-integr ]))
   t-state-event := minimum({t| t ∈ [t, t-next-integr],
                              s_t ≠ δ_state (s_t, t - tl, sx_t) } ∪ {∞} )
   t-time-event := t + ta(sd)
   tn := min {t-next-integr, t-state-event, t-time-event }
   yc_(t, tn ] := λ_c ((sc_(t, tn ], sd_t))
   send value (yc_(t, tn] , (yd, t)) in y-message to parent coordinator

when receive x-message (xc_(t, t2], (xd, t)) at time t
   if not (tl ≤ t ≤ tn) then
      error: bad synchronization
   e := t - tl
   -- construct appropriate continuous inputs
   for continuous input ports xc_i in xc = (xc_1, xc_2, ...) do
      if is-continuous-segment(xc_i) then
         sx_i (t, t2] := xc_i (t, t2]        -- store continuous segment
      elseif  is-event-segment(xc_i ) & xc_i ≠ Ø then
         sx_i (t, ∞) := xc_it -- construct piecewise constant input from event input
   if xd ≠ Ø then                         -- execute an external event
      s_t := δ_ext (s_t, e, (xd, sx_t))
      tl := t            -- elapsed time is 0
   endif
```

```
   t-next-integr := minimum{t+integr-step, t2},
            where t2 is upper bound of interval of input sx[t1, t2)
   sc(t, t-next-integr ]:= NumIntegration(f(st, sx(t, t-next-integr ]))
   t-state-event := min ({t| t ∈ [t, t-next-integr],
                        st ≠ δstate (st, e, sxt) } ∪ {∞} )

   t-time-event := t + ta(s)
   tn := min {t-next-integr, t-state-event, t-time-event }
   yc(t, tn] := λc ((sc(t, tn ], sdt))
   send value (yc(t, tn] , (Ø, t)) in y-message to parent coordinator

end devs-dess-simulator
```

Devs-analytical-dess-simulator

The devs-analytical-dess-simulator is located between a pure devs-simulator and a devs-dess-simulator discussed above. It jumps from event to event and does not use a numerical simulation algorithm, but it has to handle the state events caused by the continuous state variables. The analytical solution of the continuous states are used to compute the states at the event times and it is used to compute the times of the state events. Beside these changes the scheme is equal to the devs-dess-simulator. A devs-analytical-dess-simulator can be used as a subordinate of a devs-coordinator.

```
Devs-analytical-dess-simulator

variables:
   parent          -- parent coordinator
   tl              -- time of last state transition
   tn              -- time of next state transition
   t-state-event   -- time of the next state event
   t-time-event    -- time of the next time event
   DEVSwithAnalyticalDESS = (X, Y, S, δstate, δint, δint, λd, ta, F) -- associated model
        with total state (sc, sd, e)
   y               -- current output value of the associated model
when receive i-message (i, t) at time t
   tl := t - e
   t-time-event := tl + ta(sd)
   t-state-event := minimum({t| t ∈ [t, ∞), st ≠ δstate (st, t - tl) } ∪ {∞} )
        with st = (sd tl, sc tl + F(stl) |tl^t )
   tn := min {t-state-event, t-time-event}
```

```
when receive *-message (*, t) at time t
   if t ≠ tn then
      error: bad synchronization
   st := (sdtl, sctl + F(stl)|tl^t )

   if  t-state-event = t then              -- execute a state event
      st := δstate (st, t - tl))
   elseif t-time-event = t then                    -- execute a time event
      y := λ (st)
      send value y in y-message (y, t) to parent coordinator
      st := δint (st)
   endif
   tl := t              -- elapsed time is 0
   t-state-event := minimum({t| t ∈ [t,∞), st ≠ δstate (st, t - tl) } ∪ {∞})
          with st = (sd tl, sc tl + F(stl)|tl^t )
   t-time-event := t + ta(sd)
   tn := minimum {t-state-event, t-time-event }

when receive x-message (x, t) at time t with input value x
   if not (tl ≤ t ≤ tn) then
      error: bad synchronization
   st := (sdtl, sctl + F(stl)|tl^t )
   st := δext (st, e, xd)
   tl := t              -- elapsed time is 0
   t-state-event := minimum({t| t ∈ [t, ∞), st ≠ δstate (st, t-tl) } ∪ {∞} )
          with st = (sd tl, sc tl + F(stl)|tl^t )
   t-time-event := t + ta(s)
   tn := minimum {t-state-event, t-time-event }

end devs-analytical-dess-simulator
```

Computation of state event times

In the devs-dess-simulators and the devs-analytical-dess-simulator we specified the times of the state events as the minimum time when the application of the state event transition function results in a change of state ($t\text{-}state\text{-}event := min\{t \mid t \in [t, t\text{-}next\text{-}integr], s_t \neq \delta_{state}(s_t, t - tl, (sx_t, \emptyset))\} \cup \{\infty\}$). However, in this general form without any knowledge of the state event transition function this time is not feasible to compute. A condition employing the equality predicate for a continuous state variable

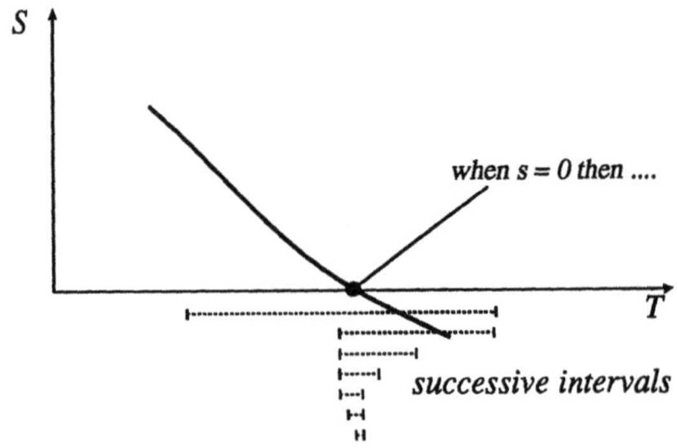

Figure 3.3.7 Bi-section method to search for state event times.

would necessitate checking for every time, and hence for a infinite number of times, if a state event occurs in the continuous interval *[t, t-next-integr]*. But the state events depend on conditions stated in the state event transition function. Having knowledge of the conditions simplifies the problem drastically. When knowing every condition like $a = b$, $a < b$, $a > b$, or a similar one, they can be transformed into condition $a - b = 0$, $a - b < 0$, or $a - b > 0$ etc. reducing the whole problem to the detection of *zero-crossings*. A lot of methods of finding zero-crossings exist and are well explored [Cellier 79, 86; Pritsker 74], from which the easiest and best known the *bi-section method* is [Cellier 79; Pritsker 74; Burden, Faires 89]. The bi-section method works the following way: In every integration step we check if there was a zero-crossing. If so, we start an binary search iteration where we subsequently halve the time interval until the time of the zero-crossing is found with sufficient accuracy. Figure 3.3.7 depicts this process.

The analytical solution for the continuous states of DEVSwithAnalyticalDESS systems facilitates the efficient determination of state event times making the usage of these system types really advantageous. If the values of the continuous state when state events occur are known, the only unknown variable in the analytical solution is the time variable Δt. The discrete states are constant until the next event. A simple root finding method like the *Newton-Raphson method* [Burden, Faires 89] can be used to determine the value Δt - the time until the next state event. In very simple cases it is even possible to transform the analytical equation so that Δt can be expressed in explicit form. In the tank system example of Section 3.2.2 the time Δt can be determined by equation $\Delta t = (level_{t+\Delta t} - level_t) / (sinflow - soutflow - sleakage)$. The values for $level_{t+\Delta t}$ are the *3* and *10* which are the values of the state variable *level* at which state events are to occur.

3.3.5 Coordinators for Coupled Multi-Formalism Models

Multi-formalism networks - couplings of systems of different types and couplings of multi-formalism systems - can be simulated using the *devs-dess-coordinator*. This coordinator is very similar in structure and operation to the devs-coordinator, with just a minor extension to handle continuous inputs and outputs.

The devs-dess-coordinator has to provide the synchronization for the subordinates which have to compute their state trajectories and outputs as required by the dynamical interpretation of combined DEV&DESS systems. We saw that the devs-dess-simulator does that for DEV&DESS systems. As DEVS systems are just special DEV&DESS systems without continuous states, inputs and outputs and hence without integration steps and state events, it can easily be seen that the devs-simulator contains the same operations as the devs-dess-simulator but without the code to handle continuous inputs and outputs, without code for integration and for state events.

However, DTSS and DESS abstract simulators cannot be used as direct subordinates of a devs-dess-coordinator, also because they do not use the same kind of messages. Therefore, we introduce special interface coordinators - *dtss-interface-* and *dess-interface-coordinators* - which are employed to translate the messages between the superior devs-dess-coordinator and its subordinate in such a way that the interface-coordinator together with its subordinate dtss- or dess-simulator computes the appropriate state and output trajectories. Figure 3.3.8 shows the structure of an abstract simulator for multi-formalism networks. In the following we will first present the coordinator for multi-formalism systems and afterwards the interface coordinators.

```
Devs-dess-Coordinator

variables:
   MFN = (X, Y, D, {Ma | a ∈ D}, EIC, EOC, IC, Select)
               -- associated multi-formalism network
   children    -- subordinate coordinators or simulators
   parent      -- parent coordinator
   tl          -- time of last event
   tn          -- time of next event
   imminent    -- list of imminent children
   a           -- selected imminent child
   sx          -- variable to store continuous inputs
   sy          -- variable to store continuous outputs
```

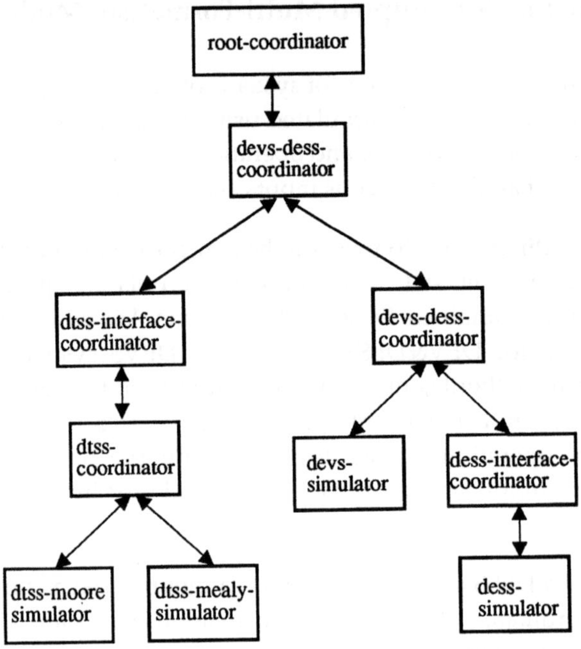

Figure 3.3.8 Multi-formalism-network abstract simulator.

```
when receive *-message (*, t) at time t
  if t ≠ tn then
    error: bad synchronization
  yd := Ø
  imminent := {a | a ∈ D, tn_a = tn}
  time-imminent := {a | a ∈ imminent , t-time-event_a = tn}
  state-imminent := {a | a ∈ imminent , t-state-event_a = tn}
  if time-imminent ≠ {} then
      a := select (time-imminent)
  elseif state-imminent ≠ {}
      a := select (state-imminent)
  else
      a := select (imminent)
  send *-message (*, t) to a
  tl := t
  tn := min {tn_a | a ∈ D}
  send value ((sy(t, tn ], (yd, t))in y-message to parent coordinator
```

```
when receive x-message (x, t) at time t with external input x = (xc(t, t2], xd)
   if not (tl ≤ t ≤ tn) then
      error: bad synchronization
   -- construct appropriate continuous inputs
   for continuous input ports xci in xc = (xc1, xc2, ...) do
      if is-continuous-segment(xci ) then
         sxi (t, t2]:= xci(t, t2]        -- store continuous segment
      elseif  is-event-segment(xci ) & xci ≠ Ø then
         sxi(t, ∞):= xci  -- construct piecewise constant input from event input
      -- consult EIC to get children influenced by the external input
   receivers := components r ∈ children influenced by external input x
   for-each r in receivers
      send x-messages (xr , t) to r with xr got from (sx, xd) through EIC
   tl := t
   tn := min {tna | a ∈ D}
   send value ((sy(t, tn ] , (Ø, t)) in y-message to parent coordinator

when receive y-message from a with output ya = (yca(t1, t2], (yda, t)
   -- check EOC if y influences external outputs
   if the output from a influences the external network output ports then
      (yc, yd) are the external outputs got from ya through EOC
      for continuous output ports yci in yc = (yc1, yc2, ...) do
         if yci is continuous segment then
            syi := yci      -- store continuous segment
         elseif  yci is event input ≠ Ø then
            syi [t, ∞):= yci -- construct piecewise constant input from event input
      endfor
   endif
   -- check IC to get children influenced by the output y of d
   receivers := components r ∈ children influenced by a
   for-each r in receivers
      send x-messages (yr , t ) to r with input value yr got from y through IC

end devs-dess-coordinator
```

Dtss-interface-coordinator

The *dtss-interface-coordinator's* main task is to save the inputs - event inputs as well as continuous inputs - for the simulation cycles scheduled at regular intervals. These inputs, saved in the state component *sx*, are used to compute the exact value of the input at the time of the state transition which are triggered by *-messages from the superior devs-dess-coordinator. After the state transition and after the output has been received in a dtss-y-message and has been stored in the *output-place*, the output for the current simulation time is sent back to the parent as an event output. Note, that through the special scheme of the dtss-moore-simulator for Moore-type DTSS systems, the output for the next simulation cycle will already reside on *output-place*. The variable *tn* indicating the time of the next simulation step is increased by *c* after every simulation step resulting in a next simulation step after *c* time units. Figure 3.3.9 shows the structure and messages received and sent by the dtss-interface-coordinator.

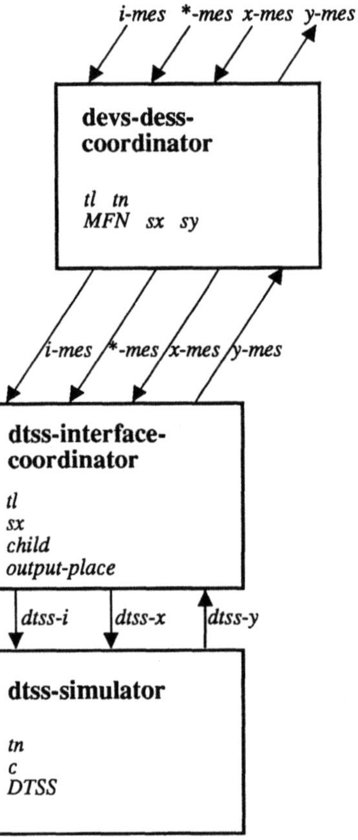

Figure 3.3.9 Dtss-interface-coordinator.

```
Dtss-interface-coordinator

variables:
   parent         -- parent coordinator
   child          -- subordinate dtss-simulator or coordinator
   tn             -- tn of child
   sx             -- variable to store input segments
   output-place   -- place to store the outputs of the DTSS model
   tl             -- time of last state transition

when receive i-message (i, t) at time t
   tl := t
   send dtss-i-message (dtss-i, t) with time t to child

when receive *-message (*, t) at time t
   if t ≠ tn then
      error: bad synchronization
   tl := t
   send dtss-x-message (sx_t, t) with input value sx_t to child
   take first value (y, t) of output-place remove it from there and
         send (y, t) in y-message to parent coordinator

when receive x-message (x, t) at time t with input value x
   if not (tl ≤ t ≤ tn) then
      error: bad synchronization
   -- construct appropriate continuous inputs sx#x
   for input ports x_i in x = (x_1, x_2, ..., x_m) do
      if is-continuous-segment(x_i) then
         sx_i (t, t2] := x_i (t, t2]          -- store continuous segment
      elseif  is-event-segment(x_i ) & x_i ≠ Ø then
         sx_i (t, ∞) := x_i    -- construct piecewise constant input from event input
 when receive dtss-y-message with output value y
   put output value (y, t) on output-place

end dtss-interface-coordinator
```

Dess-interface-coordinator

The *dess-interface-coordinator* is similar to the dtss-interface-coordinator. It is employed to store the continuous input segments in the state component *sx* and translate the messages received from the superior coordinator into messages for the subordinate child. The time *tn* of the child determines the time of the next integration step. *-messages must be sent by the superior coordinator at times *tn*.

When an input is received by an x-message at time *t*, the input is stored in the state component *sx*. In difference to the dtss-interface-coordinator, an additional simulation cycle is introduced immediately to take care of the change of the input segment. This is especially important for event inputs because they cause a discontinuous jump from one input value to the next.

3.3.6 Root-Coordinator

A *root-coordinator* is attached to the coordinator (or simulator) of the overall model. This root-coordinator has the task of initiating the simulation cycles. Given an initial time *t0*, it first sends an i-, dess-i- or dtss-i-message to the subordinate coordinator or simulator depending on the type of the simulator or coordinator. Then it executes a loop where it sends *-messages, dtss-x-messages, or dess-x-messages to its subordinate with the time of the next internal event *tn*. When dtss-x- or dess-x-messages are sent the input value is Ø because the overall network has to be autonomous, i.e., it must not have inputs.

```
Root-Coordinator
variables:
   child      -- coordinator (or simulator) of overall model
   tn      -- time of next event
case of abstract simulator type of child
   (devs, devs-dess): send i-message (i, t0) with initial time t0 to child
   dtss:        send dtss-i-message (dtss-i, t0) with initial time t0 to child
   dess:        send dess-i-message (dess-i, t0) with initial time t0 to child
end case
tn := tn of child
while tn ≠ ∞ & simulation should not be terminated do
   case of abstract simulator type of child
      (devs, devs-dess): send *-message (i, tn) with time tn to child
      dtss:        send dtss-x-message (Ø, tn) with input value Ø to child
      dess:        send dess-x-message (Ø, tn) with input value Ø to child
   end case
   tn := tn of child
end while
end Root-Coordinator
```

3.4 IMPLEMENTATIONS OF MODELLING AND SIMULATION ENVIRONMENTS

Two implementations have been done based on the modelling and simulation concepts outlined in Sections 3.2 and 3.3. The first one [Praehofer 89; Praehofer, Zeigler 90] is an extension of *DEVS-Scheme* [Zeigler 87, 90a] - Zeigler's implementation of a discrete event modelling and simulation environment in the LISP dialect *PC-Scheme*. In this first implementation we implemented the abstract simulator for discrete time systems. But only a rudimentary approach to continuous and combined modelling and simulation is used which involves a fixed step numerical integration of continuous models and approximates continuous trajectories by discrete sequences. Only one type of multi-formalism model, viz. couplings of DEVS, DTSS, and DESS systems, is available.

The second is an implementation in *Interlisp-D* and its object-oriented superset *LOOPS* called *STIMS - Systems Theory Instrumented Modelling and Simulation* [Spalt 88; Praehofer 90b]. The concepts used for this implementation are almost equivalent to the concepts outlined in Sections 3.2 and 3.3, only with minor differences in atomic multi-formalism models. The STIMS modelling and simulation environment also incorporates a convenient user interface for interactive modelling and simulation. Nevertheless, it is a prototypic implementation which is mainly used to develop and validate the theoretical concepts.

3.4.1 Implementation in PC-Scheme

Modelling and simulation concepts for discrete time systems, differential equation specified systems, as well as multi-formalism coupled systems have been implemented in PC Scheme and *SCOOPS* as a extension of DEVS-Scheme. In that implementation, however, differential equation specified systems work with fixed integration steps like discrete time systems. They are not able to treat events instantly and they put out discrete values at the times of the integration steps. Consequently, the precision of the event times depends on the integration intervals.

DEVS-Scheme is an implementation of the DEVS formalism. It has been developed by Zeigler at the University of Arizona [Zeigler 87, 90a]. It is a general purpose modelling and simulation environment for constructing hierarchical discrete event models. PC Scheme [Texas Instruments 85] is a powerful LISP dialect for PC-compatible microcomputers. SCOOPS is the object-oriented superset of PC-Scheme. DEVS-Scheme is implemented as a shell which sits upon PC-Scheme in such a way that all of the underlying LISP based and object-oriented programming language features are available to the user. The result is a powerful basis for combining AI and simulation techniques.

The DEVS-Scheme environment also employs a second layer, *ESP-Scheme*, which implements a knowledge representation scheme, called the *system entity structure* [Zeigler 84a, 90a; Rozenblit, Zeigler 88]. It is a means of representing static system structure knowledge of a variant family of models. These features have not been used in our extension and so are not available in our extended environment although such an extension would be straightforward.

DEVS-Scheme is principally coded in SCOOPS. System types as well as the abstract simulator concepts are implemented by classes. Figure 3.4.1 shows this class system of DEVS-Scheme. The main idea of our extension was to extend this class system and so to integrate our new system types and simulation concepts. Most of the features already available in DEVS-Scheme have been left unchanged and used in the new parts. In the following we will briefly describe this extended class system. Main differences of the original and extended environments will be worked out. The reader is referred to [Zeigler 87, 90a] for a detailed description of DEVS-Scheme.

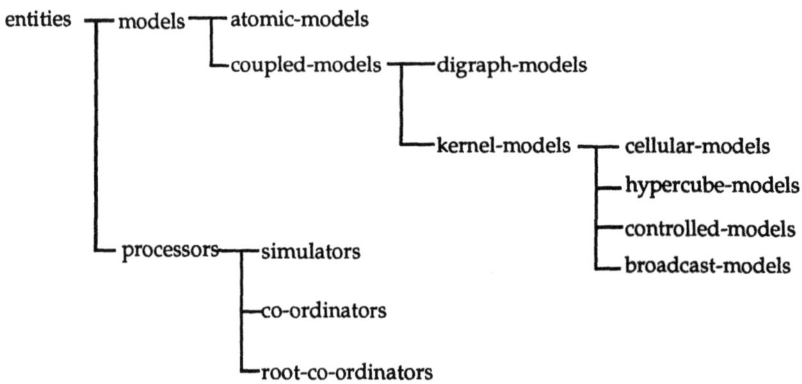

Figure 3.4.1 Original class systems of DEVS-Scheme.

System types implementation

Figure 3.4.2 shows our extended class system for system types implementation which implements atomic and coupled discrete event, discrete time and differential equation specified systems. Most of the parts of the original class system of DEVS-Scheme have been reused.

The class *devs-atomic-models* corresponds to the class atomic-models of the old class system and implements atomic discrete event systems. *Dts-atomic-models, dts-mealy-models, dess-atomic-models* and *memoryless-models* represent the other respective system types which are new in the implementation. As these classes share common features among each other, we tried to extract these common features and summarize them in superclasses. *Atomic-models* and *memory-models* are such classes. Atomic-models implement variables and methods which are shared by all atomic models. In memory-models, all the features related to states and state variables are summarized.

The class *coupled-models* has been left unchanged from the old implementation but subclasses of it were changed. Here the object-oriented feature with multiple inheritance proved its usefulness. Under coupled-models two classes were introduced to represent the two types of classifications of networks. *Coupled-models-type-spec* represents the first possible classification of networks which is done according to the type of their components. The subclass *devs-coupled-models* implements the features of DEVS networks and multi-formalism networks.

These are the features which have to do with tie breaking. The subclass *dts-coupled-models* implements DTSS and DESS networks and incorporates mainly methods to set the time-base or integration interval.

The second subclass of coupled-models, *coupling-spec,* represents the classification of networks according to the coupling scheme they use. We only employed *digraph-models* in our implementation. Digraph model implements this kind of coupled model where each coupling has to be specified explicitly. In Figure 3.4.2 it is indicated how other types of coupling schemes could be incorporated in the environment.

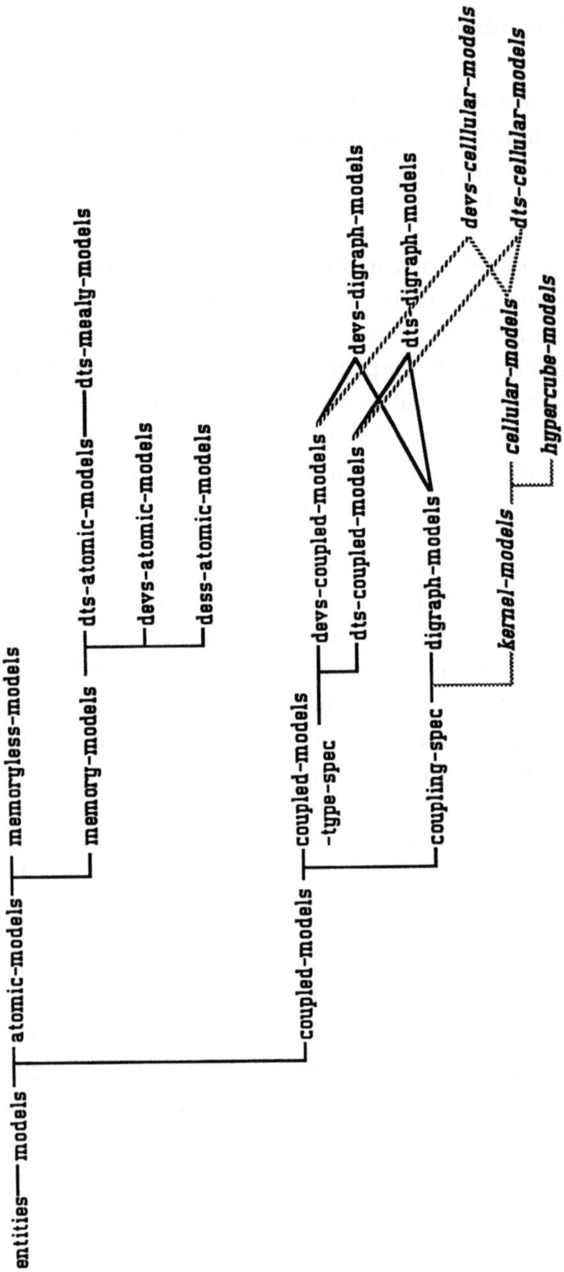

Figure 3.4.2 Extended class system for DEVS, DTSS and DESS modelling.

Specific types of coupled models now can be simply derived by summarizing the features of one subclass of coupling-spec and one subclass of coupled-system-type-spec through multiple inheritance. Figure 3.4.2 shows the classes *devs-digraph-models* and *dts-digraph-models* and it is indicated how it would work for cellular models. The implementation of the new atomic-models was straightforward since most of the features dealing with states, transition functions etc. have been reused. One extension is that the input and output ports for all models have to be specified explicitly now. The methods to support these are implemented in the class *models.*

Abstract simulator implementation

Figure 3.4.3 shows the class system which implements all the abstract simulator concepts. *Devs-simulator* and *devs-coordinator* are left unchanged except for the addition of methods for initialization. The *root-coordinator* has also been left unchanged except for the addition of methods for initialization. *Dts-simulator, mealy-simulator, fnss-simulator* and *dts-coordinator* are straightforward implementations of the algorithms for discrete time simulation. Dts-simulator implements the simulation scheme for Moore-type DTSS. The dtss-interface-coordinator is implemented in the class *dts-processor* and so inherited by every dts-simulator or dts-coordinator. Hence each of them also has the capability to serve as an interface coordinator. This has the advantage that no interface coordinator has to be introduced explicitly but is implicitly available through the overall dts-coordinator or simulator.

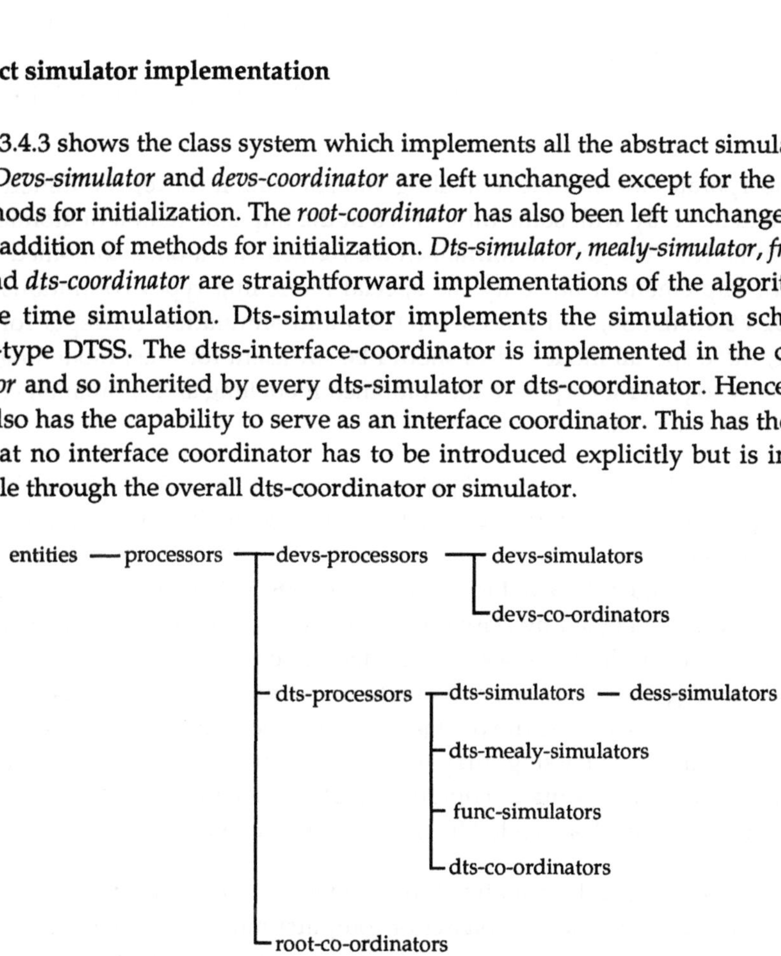

Figure 3.4.3 Extended class system for DEVS, DTSS and DESS simulation.

3.4.2 The STIMS Modelling and Simulation Environment

The *STIMS - Systems Theory Instrumented Modelling and Simulation* - modelling and simulation environment [Spalt 88; Praehofer 90b] is an implementation of the concepts outlined in Sections 3.2 and 3.3 in Interlisp-D / LOOPS. Interlisp-D [Xerox 83] is a LISP dialect which emphasizes interactive and display oriented programming. LOOPS [Bobrow, Stefik 83] is the object-oriented superset of Interlisp-D. Interlisp-D and consequently STIMS runs on *SIEMENS 5815* workstations which are equivalent to the *XEROX 1108* workstations. From this rather exotic hardware and software environment it is already obvious that STIMS shall not serve as a real product but it was used to develop and validate theoretical concepts by a prototypic implementation.

STIMS incorporates most of the modelling and simulation concepts outlined in Sections 3.2 and 3.3, only with some differences in the implementation of atomic multiformalism models. The general formalism DEV&DESS for atomic combined modelling is not implemented but only the special modelling types continuous-input-DEVS, DEVSwithDESS, and DESSwithDEVS and simulation concepts for them. However, these special model types should be sufficient for most modelling examples and they together incorporate the combined modelling and simulation concepts outlined in Section 3.2.2.

STIMS is principally coded in the object-oriented language LOOPS. That object-oriented programming [Meyer 88; Keene 89; Stroustrup 86; Blaschek et al. 89] is well suited for the implementation of systems theoretical concepts has been proved by several projects [Zeigler 87, 90a; Pichler, Praehofer 88; Mittelmann 88]. System types can be readily implemented by class definitions, systems of a particular type correspond to instances of these classes. The modelling formalisms in STIMS are implemented by setting up a complex class system and hierarchical models are implemented by a hierarchical structure of objects. As abstract simulators consist of objects and work with message passing, they would be readily implementable in LOOPS. However, as message passing is computational inefficient in LOOPS, the abstract simulator concepts have been coded by Interlisp-D functions.

The STIMS modelling and simulation environment is organized around four layers which support the user in several phases of computer modelling and simulation. The uppermost layer - the *file layer* - is used for file handling and model base handling. It provides operations to create model bases and save and retrieve models and model bases from files. The second layer - the *model base layer* - is used to set up model bases for different areas of applications. Model bases in STIMS are implemented by class systems whose classes are based on the classes implementing the system formalisms. Such classes in model bases define generic atomic models which can be used as atomic building blocks in complex modular, hierarchical models. The third layer is the *modelling layer*. The modular hierarchical models are built here. Atomic components are retrieved from model bases or can also be set up from scratch and then coupled together to form complex hierarchical, coupled models. The last layer then is the *simulation layer*. After setting initial states and setting up the experimental conditions of the simulation run, the hierarchical models can be simulated by the abstract simulator.

STIMS is an interactive modelling and simulation environment. The user is supported in each of his works by modern, interactive, mouse and menu oriented user interfaces. User interface modules have been implemented for each of the four layers under strong utilization of the interface building blocks already provided by LOOPS. Of special importance here are the so-called *browsers* which are used to represent objects and interrelation of these objects by a graph and to facilitate the selection and application of operations on objects. E.g., the *LOOPS class browser* is used to represent classes, which are only special objects in LOOPS, and their super-subclass relations and provides operations for class and method definition. The *LOOPS instance browser* is used to represent arbitrary objects and user defined interrelations and operations. An extension of the LOOPS instance browser, conveniently implemented as an specialization of the class implementing standard LOOPS browsers, is used as the basic command unit for modelling and simulation. To observe the simulation runs and put out the simulation results, display oriented elements - *gauges* and *trajectories* - have been implemented.

System types implementation in STIMS

Analogous to the extended DEVS-Scheme environment and other implementations of system theory concepts like CAST.FSM [Pichler 88; Pichler, Praehofer 88; Pichler, Schwärtzel 90], system formalisms have been implemented by setting up a class system. Figure 3.4.4 shows this class system which is much more complex than the extended DEVS-Scheme class system since it provides much more modelling facilities. Beside providing more concepts for combined modelling, this class system also provides formalisms for linear system modelling. The class system is also constructed in such a way that, where it is feasible, common features of classes are summarized in a common superclass. Classes define the variables for storing the parts of the system formalism and other data, and they define the operations required for modelling and simulation in form of methods.

Looking at the class system presented in Figure 3.4.4 we see that the overall class is class *STIMSSystem* which is a direct descendant of the most general LOOPS class *Object*. In this class all the features common to all system formalisms are implemented, that is everything which has to do with input and output interfaces. Below *STIMSSystem* a first distinction is made into *AtomicSystem* and *CoupledSystem*. The class *AtomicSystem* summarizes everything which is common to every atomic model, viz. every atomic model has to have an output function. The direct descendant *FunctionSystem* then implements the function specified system type.

MemorySystem is employed to implement everything which has to do with states. Hence, all atomic formalism implementations, except the *FunctionSystem*, has it as superclass. Subclass *AtomicDEVS* systems implements DEVS systems. Subclass *AtomicDTS/DESS* implements features common to every discrete time or differential equation system type. The first branch under *AtomicDTS/DESS* separates discrete time and differential equation specified systems and then further distinguishes Mealy and Moore discrete time systems.

The second branch under *AtomicDTS/DESS* - *LinearDTS/DESS* - implements the features of linear discrete time and differential equation specified systems. However, the classes *LinearSeqMach*, *LinearDifferenceEqu*, and *LinearDESSAbstract* do not implement linear systems themselves but only the features of linearity. Usable linear system formalisms (*LinearMooreSeqMach*, *LinearMealySeqMach*, *LinearMooreDifferenceEqu*, *LinearMealyDifferenceEqu*, *LinearDESS*) are gained by combining by multiple inheritance all the features of ordinary discrete time and differential equation specified systems and the features of linearity.

Atomic combined modelling formalisms are implemented in three classes. First of all continuous-input-DEVS are available through class *AtomicDEVS* also implementing ordinary DEVS. Then DEVSwithDESS are implemented by class *AtomicDEVS&DESS* and DESSwithDEVS are implemented by class *AtomicDESS&DEVS*. Both classes are implemented by multiple inheritance by combining the features of *AtomicDEVS* and *AtomicDESS* in such a way that in *AtomicDEVS&DESS* the class *AtomicDEVS* has precedence and in the class *AtomicDESS&DEVS* the class *AtomicDESS* has precedence.

The implementation of coupled systems is similar to the implementation in the extended DEVS-Scheme environment (see Section 3.4.1), the only difference being that we have separate classes for differential equation specified coupled systems (*DESSCoupledSystem*, *DESSNetwork*). Again, multi-formalism networks are available through class *DEVSNetwork*.

Only some of the classes of the class system implement system formalisms from which models can be built, the others only are abstract classes implementing some common features. The classes finally implementing modelling formalisms are *FunctionSystem, AtomicDEVS, AtomicDESS, AtomicMooreDTS, AtomicMealyDTS, AtomicDEVS&DESS, AtomicDESS&-DEVS, LinearMooreSeqMach, LinearMealySeqMach, LinearMooreDifferenceEqu, LinearMealyDifferenceEqu, LinearDESS, DEVSNetwork, DESSNetwork and DTSNetwork.* These classes can be further specialized to realize more elaborate modelling concepts. Model bases in STIMS are based on this possibility and are set up by further specializing the basic system formalism classes.

Models in STIMS then are instances of these classes. A hierarchical model is implemented by a hierarchical, tree-like object structure where the inner nodes are of coupled system classes and the leaves have to be instances of classes implementing atomic modelling formalisms.

The network classes own a variable *components* holding the pointers to the subordinate system objects. Each system object has a variable *parent* holding a pointer to its superior network in the hierarchical model. Besides the components, the inner network objects have to define the couplings. Therefore, every network object owns variables *extInpCoupling, extOutCoupling,* and *IntCoupling* where the couplings are stored in form of list structures.

The leaves in the hierarchical objects structure have to be instances of an atomic system formalism class. They own variables to store the state variables and also the current state. A state in STIMS is represented by an association list where the keys in the association list define the state variables and the values define the state variable values. Besides a state s we also use parameters p. These parameters belong to the set of states but they have the special characteristic that they do not change their values during a simulation run. They are set before a simulation experiment to set up a special experimental condition. Parameters are implemented in the same way as states. Also input and output vectors are implemented by similar structures.

Each atomic model then owns variables to hold the state transition and output function definitions. In STIMS, state transitions and output functions are defined by Interlisp-D functions and hence the variable of the system object holds the name of the LISP function implementing the state or output function. Figure 3.4.5 gives an example of LISP functions implementing the state transition and output functions of a STIMS atomic DEVS model of the FIFOStation outlined in Section 3.2.1.

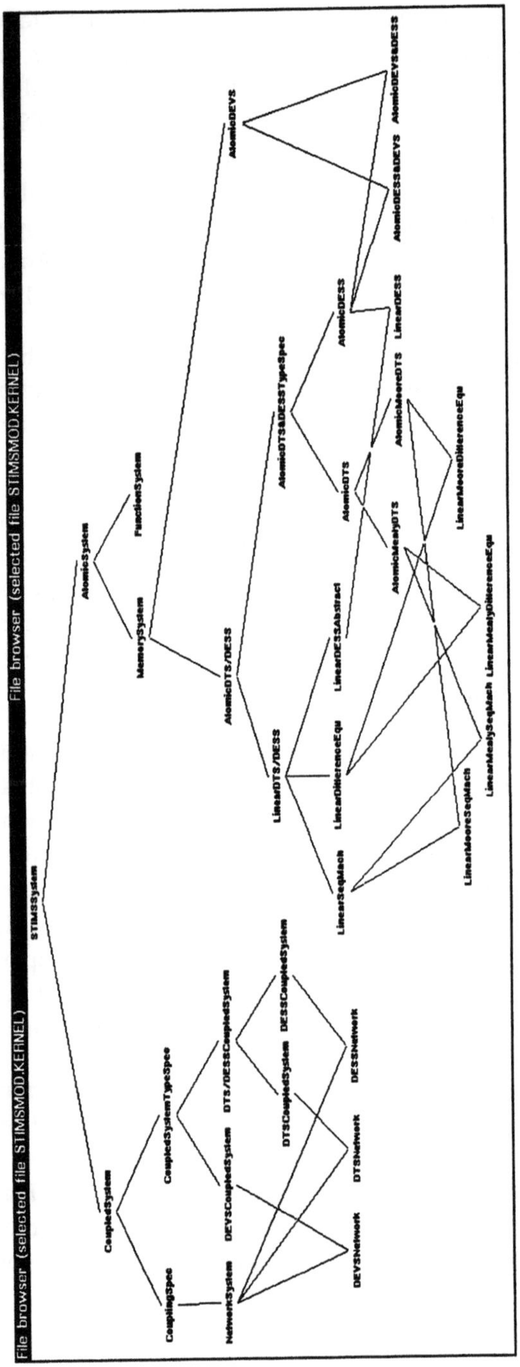

Figure 3.4.4 Class system implementing modelling formalisms in STIMS.

DEdit of function FIFOStation-E

```
(LAMBDA (s p e x)                              (* hp "17-Jan-90 15:35")

        (* * external transition fn)

  (←s
   (QUOTE queue)
   (FIFOInsert (↑x (QUOTE in))
         (↑s (QUOTE queue))))
  (model-in-phase
     (idle (Hold-in (QUOTE busy)
                (↑p (QUOTE procTime))))
     (busy (Continue))))
```

DEdit of function FIFOStation-O

```
(LAMBDA (s p y)                                (* hp "17-Jan-90 15:36")

        (* * output fn)

  (←y
   (QUOTE out)
   (FIFOFirst (↑s (QUOTE queue))))
```

DEdit of function FIFOStation-I

```
(LAMBDA (s p)                                  (* hp "17-Jan-90 15:35")

        (* * internal transition fn)

  (←s
   (QUOTE queue)
   (FIFORest (↑s (QUOTE queue))))
  (if (FIFOEmpty (↑s (QUOTE queue)))
      then (Passivate-in (QUOTE idle))
    else (Hold-in (QUOTE busy)
                (↑p (QUOTE procTime)))))
```

Figure 3.4.5 State transition and output function of the STIMS model FIFOStation.

Arguments to the LISP functions correspond to the arguments of the mathematical definition of the transition function. Only the output function has an additional argument *y* for the output. New state values and output values are not defined by the return value of the function but the new values for the state variables and the output ports are put into the association list.

Special access functions are used in the function definition to retrieve values of state variables, parameters, input ports (*↑s, ↑x, ↑p, ↑y*) and to define new values for state variables, derivatives, and output ports (*←s, ←d, ←y*). Several other pseudo-code elements beside these access functions have been implemented by LISP macros.

They primarily support discrete event modelling and basically are used to set the two state variables *phase* and *sigma*. E.g. the pseudo-code statement *(Hold-in 'busy (p 'procTime)* sets the phase variable *phase* to busy and the variable to *sigma* to the value of the parameter *procTime*. The statement *(Continue)* sets *sigma* to *sigma - e* and leaves phase unchanged. Other macros are used to distinguish the *phase* variable (*model-in-phase*), to determine the input port with a value unequal to the non-event (*when-x-on-port*) and to specify state events (*when-event*).

Abstract simulator implementation in STIMS

The abstract simulators have not been implemented using the object-oriented paradigm of LOOPS, as would be most straightforward, but have been implemented by LISP functions because this is computationally much more efficient. Each message type of the abstract simulators has been implemented by one unique LISP function. The different simulator and coordinator types depending on the type of the associated model are distinguished by a case statement in the LISP function. Message passing is implemented by recursive function calls. The data of the simulators and coordinators are stored within the associated model object. The following piece of Interlisp-D code outlines the implementation of *-messages by the LISP function *STIMSim*DEVS*.

```
(STIMSim*DEVS (LAMBDA  (model sender t mode)
...
   (SELECTQ (@ model proctype)
      (DEVSCoord
         (* * code for devs-coordinator)
         ...
         (starChild ← (DEVSSelect ...)
         (STIMSim*DEVS starChild model  t  mode)
         ... )
      (DEVSSimu
         (* * code for devs-simulator)
         ...
         (APPLY*  (@ model outputFn)  (@ model s)  ...)
         (STIMSimOutDEVS sender model t  mode)
         (APPLY*  (@ model intTransFn)  (@ model s)  (@ model parameters) )
         ... )
```

```
((DTSSCoord MEALYCoord DTSSSimu MEALYSimu)
   (* * code for dtss-interface-coordinator)
   ...
   (STIMSimTransDTSS sender model t  mode)
   (STIMSimOutDEVS sender model t  mode)
   ... )
((DESSCoord DESSMEALYCoord DESSSimu)
   (* * code for dess- interface-coordinator)
   ... )
((DEVS&DESSSimu
   ... ) )
((DEVS&AnalyticalDESSSimu
   ... ) )
...)
```

The first argument *model* of the LISP functions holds the pointer of the associated model object. The second argument *sender* holds the pointer to the model object from which the message has been received. When sending back a y-message (function *STIMSimOutDEVS*) the value of sender defines the value of argument *model* in the function call. The argument *t* gives the current simulation time and argument *mode* is of minor importance, it is used to control the simulation output. The processor type required by the associated model is given by its instance variable *procType*. The overall *SELECTQ* statement (is a case statement) reads this variable and, based on its value, selects the piece of code implementing that particular processor type. The first piece of code implements the devs-coordinator (which is also the devs-dess-coordinator implementation). We see that *starChild* is determined and a recursive function call of *STIMSim*DEVS* with arguments *starChild* for model and current model for sender implements the sending of the *-message to the imminent component. The next piece of code is for the devs-simulator. Output function and internal transition functions are retrieved from the model and applied to the required arguments. Other cases in the case statement implement #, dess-interface-coordinators and devs-dess-simulators.

The state events in the state event transition functions must be specified using the *when-event* macro. This macro is like an if statement without an else part. Before the simulation run, the state event transition function of combined models is searched for *when-event* statements and the conditions specifying the state events are gathered. LISP now facilitates the transforming of arbitrary conditions in such a way that the detection of state events is reduced to the detection of zero-crossings. In the simulation runs a simple bi-section algorithm is used to search for state events.

The STIMS user interface

STIMS is an interactive modelling and simulation environment. The facilities of Interlisp-D and LOOPS have been used to set up a convenient mouse and menu-driven user interface which supports the "simulationist" in all phases of his work with STIMS. As STIMS is organized around four layers, the user interface of STIMS also is organized around four layers. For each layer there is an interactive command unit which provides the respective operations through pop-up menus. The first layer for file and model base handling is represented by the *STIMSIcon* (Figure 3.4.6). A menu, which is popped up when clicking at the icon, provides operations for file and model base handling.

The command unit for the model base layer is a specialized LOOPS class browser. The class hierarchy implementing the generic models is presented and operations to define new generic models or alter existing ones are provided. Also operations for basic model handling like model creation, browsing etc. are done in this browser (Figure 3.4.7).

The command unit for modelling is implemented by a special LOOPS instance browser. The instance browser shows the composition tree structure of the hierarchical model (Figure 3.4.8). When objects in the tree are selected, a menu is popped up providing the operation for model building, viz. definition of input ports, output ports, state variables, state transition and output functions, or components and couplings. In distinction to standard LOOPS instance browsers, the pop-up menus are object specific, i.e., different menus are generated depending on the class of the object [Müller-W. 86]. When selecting a network model, there is an operation to add new components. This new component has to be selected from a model base by selecting the appropriate class in the class browser. State transitions and output

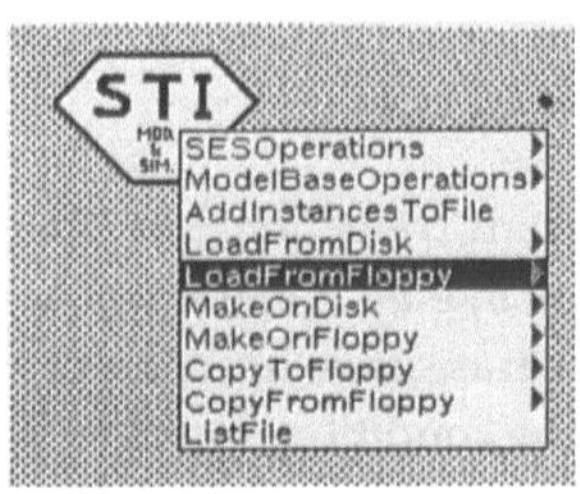

Figure 3.4.6 The STIMS icon.

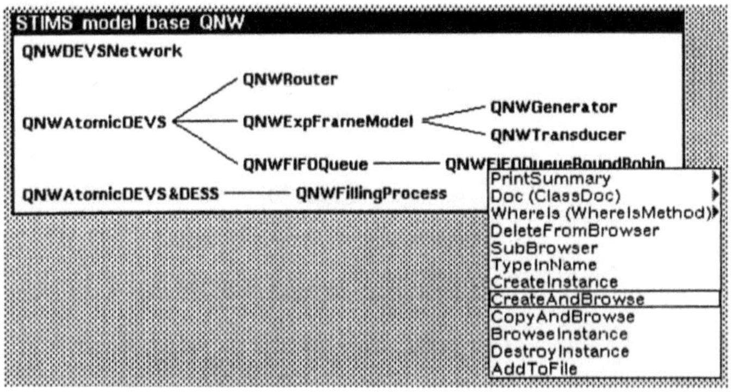

Figure 3.4.7 A model base browser.

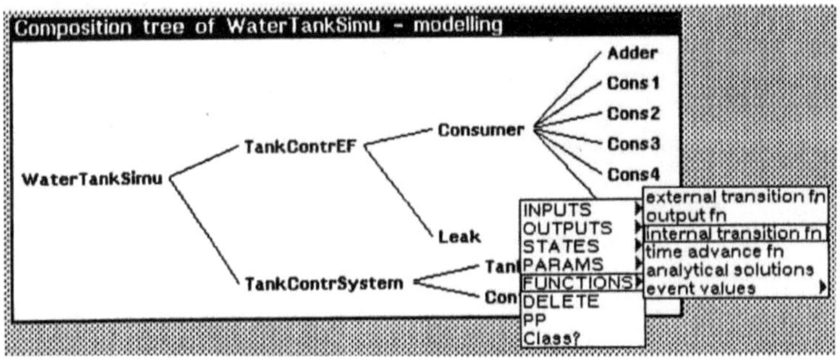

Figure 3.4.8 An instance browser of a hierarchical model.

functions of atomic components can be edited using the syntax editor of Interlisp-D (see Figure 3.4.5). When selecting the menu item to edit a function, a syntax editor window is opened and the function can be defined.

The instance browser is also used as the command unit for the simulation layer. The mode of the instance browser can be changed to *simulation*. New menus are provided in this new mode to set initial states, to set up experimental conditions and to start simulation runs. There are several different modes of simulation, viz. a *stepwise mode*, a *trace mode*, where all the events are put out in trace windows, and a normal mode.

The simulation runs can be observed and simulation results can be put out on the screens by gauges and trajectories. LOOPS provides several classes implementing gauges to put out number values in a pictorial way. We have implemented a set of

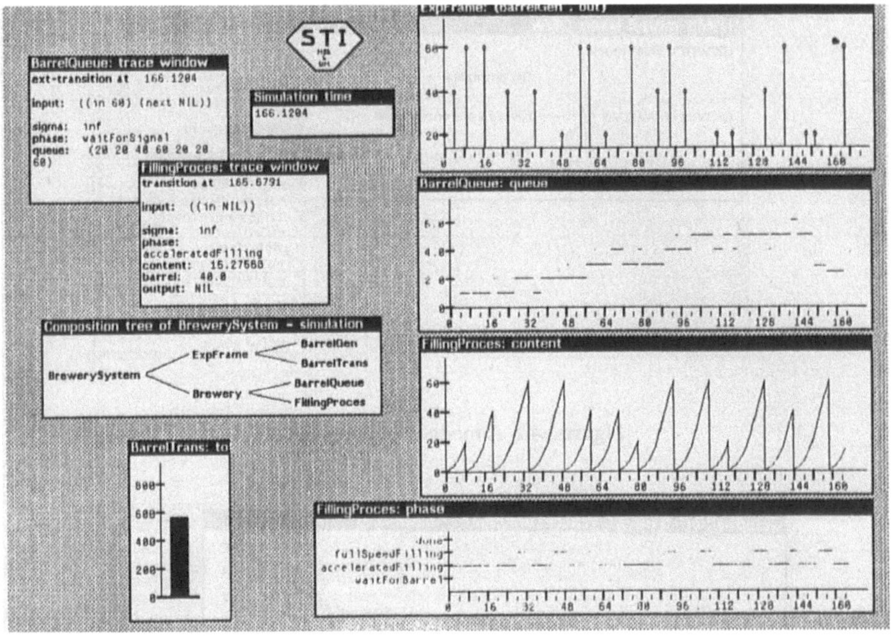

Figure 3.4.9 A typical screen of a STIMS session.

classes to implement gauges and trajectories for arbitrary values (beside numbers also symbols and boolean vectors). Figure 3.4.9 shows a screen of a typical STIMS session. A detailed description of the user interface of STIMS can be found in [Praehofer 90d].

3.5 MULTIFACETTED MODELLING

3.5.1 Introduction to Multifacetted Modelling

In the first section, when discussing the general model-based problem solving process, we stated that the modelling process should be guided by the objectives of the problem solving enterprise. We also stated that through the complexity of real systems, a fully comprehensive model of a real system is infeasible to build [Zeigler 84c]. Modelling always means idealization, abstraction and simplification. Only some aspects, the aspects of the real system relevant for the objectives in hand, may be incorporated into the model.

In an objectives-driven methodology, the multiplicities of objectives result in multiplicities of models representing different aspects of the same real system. When the goal of an computer-aided environment is not the building and usage of one single model for one single objective, but when a set of different models emerged from different objectives should be built and maintained as whole, new concepts and techniques have to be introduced. Zeigler summarized the concepts introduced for that purpose by the term *multifacetted modelling methodology* [Zeigler 84a, 84c].

The heart of the multifacetted modelling methodology is the *system entity structure* [Zeigler 84a] knowledge representation scheme. This knowledge representation scheme has its origins in the Systems Theory context and in artificial intelligence [Zeigler 90a]. Systems Theory provides the concepts of *decompositions* and *couplings* while it does not provide a representation of *taxonomic* knowledge known from the frame concept of artificial intelligence and the object-oriented programming paradigm. The system entity structure incorporates these concept to represent a system in its multiplicities of decompositions and possible variants.

In the context of analysis of real, existing systems, the system entity structure mainly is used to integrate the various models representing the different aspects of the real system into a coherent whole. Here the concept of decomposition plays a dominant role and taxonomy is less important. In the context of system design however, the system entity structure is used to represent all feasible design configurations. Here taxonomic knowledge is much more important and is used to represent the different design alternatives.

The system entity structure only represents the static structures of the possible models. A model base contains the dynamical structures of the models in form of system specifications. Together with a base of system entity structures, it forms a knowledge base representing static as well as dynamical knowledge about real systems. Employing a system entity structure / model base, the modelling process starts by first defining the objectives of the problem solving enterprise. Then driven by the objectives, the entity structure base is consulted to retrieve the static structure of a model which is appropriate for the objectives. This process is denoted *pruning* the entity structure [Zeigler 84a; Rozenblit 85] and means selecting an appropriate entity structure and the appropriate decompositions and variants. Based on the static structure, a hierarchical model can be built by retrieving the dynamical knowledge from the model base. When no appropriate model can be found in the entity structure / model base, a new model has to be built from scratch which can be "recycled" by putting it into the entity structure / model base [Zeigler 84c].

3.5.2 System Entity Structure Definition

The system entity structure is used to manage a family of models and has to support the representation of *decompositions, couplings* and *taxonomy*. Decomposition is used to represent the way real systems are composed into components in a hierarchical manner. Coupling specifications define how these components can be coupled together. Taxonomic knowledge means the representation of the different kinds of an object, i.e., how they are categorized and subclassified [Zeigler 90a].

Zeigler [Zeigler 84a, 90a] defines the system entity structure as a labeled tree with attached variable types which has to satisfy the following axioms:

1. *uniformity*: Any two nodes which have the same name have identical attached variable types and isomorphic subtrees.
2. *strict hierarchy*: No label appears more than once down any path of the tree.
3. *alternating mode*: Each node has a mode which is either *entity, aspect* or *specialization*; if the mode of a node is entity, then the modes of its successors are aspect or specialization; if the mode of a node is aspect or specialization, then the modes of its successors are entity. The mode of the root is entity.
4. *valid brothers*: No two brothers have the same label.
5. *attached variables*: No two variable types attached to the same item have the same name.
6. *inheritance*: Every entity in a specialization inherits all the variables, aspects and specializations from the parent of the specialization.

While the system entity structure is completely defined by the axioms, the interpretation is open to the user. However, to be meaningful in the modelling context the interpretation should be as follows [Zeigler 84a; Zeigler, Rozenblit 88]:

An entity is intended to represent a real world object. The object can be independently identified or can represent a component of one or more decompositions. An aspect represents one out of many possible decompositions of an entity. The alternating mode axiom requires that the children of an aspect are entities which are the components of the decomposition. The coupling constraints are attached to the aspects and represent the way the entities of an aspect can be coupled together. The variable types attached to entities define attributes of the entity. Variable types attached to an aspect belong the parent entity but only in the context of the aspect.

Specializations are used to codify taxonomic knowledge of an entity. Several specializations of an entity may exist, each classifying the entity according to a different classification criteria. The children of a specialization are entities representing the different variants of the general entity according to the particular specialization criteria.

A special form of an aspect used in entity structures is the *multiple decomposition*. By that means it is possible to represent a decomposition of an entity into an arbitrary and varying number of identical entities. As a typical example of such a multiple decomposition we can think of the entity *group* which can be decomposed into an arbitrary number of *individuals*.

The best way to represent an entity structure is by a pictorial representation of the tree structure. The nodes of the tree represent entities, aspects and specializations. The arcs of the tree represent the relations between parents and their children. As, depending on the type of the node, we can identify different relations between the nodes, we use different arcs to represent these relations. An aspect relation is depicted by a single line, an multiple decomposition relation by three parallel lines and a specialization relation by a double line arrow.

Figure 3.5.1 shows a system entity structure of a multi-processor architecture model. The root *MPS-EF* represents the multi-processor - experimental frame coupling. The first descendant *mp-ef-dec* of the root is an aspect node and stands for the decomposition of the root into the multi-processor system and the experimental frame. The experimental frame *EF* is further decomposed into a transducer *Transd* and an acceptor *Accept* entity which are not further decomposed and stand for atomic components.

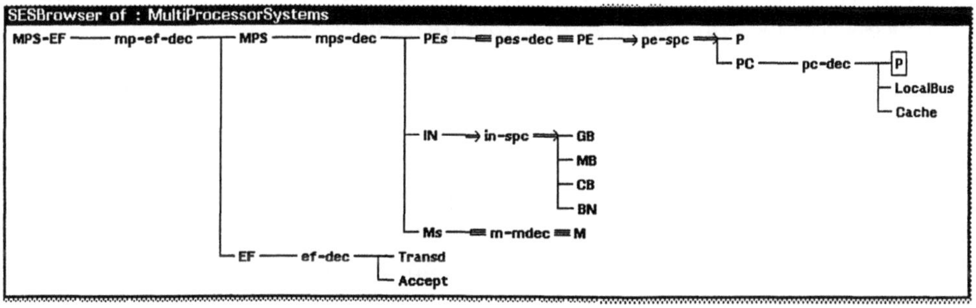

Figure 3.5.1 System entity structure for multi-processor system architectures.

The tree under the multi-processor node *MPS* codifies all possible design configurations of the multi-processor system architecture. The multi-processor architecture is first of all composed into a set of processor elements *PEs,* interconnection network *IN,* and a set of memory elements *Ms.* The set of processor elements and the set of memory elements consist of an arbitrary number of processor elements and memory modules respectively. This is represented by multiple decompositions *pes-dec* and *m-dec*. *PE* denotes an individual processor element and *M* an individual memory module.

Several different types of interconnection networks can be used in the multi-processor architecture. This is signified by an specialization *in-spc*. Possible variants of interconnection networks are the global bus *GB,* the multi-bus *MB,* the cross bar *CB,* and the Banyan network *BN.*

Processor elements can be of different variants and hence are specialized by the specialization *pe-spc*. A processor element can be a single processor *P* or can be an element *PC* which is decomposed into processor *P*, a cache *Cache,* and a local bus *LocalBus*. It should be recognized that the processor *P* in the *PC* decomposition and the processor *P* as an variant of the processor element *PE* have identical names and hence are identical entities.

3.5.3 Pruning the System Entity Structure

The system entity structure is employed to codify possible models or configurations of a system in a compact manner. E.g., the multi-processor system entity structure codifies design configurations of a multi-processor system architecture. From the system entity structure then, different hierarchical model structures can be extracted. This process is called *pruning* the entity structure [Zeigler 84a; Rozenblit 85].

The pruning process is guided by the objectives of the problem solving process. Rozenblit [Rozenblit 85; Rozenblit, Huang 91] defined procedures for automatic pruning, which are based on the formulation objectives in form of *generic experimental frames* and the formulation of *design constraints* using a rule-based approach. Pruning also can be done interactively by which the user has to decide for the different variants to get a model structure which is appropriate to solve his problem. We define a technique for interactive pruning by the following procedure (this simplified procedure is based on the procedure found in [Zeigler 90a]; a more elaborated pruning process can be found in [Zeigler 84a]):

```
Procedure Interactive Pruning (SES):
  Select a root entity in the entity structure SES
  to become the root of the hierarchical model

  loop Starting with the root entity traverse the system entity structure SES
  in a depth-first manner to visit all entities until all have been visited do:
    Decide if the currently visited entity shall become a leaf
          in the hierarchical model
    if yes then abort subsequent pruning of its subtree
    if no then
      Select one (or none) aspect from those available under the entity
      In a multiple decomposition define the number of entities
          of the multiple decomposition
      Select one (or none) specialization from those available
      Select one entity of the selected specialization

end Interactive Pruning.
```

The resultant of the procedure is a *pruned entity structure* where at most one aspect and at most one specialization follows an entity and at most one entity follows each specialization node. From this pruned entity structure we derive a composition tree by replacing the entities which have specializations by the specialized entity and the specialization nodes are discarded. Afterwards, the attributes and the coupling specifications and the children of the aspects are moved to their parent entities and the aspect nodes are discarded. The result is the composition tree of the hierarchical model. A hierarchical model can be constructed from the composition tree when dynamical models for all the leaf entities reside in the model base.

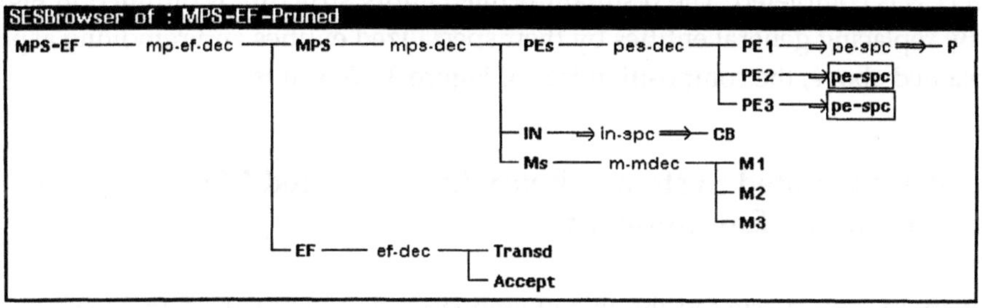

Figure 3.5.2 Pruned entity structure.

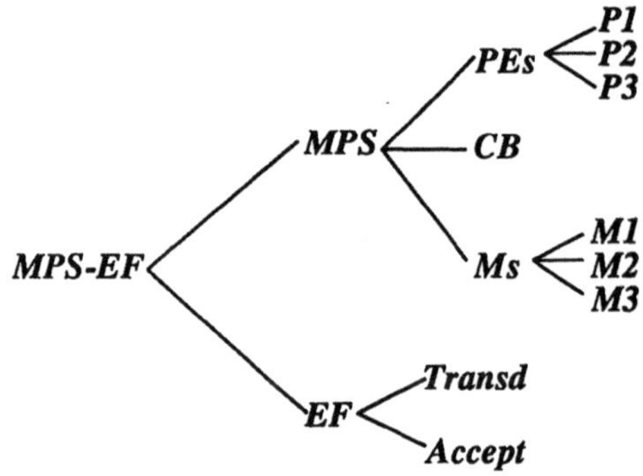

Figure 3.5.3 Resulting composition tree.

In our example of the multi-processor system entity structure, a pruning process could look the following: we choose the entity node *MPS-EF* as the root . This shall not be a leaf hence we choose the unique aspect *mp-ef-dec.* Following the depth first scheme, we go to the node *EF* and decompose this node by choosing aspect *ef-dec. Transd* and *Accept* of course have to be leaves.The next entity in the depth-first traversal is the entity *MPS* which shall be further decomposed by aspect *mps-dec*. The next entity is *Ms* with multiple decomposition *m-mdec*. For the multiple decomposition *m-mdec* we decide that the number of memory elements *M* should be three. The entity *IN* owns a specialization *in-spc*. For this specialization we choose the entity cross-bar *CB* to be our interconnection scheme. The set of processor elements *PEs* has a multiple decomposition node *pes-dec* and we decide to take three processor elements *PE*. From the specialization under the processor elements we then decide to take the single processor *P*. Now all the leaf entities have been visited and the pruning process is completed. The resultant pruned entity structure is outlined in Figure 3.5.2. By replacing general entities by their specialized entities and removing aspects as explained above, the composition tree of Figure 3.5.3 results.

3.5.4 An Integrated Systems Theory Instrumented Modelling and Simulation Environment

In Section 3.1.1 we discussed the role of simulation in the scientific problem solving process from a very general point of view. In Sections 3.1.2 and 3.2 we presented how Systems Theory can help and how modelling and simulation is done following a rigorous systems theoretic approach. In Section 3.3 we developed the different abstract simulator concepts for simulation of discrete, continuous and combined modular, hierarchical models. Above we described a new knowledge representation scheme which is employed to support the management of families of related models in model/knowledge bases and its role in the multifacetted modelling methodology. Now we will outline how all these concepts can be employed to set up an integrated Systems Theory instrumented modelling and simulation environment. Such an environment should support modelling and simulation of various kinds of models - discrete, continuous as well as combined - and all the model management capabilities described above.

Figure 3.5.4 sketches the architecture of such an integrated environment. The environment is built around two bases. One is the SES base containing the system entity structures, viz. all the possible static model structures of several systems. The second is the model base containing all the dynamical behavior descriptions in form of system specifications.

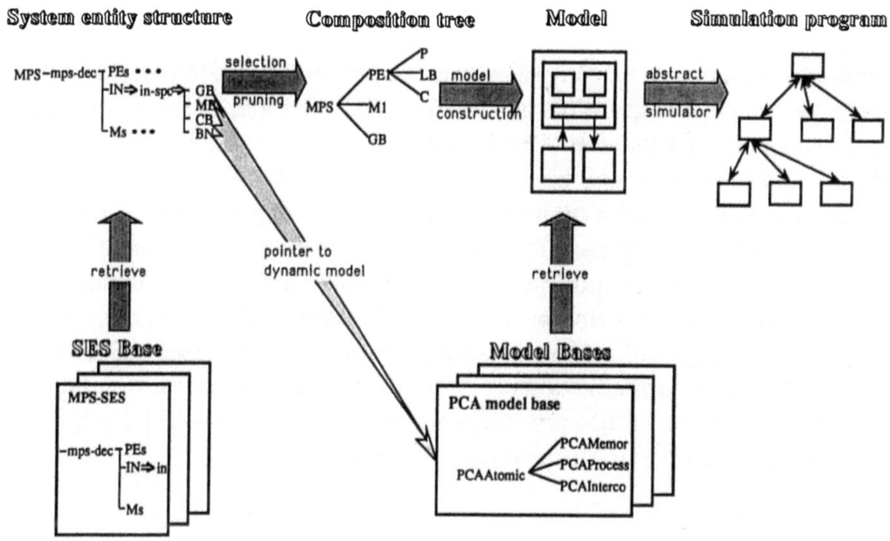

Figure 3.5.4 Systems Theory instrumented modelling and simulation architecture.

The modelling and simulation process starts by retrieving the appropriate system entity structures from the SES base. The next step is pruning the selected entity structure. The result of this process is a composition tree. With the leaves of the pruned entity structure, pointers to the dynamical models in the model base have to be associated. Using the hierarchical structure as well as the coupling specification encoded in the composition tree as well as the dynamical structure of the leaves retrieved from the model base, a modular hierarchical model can be constructed. Herein, the dynamical descriptions associated with the leaves can be of arbitrary type. The system theoretic concepts support discrete, continuous as well as all kind of combined models. This modular, hierarchical model now readily can be transferred into an executable simulation program using the abstract simulator concepts.

3.6 TOWARDS VARIABLE STRUCTURE MODELLING AND SIMULATION*

3.6.1 Introduction and Motivation

Conventional Systems Theory and simulation systems adequately support only a single level at which change occurs in a model, that of changes in the model descriptive variables, viz., its behavior. Although changes in the model structure may be introduced, and to some extent, tracked in such systems, specific and powerful support is not provided for such activities. Many kinds of real systems are most readily perceived as exhibiting changes simultaneously at structural and behavioral levels. To name a few problems requiring variable structure modelling: vulnerability of computer networks and distributed systems to local failures, automated management of failure responses of complex systems, adaptive systems and self-reconfiguring computer/robotic organizations. A new paradigm, structural simulation (as opposed to conventional "trajectory" simulation) is needed to avoid having to force structural changes down to the same level as behavioral ones.

---

* This section is based on the article "System Theory Challenges in the Simulation of Variable Structure and Intelligent Systems" by B. P. Zeigler and H. Praehofer [Zeigler, Praehofer 90]. We gratefully acknowledge the contributions of Bernard P. Zeigler in writing this section.

In previous work we have defined variant family as a set of models that can be generated by applying a sequence of admissible transformations to an initially specified skeleton. Such a variant family may arise for example, in a design environment in which alternative design models are to be evaluated. A variable structure model contains within itself, the ability to transform, in appropriate circumstances, to any of the models in a variant family.

The problem of facilitating the model-controlled variation of model structure introduces new elements into a simulation environment research that have not been previously considered among the standard set of issues in Systems Theory. In variable structure modelling methodology, a model is able to change its structure, i.e., to transform from one structure state (synthesis template) to a successor. We have proposed that the system theoretic representation scheme, the system entity structure, provides an appropriate formal tool for describing the structure state of a model. The entity structure specifies the family of structure state variants that a model can assume. We will show how control of the entity structure synthesis must be embedded in the model so that it puts into effect a structure state transition over an extended time interval.

A host of system-theoretic, modelling methodological, and simulation environment implementation issues are raised by this approach. One we have considered previously is that of distributed maintenance of structure knowledge. In a variable structure system containing modular components there can be no sharing of, nor direct access to global or other components' variables. Thus knowledge of current structure must be distributed among the components and co-operatively maintained. Each component must have its own knowledge - a dynamical model - of just that projection of the structure state necessary for it to carry out its intended behavior. The state of this model must be maintained in homomorphic correspondence with the structure state of the system via communication with other components. Such a conceptual framework based on system concepts helps in understanding how to design such systems and should even provide a way to prove design correctness.

In this paper we continue to explore some of the fundamental issues in the variable structure system paradigm. We develop a ranking of difficulty for variable structure models. For example, the simplest case is that of a single component which changes from one state transition structure to a second under some conditions (e.g., fall before and after the opening of a parachute). Much more complex is the handling of system synthesis operations, diagnosis and repair which calls for assembling and disassembling components to form new ones during a simulation run. Yet more complex is the representation of adaptive systems or intelligent agents that employ a modelling methodology to develop and alter models to learn a complex environment.

The ranking typology should be useful in guiding future developments of concepts and environments for variable structure methodology. We use it to discuss issues in the implementation of variable structure models in the DEVS-Scheme object-oriented simulation environment. Since this environment is based on system-theoretic formalisms, such implementation should provide the basis for system-theoretic considerations of variable structure systems. Can Systems Theory develop useful formalisms to deal with such variable structure models or does the complexity involved force reliance on computer implementations? As we will see, one helpful orientation can be found in the theory of multi-level systems developed by Mesarovic, Macko and Takahara [Mesarovic et al. 1970]. Much more in the way of CAST support is needed.

The basis for variable structure modelling and simulation provides the multifacetted modelling methodology developed by Zeigler [Zeigler 84a] and reviewed in Section 3.5. The multifacetted methodology denotes an approach which recognizes the existence of multiplicities of objectives and models in a simulation project. It provides formal representation schemes that support the modeler in organizing the model construction process [Rozenblit 85].

3.6.2 Ranking Typography of Variable Structure Models

Many real world systems and problems demand variable structure modelling and simulation. A ranking of difficulties in modelling such real systems can help to guide the development of concepts and supportive environments for variable structure modelling. In the following we classify the problems according to the actions to be undertaken when making a structure state transition and the elements influencing this decision.

Figure 3.6.1 Change of one single model component.

The simplest case exhibiting a variable structure occurs when a single component changes from one structure state to another. This is called a multi-model by Ören [Ören 89b]. As an example we can think of the well known parachute problem where the dynamics change from free fall to retarded fall after opening the parachute. More generally, structure change of this kind involves resetting of parameter values and replacement of one component by another (Figure 3.6.1). This situation is represented in multifacetted modelling methodology in that one specialization of an entity is replaced by another one (E1 specialization of E is replaced by E2 specialization). It must be emphasized that variables of the general entity (E) are common to all specialized entities (E1, E2). In replacing the component represented by one specialized entity by the component represented by a second, the values of the common variables have to be transferred as well. For example, the variable representing the height of the falling man has to be transferred from the model of the free fall to that of retarded fall.

Already a much more complex situation is given when the problem calls for creation and destruction (or modification) of components during the simulation runs. Vulnerability of distributed systems to local failures is an example of such a problem. We include this class within a wider one in which components within a model may be assembled and disassembled. Possible actions are adding, deleting and exchanging of components and adding and deleting couplings. To distinguish this rank from still more complex problems we require that the structure change depends only on processes inside the model and cannot be directly influenced by decisions made outside. In other words, decisions causing structure changes and actions to be undertaken are locally made. Figure 3.6.2 exhibits such a complex structure change which is governed by the possible structure states represented by the system entity structure ES. As in the first case, when exchanging components, the values of variables of the general entity have to be transferred to the new component.

Even more complex is the situation where model changes are decided by components outside the model. The area of adaptive systems belongs to this class. Here the problem is the design of an optimal scheme for adapting structure so as to maintain performance objectives in the face of changing environments. Examples of adaptive systems are adaptive computer architectures [Zeigler 86], flexible manufacturing systems, self-reconfiguring multi-robot organizations and economic organizations.

In addition to the actions which have already been discussed in the first two cases, this class requires us to facilitate the exchange of components among different networks. It should be possible to transfer a component representing a resource to a new "place of work" where it is needed more urgently. Examples of problems requir-

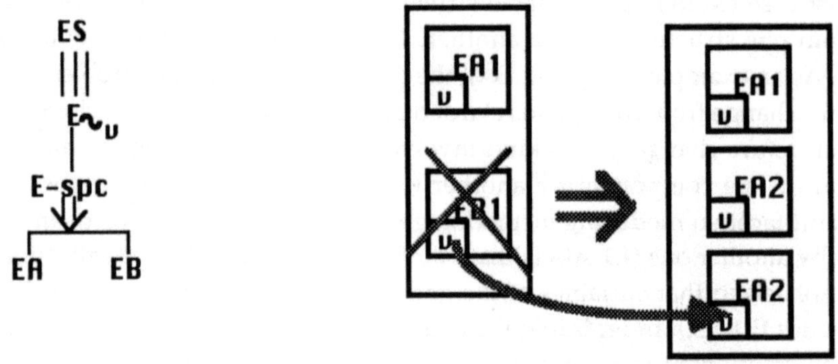

Figure 3.6.2 Adding and exchanging components.

ing such operations are exchange of workers between different departments or transfer of fertile breeders from one cattle herd to another to enhance reproduction.

Another problem arising here is the modelling of intelligent agents which make decisions causing structure changes. As the human intelligent agent does, the model counterpart should have an image (= model) of the prevailing world situation upon which his decisions are based. To represent knowledge of real world situations faithfully, this internal model must necessarily be imprecise and uncertain. We will discuss this problem in detail in Section 3.6.5.

3.6.3 A Multi-Level System Approach to Variable Structure Modelling

The theory of of hierarchical, multi-level systems developed by Mesarovic, Macko and Takahara [Mesarovic et al. 1970] in the context of control of large scale systems can serve as a guideline in the development of concepts for the simulation of variable structure models. Their book "Theory of Hierarchical, Multi-level Systems" introduced a classification of multi-level systems into multi-strata, multi-layer and multi-echelon systems. In a multi-strata system, on each level the system is described as a whole but with different degree of abstraction and/or from different points of view. In a multi-layer system the whole system is divided into different levels according to different complexities of the decision making process. In distinction to the multi-strata system, all layers must be taken together to represent the whole system. A multi-echelon system is used to represent organizational hierarchies of decision making units in the sense that the infimal units are controlled by the supremal units.

Although this theory is not transferable in unmodified form to the problem of variable structure modelling, we will use the notions for a general orientation. We state that a variable structure model should be organized like a multi-echelon system where the supremal levels have control over the structure of the infimal levels. The supremal units base their decision upon the state transitions of the infimal units.

Inter-level dependencies are of primary concern in the expositions of Mesarovic and Takahara as they are here. In variable structure modelling we observe two types of dependencies: first the supremal units have to observe the dynamical processes, i.e., the changes of the state variable values, of the infimal units and second, according to the decisions based upon these changes, they have to change the structure states of the infimal units. Figure 3.6.3 shows this general structure of a multi-level, variable structure system.

For most cases of variable structure models a two level system will be sufficient. This is especially true for the first two classes of our ranking typography. The first level will always represent the behavioral model where the normal behavioral simulation takes place. The second level then is used to control the structure state of the first one. But looking at class three of our typology, one can easily imagine that the second level, which is a model of the intelligent agents responsible for the adaptive control, is itself subject to change by a higher control unit, i.e. a third level. As an example we can think of a industrial organization. The intermediate management controls the structure state of the manufacturing system while the top management also controls the structure state of the intermediate management.

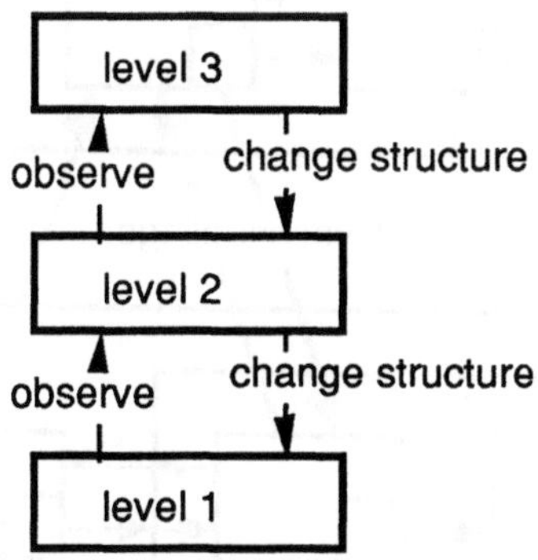

Figure 3.6.3 General scheme of a multi-level, variable structure system.

3.6.4 DEVS-Scheme Implementation of Multi-Level, Variable Structure Models

The DEVS-Scheme simulation environment (see Section 3.4) provides a strong basis on which an environment for variable structure modelling and simulation can be built. A multi-level variable structure system can be built in DEVS-Scheme in such a way that each level is a modular, hierarchical DEVS system. An atomic model at a supremal level would own state variables which represent the structure state of the infimal unit it is controlling. With such knowledge, the supremal unit can direct a change in structure state in the infimal one.

In addition to couplings on each level, to facilitate communication between the different levels, inter-level couplings have to be introduced. These couplings are used to observe the changes of the state variables' values in the atomic models in the infimal units or for controlled interrogations into their states. Hence we have to erect inter-level couplings between the atomic models of the different levels (Figure 3.6.4).

The development of simulation concepts to simulate such multi-level systems will be a challenge for further developments. However, the current status of DEVS-Scheme already provides some support to implement a multi-level, variable structure system.

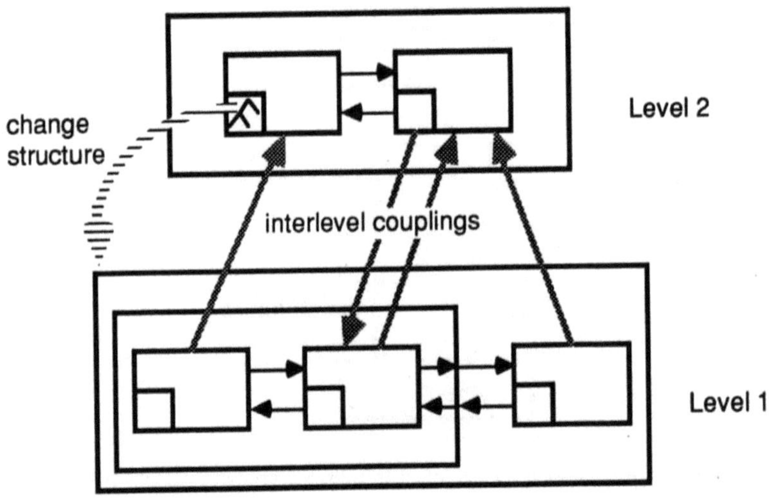

Figure 3.6.4 Multi-level, variable structure system implementation.

For observing the processes of the infimal units, DEVS-Scheme provides the concepts of event-observation and introspection. Event-observation allows one to instrument any atomic model so that the value of any expression computable in the environment of the atomic model can be output prior to the internal or external events. Such outputs appear at special ports: *%ext-event%* and *%int-event%* , for external and internal events respectively. These ports should be coupled to ports of atomic models of the supremal level which, by this means, observe the dynamical behavior of the infimal level.

The similar concept of introspection can be used for controlled interrogation of a atomic model in the infimal unit by a model of the supremal unit. The model of the infimal level is augmented with a special input port *%interrogate%* and a special output port *%report%*. By couplings of the supremal model to infimal model, the supremal can ask questions through the input port %interrogate% and receive answers through the output port %report%.

Equipped with these concepts we are in the position to implement multi-level, variable structure systems, particularly the inter-level couplings as shown in Figure 3.6.5. The multi-level system would be modeled by a coupled system where each component represents one level. The higher levels hold the structure state of components of the infimal levels in their state and so are in the position to change the structure of these components. Inter-level couplings - where required - are done through the special ports *%ext-event%*, *%int-event%*, *%interrogate%*, and *%report%* which have to be brought out to the outermost coupled model of one level to couple them to the coupled model implementing the other levels. A multi-level system implemented in such a way can readily be simulated using the abstract simulator concept of DEVS-Scheme.

3.6.5 Models of Intelligent Agents

The DEVS formalism expressed in the atomic-model class of DEVS-Scheme serves as a basis on which higher level formalisms can be implemented. In particular, a convenient rule-based paradigm is supported by *forward-models*, a specialized class of atomic models. Forward-models is based on the forward chaining paradigm and is an appropriate means to model the decision making capabilities of intelligent agents. To develop a model of type forward-models, we write rules which together specify the internal, external and output functions of a DEVS model. The inference engine underlying forward-models evaluates the rules in the order they are added to the model.

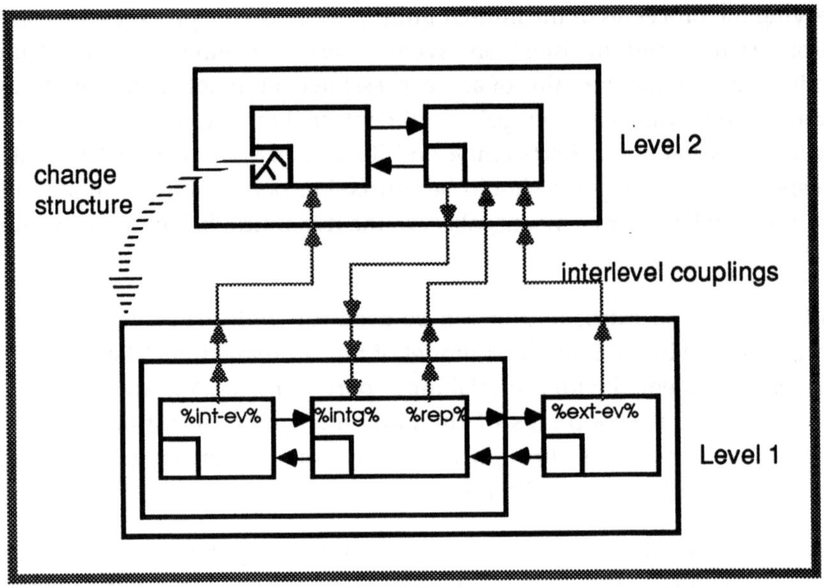

Figure 3.6.5 Multi-level, variable structure model implementation in DEVS-Scheme.

But to achieve realism, models of intelligent agents must be able to represent not only their decision making capabilities but also the models on which such capabilities are based. In the context of event based control [Zeigler 90a; Zeigler et al. 88] the situation is depicted in the following way: the simulation model consists of the model of the system to be controlled and the model of the intelligent agent which in turn consists of the model of the decision making component and of its internal model of the system to be controlled (used by the intelligent agent to picture of the real world) (Figure 3.6.6). Table-models, another specialization of atomic models can be used to abstract internal models from models of the external system.

An interesting parallelism can be observed between the above scheme of event based control and the problem of controlling the structure state of variable structure model. While in event based control the decisions of the control device are based on the internal model of the system to be controlled, in variable structure modelling the "internal model" has to represent the possible structure states, and hence the system entity structure, of the variable structure model. In normal simulation, the entity structure is pruned in the modelling process only between distinct simulation runs. This pruning is driven by the objectives of the human or artificial modeler [Rozenblit 85; Rozenblit, Huang 91; Rozenblit et al. 90]. In variable structure simulation, this

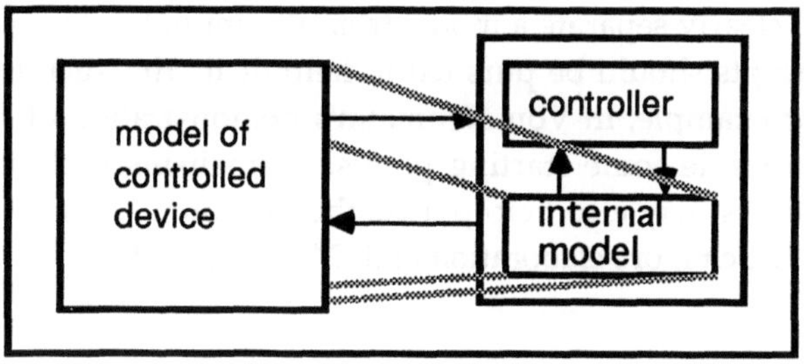

Figure 3.6.6 Event based control.

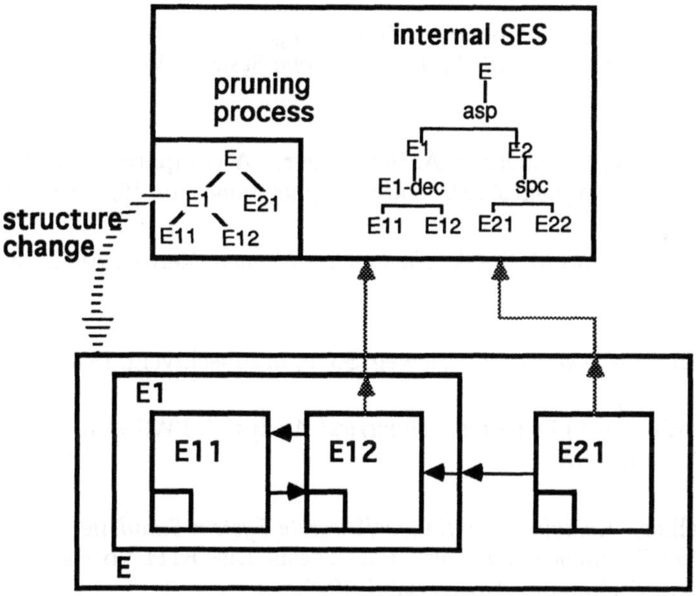

Figure 3.6.7 Modelling the pruning process for variable structure simulation.

pruning process has to be done by the (intelligent) model representing the objective driven pruning process and it uses a system entity structure as internal knowledge of possible structure states (Figure 3.6.7).

The development of models which incorporate internal system entity structures will pose a major challenge to the development of DEVS-Scheme. More generally stated, how much of a simulation environment can be incorporated within its models? It should be noted that classical simulation languages cannot even state this problem since they do not clearly separate a model from its simulator. Developments in Systems Theory concepts should be pursued to help deal with such issues. Results on self-reference, for example, in Von Neumann's demonstration of self-reproducing automata, may provide some starting points. Formalisms such as Takahara's for time-varying systems may also help phrase the questions. However, practical systems will require much more elaboration and CAST support.

REFERENCES – CHAPTER 3

[Antsaklis et al. 89]
P. J. Antsaklis, K. M. Passino and S. J. Wang, "Towards Intelligent Autonomous Control Systems". *J. of Intelligent and Robotic Systems,* Vol. 1, No. 4, pp. 315–342, 1989.

[Blaschek et al. 89]
G. Blaschek, G. Pomberger, A. Stritzinger, "A comparison of object-oriented programming languages". *Structured Programming,* Vol. 10, No. 4, 1989.

[Bobrow, Stefik 83]
D. G. Bobrow, and M. Stefik, *The LOOPS Manual,* Xerox PARC, Palo Alto, CA, December 1983.

[Bossel 87]
H. Bossel, "*Systemdynamik*". Vieweg, Sohn, Braunschweig, 1987.

[Burden, Faires 89]
R. L. Burden and J.D. Faires, "*Numerical Analysis*". PWS-Kent Publishing Company, Boston, 1989.

[Cellier 79]
F. E. Cellier, "*Combined Continuous/Discrete System Simulation by Use of Digital Computers. Techniques and Tools*". PhD Thesis, Diss ETH No. 6483, Swiss Federal Institute of Technology, Switzerland, 1979.

[Cellier 86]
F. E. Cellier, "Combined Continuous/Discrete Simulation - Application, Techniques, and Tools". *Proceedings 1986 Winter Simulation Conference,* Washington D.C., pp.24–33, 1986.

[Fahrland 70]
D. Fahrland, "Combined discrete event continuous system simulation". *SIMULATION,* February, pp. 61–72, 1970.

[Garzia et al. 86]
R. F. Garzia, M . R. Garzia, B. P. Zeigler, "Discrete-event simulation". *IEEE Spectrum,* December, pp. 32–36, 1986.

[Golden, Schoeffler 73]
D.G. Golden, J.D. Schoeffler, "GSL - a combined continuous and discrete simulation language". *SIMULATION*, January, pp. 1–8, 1973.

[Keene 89]
S. E. Keene, *"Object-Oriented Programming in COMMON LISP"*, Addison Wesley, Reading, Massachusetts, 1989.

[Klir 85]
G. J. Klir, *"Architecture of System Problem Solving"*. Plenum Press, New York, 1985.

[Laux 81]
G. Laux, *Kybernetik Lexikon*, Akademie Verlag, Berlin, 1981.

[Lerner 67]
A. Ja. Lerner, *"Kybernetik"*. VEB Verlag Technik, Berlin, 1967.

[Mesarovic et al. 70]
M.D. Mesarovic, D. Macko, and Y. Takahara (1970), *"Theory of Hierarchical, Multilevel, Systems"*, Academic Press, New York, 1970.

[Meyer 88]
B. Meyer, *"Object Oriented Software Construction"*. Prentice Hall, New York, 1988.

[Mittelmann 88]
R. Mittelmann, "Object Oriented Implementation of Petri Nets Concepts".
in: *Cybernetics and Systems'88* (ed. R. Trappl), Kluwer Academic Publisher, pp. 759-766, 1988.

[Müller-W. 86]
T. Müller-W. *"LISAS – Linzer Systolic Array Simulator"*. Master Thesis, Department of Systems Theory, University Linz, Austria, 1986 (in German).

[Nasa 85]
Nasa, *The Space Station Program*. Nasa Publication, 1985.

[Ören 85]
T. I. Ören, "GEST - A Modelling and Simulation Language Based on System Theoretic Concepts". In: *Simulation and Model-Based Methodologies: An Integrative View*, eds. T.I Ören, B.P. Zeigler and M.S. Elzas, NATO ASI Series, Series F: Computer and System Sciences, Vol. 10, Springer Verlag, pp. 281–335, 1985.

[Ören 89a]
T.I. Ören, "Simulation Models: Taxonomy". In: *Encyclopedia of Systems and Control*, ed. M. Singh, New York, NY, Pergamon Press, 1989.

[Ören 89b]
T. I. Ören, Dynamic Templates and Semantic Rules for Simulation Advisers and Certifiers". In: *Knowledge-Based Simulation: Methodology and Application*, eds. P.A. Fishwick and R.B. Modjeski, Springer-Verlag, Berlin, Heidelberg, New York, 1989.

[Pegden 85]
C. D. Pegden, *"Introduction to SIMAN"*. System Modelling Corporation, State College, Pennsylvania, 1985.

[Pichler, Praehofer 88]
F. Pichler and H. Praehofer, "Computer Aided Systems Theory: Finite State Machines", *Cybernetics and Systems*, pp. 737–742, 1988.

[Pichler, Schwärtzel 90]
F. Pichler, H. Schwärtzel, *"CAST– Computerunterstuetzte Systemtheorie"*, Springer Verlag, Berlin, 1990.

[Pichler 75]
F. Pichler, *"Mathematische Systemtheorie"*. Walter de Gruyter, Berlin, 1975.

[Pichler 88]
F. Pichler, "CAST - Computer Aided Systems Theory: A Framework for Interactive Method Banks", *Cybernetics and Systems*, pp. 731–736, 1988.

[Praehofer, Spalt 88]
H. Praehofer and A. Spalt "An Interactive Simulation Environment Based on Systems Theory Concepts and Object Oriented Programming". *Proc. of the SCS European Simulation Multiconference '88*, SCS Press, Nice, France, pp. 199-203, 1988.

[Praehofer, Zeigler 90]
H. Praehofer and B. P. Zeigler,"Modelling and Simulation of Non-Homogeneous Models". In: *Computer Aided Systems Theory - EUROCAST '89*, eds. F. Pichler and R. Moreno-Diaz, Lecture Notes in Computer Science 410, Springer Verlag, Berlin, pp. 200–211, 1990.

[Praehofer 89]
H. Praehofer, *"Modelling and Simulation of Non-Homogeneous Specified Models"*, Technical Report, Department of Systems Theory, University Linz, Austria, 1989.

[Praehofer 90a]
H. Praehofer, "Simulation of Event-Based Control of Continuous Systems". In: *AI, Simulation and Planning in High Autonomy Systems*, eds. B. P. Zeigler and J. W. Rozenblit, IEEE Computer Society Press, pp. 89–96, 1990.

[Praehofer 90b]
H. Praehofer, "STIMS - A prototypical implementation of simulation environment based on systems theoretical concepts". *Proc. of the SCS European Simulation Multiconference '90*, SCS Press, Nuremberg, FRG, pp. 75-80, 1990.

[Praehofer 90c]
H. Praehofer "Systems Theoretic Formalisms for Combined Discrete Continuous System Simulation". *International Journal of General Systems*, Vol. 19, No. 3, pp.. 219-240, 1991.

[Praehofer 90d]
H. Praehofer, *"STIMS User's Guide"*. Technical report, Department of Systems Theory, University Linz, Austria, 1990.

[Praehofer 91]
H. Praehofer *"Systems Theoretic Foundations for Combined Discrete Continuous System Simulation"*. PhD Thesis, Department of Systems Theory, University Linz, Austria, 1991.

[Pritsker, Pegden 79]
A. B. Pritsker and C. D: Pegden, *"Introduction to Simulation and SLAM"*. Halsted Press, New York, 1979.

[Pritsker 74]
A. A. B. Pritsker, *"The GASP IV Simulation Language"*. John Wiley, Intersience, 1974.

[Reisig 86]
W. Reisig, *"Petrinetze - Eine Einführung"*. Springer Verlag, Berlin, 1986.

[Rozenblit, Zeigler 88]
J.W. Rozenblit and B.P. Zeigler, "Design and Modelling Concepts". In: *Encyclopedia of Robotics*, pp. 308 –322, John Wiley, New York, 1988.

[Rozenblit 85]
J. W. Rozenblit, *"A Conceptual Basis for Integrated, Model-based System Design"*. PhD Thesis, Wayne State University, Detroit, Michigan, 1985.

[Rozenblit, Huang 91]
J.W. Rozenblit, Y.M. Huang, "Rule-Based Generation of Model Structure in Multifacetted Modelling". *ORSA Journal of Computing*, 4(3), 1991.

[Rozenblit et al. 90]
J. W. Rozenblit, J. Hu, T. G. Kim, and B. P. Zeigler, "Knowledge Based Design and Simulation Environments (KBDSA): Foundational Concepts and Implementation". *J. Operational Research Society*, Vol. 41, No. 6, pp. 475–489, 1990.

[Schmidt 85]
B. Schmidt, *"Systemanalyse und Modellbau"*. Springer Verlag, Berlin, 1985.

[Schmidt 87]
B. Schmidt, *"Model Construction with GPSS-FORTRAN Version 3"*. Springer Verlag, Berlin, 1987.

[Simon 62]
H. A. Simon, "The Architecture of Complexity". *Proc. Am. Philos. Soc. 106*, pp. 467–482, 1962.

[Spalt 88]
A. Spalt, *"Systems Theory Instrumented Discrete Event Simulation"*, Master Thesis, University of Linz, Austria, 1988 (in German).

[Stachowiak 73]
H. Stachowiak, *"Allgemeine Modelltheorie"*. Springer Verlag, Wien, 1973.

[Stroustrup 86]
B. Stroustrup, *"The C++ Programming Language"*. Addison Wesley, 1986.

[Texas Instruments 85]
Texas Instruments, *"TI Scheme Reference Manual"*. Texas Instrument Incorporated, 1985.

[Wymore 76]
A.W. Wymore, *"Systems Engineering Methodology for Interdisciplinary Teams"*, John Wiley, New York, 1976.

[Xerox 83]
Xerox, "*Interlisp Reference Manual*". Xerox Palo Alto Research Center, October 1983.

[Zeigler, Praehofer 90]
B. P. Zeigler and H. Praehofer ,"System Theory Challenges in the Simulation of Variable Structure and Intelligent Systems". In:*Computer Aided Systems Theory - EUROCAST '89*, eds. F. Pichler and R. Moreno-Diaz, Lecture Notes in Computer Science 410, Springer Verlag, Berlin, pp. 41–51, 1990.

[Zeigler 76]
B. P. Zeigler, *"Theory of Modelling and Simulation"*. John Wiley, New York, 1976.

[Zeigler 84a]
B. P. Zeigler, *"Multifacetted Modelling and Discrete Event Simulation."*. Academic Press, London, 1984.

[Zeigler 84b]
B. P. Zeigler, "System-Theoretic Representation of Simulation Models". *IIE Transactions*, March, pp. 19–34, 1984.

[Zeigler 84c]
B. P. Zeigler, "Multifacetted Modelling Methodology: Grappling with the Irreducible Complexity of Systems". *Behavioral Science*, Vol. 29, pp. 169-178, 1984.

[Zeigler 86]
B. P. Zeigler, "Toward a Simulation Methodology for Variable Structure Modelling", in: *Modelling and Simulation Methodology in the Artificial Intelligence Era*, eds.: Elzas, M.S., T.I. Oren and B. P. Zeigler, North Holland, Amsterdam 1986.

[Zeigler 87]
B. P. Zeigler, "Hierarchical, Modular Discrete Event Simulation in an Object Oriented Environment". *Simulation Journal*, Vol. 49:5, pp. 219–230, 1987.

[Zeigler et al. 88]
B. P. Zeigler, F.E. Cellier, J. W. Rozenblit, "Design of a Simulation Environment for Laboratory Management by Robot Organizations". *Journal of Intelligent and Robotic Systems*, Vol. 1, pp. 299–309, 1988.

[Zeigler 89]
B. P. Zeigler, "DEVS Representation of Dynamical Systems: Event-Based Intelligent Control". *Proceedings of the IEEE*, Vol. 77, No. 1, pp. 72–80, Jan. 1989.

[Zeigler 90a]
B. P. Zeigler, *"Object-Oriented Simulation with Hierarchical, Modular Models"*. Academic Press, London, 1990.

[Zeigler 90b]
B. P. Zeigler, "High Autonomy Systems: Concepts and Models". In: *AI, Simulation and Planning in High Autonomy Systems*, eds. B. P. Zeigler and J. W. Rozenblit, IEEE Computer Society Press, pp. 2–7, 1990.

4 MACHINE VISION

4.1 INTRODUCTION

The goal of machine vision, in very general terms, is the construction of artificial mechanisms that are capable of "seeing" similar to the visual systems found in many natural creatures. There are two major motivations for pursuing work in this direction. *First,* there exist numerous practical applications for an artificial vision technology, e.g., in the areas of robotics, space exploration, remote sensing, security, surveillance systems, and many others. *Secondly,* it is a widely held view that research in machine vision may eventually provide us with a deeper understanding of biological vision systems, the human visual system in particular.

To build machines that "see" is a complex systems engineering task. While, at present, universal vision machines (with performance comparable to the human visual system) appear to be impractical, special purpose machines have been devised successfully to operate in restricted application domains. Designing such a dedicated system usually requires detailed knowledge of the application domain, experience with available processing techniques, plus a large experimentation effort. Clearly, this calls for computerized support tools in the sense of CAD and CAST in order to make the development process for vision systems less time-consuming and costly.

While machine vision is not in the mainstream of contemporary CAST research, there are many common aspects that are worth to be looked upon. CAST originated from Systems Theory with the goal of using the computer to make that theoretical body more accessible to engineers. In machine vision, computers have been used from the very beginnings, however, with little support from theory. No need to say that both areas are quite different, at least in their present state. How both fields could profit from each other, despite these obvious differences, is the main theme of Section 4.2.

Problems in machine vision are of a large variety. The area we have focussed on within the last few years is *object recognition*, which is a central and traditional problem in machine vision and artificial intelligence (AI). As a general guideline for this research, we have defined a very ambitious (perhaps even illusionary) long-term goal which is called the *Visual Modelling Tool* and is described in Section 4.3. This super-project, while not intended to be ever completed, provides numerous hooks for the development and implementation of individual components that can be attacked as projects at the diploma or dissertation level.

One of these components, in fact a most essential one, is an effective method for structural object recognition. As described in Section 4.4, structural object recognition has become a key paradigm in computer vision. Unfortunately, practical approaches are usually computationally expensive and intolerant against variation in object appearance. Recognition time grows at least linearly with the number of stored object models, while the number of known objects is is often assumed to very small. We argue that a more *spontaneous* performance cannot be achieved with traditional, combinatorial matching methods, but that other representations and concepts of similarity are required. In particular, we consider probabilistic matching techniques that have been applied successfully in speech recognition and relate this to the sequential scan paths exercised by the human visual system. From this, a framework for structural recognition is proposed that involves perceptual grouping as well as the use of contextual information.

Machine vision, particularly knowledge-based vision, creates a large variety of different design tasks including low-level image processing, components for symbol manipulation, and complex cognitive processes. The diversity of requirements of these tasks calls for radically new concepts in terms of hardware and software structures. Classical design tools, such as image processing systems, support only small portions of this large task set. In Section 4.5 we describe an interactive and homogeneous software environment termed the *Vision Kernel System* (VKS) that aims at supporting the entire spectrum of problems encountered in knowledge-based vision. VKS can be considered a first attempt towards a CAST tool for machine vision development.

4.2 CAST AND MACHINE VISION

4.2.1 Building Machines That "See"

The goal of machine vision (MV) is to derive from a visual input some information that is useful in performing a given task. Recognizing components on an assembly line, automatic traffic control, surveillance, repair tasks in hazardous environments, or autonomous navigation on Mars are typical examples. Since its beginnings, when the problems related to mechanizing vision tasks were grossly underestimated, the field of MV has undergone exciting developments. MV can be considered as one of the main roots of Artificial Intelligence (AI) and also as one of its most demanding applications. Consequently, MV has had a strong influence upon the directions taken in AI itself. However, since two-dimensional images of real, physical scenes are the typical input to MV systems, many specific techniques have been developed to operate on this particular form of input data. These include many image processing algorithms, such as filtering, thresholding, convolution, correlation, etc.

One important direction in MV is the reconstruction of the three-dimensional spatial layout of a scene from one or more two-dimensional images. This problem (sometimes referred to as the "inverse optics" problem) is ill defined, since the information contained in a 2-D projection is not sufficient to reconstruct a 3-D scene without additional constraints. A whole class of algorithms has been developed that utilize information contained in shading, motion, stereo, etc., to compute 3-D structures in a bottom-up fashion, often referred to as "shape-from-X" approaches. While the analytical (i.e., numerical) nature of these algorithms appears to be somewhat off the mainstream of AI research, they have become well established in MV. Unfortunately, most of these reconstructive techniques rely upon restrictive assumptions about the physical properties of the scene (e.g., the type of illumination, reflection properties) and are often unstable, thus limiting their applicability under real conditions.

As far as implementing "intelligent" behavior is concerned, MV has so far fallen short behind the enthusiastic expectations of its beginnings. Probably the most important result of MV research was to recognize the real difficulty of the problem itself. In particular, the task of constructing universal vision mechanisms that could even remotely match the performance of the human visual system, still appears to be an impossible dream. There has been, however, some impressive success in limited task domains, such as character recognition, printed circuit board inspection, or automatic analysis of blood cells. Typically these vision systems are highly tuned to their specific purposes, the properties of the imagery encountered, and representations used. However, the development of dedicated vision systems is time-consuming and expensive and thus calls for assistance by computerized tools.

The aim of *Computer-Aided Systems Theory* (CAST) [Pichler, Schwärtzel 90] is to support human engineers working in complex systems domains with suitable computerized tools. Easy access to a large theoretic framework, decision support, and interactive experimentation are among the key paradigms of the CAST methodology. Obviously, CAST tools pursue very similar goals to those that would be useful for vision systems design. In the remaining part of this section, we focus upon these similarities and the applicability of CAST for machine vision. First, the process of developing dedicated vision systems is described in more detail. Then, we take a closer look at the present state of CAST research and discuss its direct applicability to the vision domain, followed by a concrete proposal for a vision support tool.

4.2.2 Dedicated Vision Systems

Building a vision system for a particular application requires extensive knowledge about the application domain, the suitability of existing processing methods, plus a large amount of experimentation. Knowledge accumulated by individual experts is hardly accessible or documented, resulting in little reusability for other, even similar, applications. Depending upon the required performance, the choice (or development) of a computational architecture may be crucial. All this makes the design process time-consuming, often unpredictable, and consequently very expensive. For many applications, the high development costs for dedicated vision systems are still prohibitively high, although solutions may be feasible from the technological point of view. In the following, we concentrate on applications in the wide area of *image understanding*, where problems of object recognition, classification, scene labeling, etc. are of central importance.

A Vision System as a Black Box

A vision machine can be viewed as a system **S** that transforms an image **I** (or a sequence of images) into some other representation **R** that is a useful description of that image or the scene that produced it (Figure 4.2.1). The type of representation (i.e., what is considered "useful") depends upon the particular application domain.

In all but the simplest domains there are no simple mathematical rules, formulae, or algorithms to define **S**. A direct approach would be to take **S** as a function (black box) from the set of all possible images **I*** to the set of useful representations **R*** and specify the behavior of that system by enumerating all possible pairs (**I′**,**R′**), with **I′**∈ **I***, **R′**∈ **R***. This approach is not feasible except for trivial applications because of the large number of possible images. Consequently, more sophisticated ways for defining **S** must be found. Very generally speaking machine vision can be defined as the task of specifying and realizing **S** on some computational device.

It is widely accepted that the necessary transformation between the *iconic* representation **I** and the target representation **R** cannot be accomplished in a single step, but that it is necessary to introduce *intermediate* representations and transformations between these representations. The central goal is to decompose the transformation **S** into smaller pieces that can be specified algorithmically and realized efficiently with available computational means. E.g., Marr [Marr 82] proposed the *primal sketch* and the *2½-D sketch* as universal intermediate representations.

Vision Processes

An alternative view is to separate vision coarsely into 3 layers of representations and related processes: *low-level, mediate-level,* and *high-level* vision. *Low-level vision* relates to pixel-oriented processes to extract primitive image features (e.g., contours and regions), including many image processing techniques. The *mediate level* is primarily concerned with detecting salient feature groupings, while the final (symbolic) descriptions are manipulated at the *high level.* This architecture is characterized by a stepwise reduction of the amount of data from the low to the high level, while the representations become increasingly expressive and abstract towards the top. The proper choice of the intermediate representations in combination with suitable transformations is a key issue in multi-layer vision systems design.

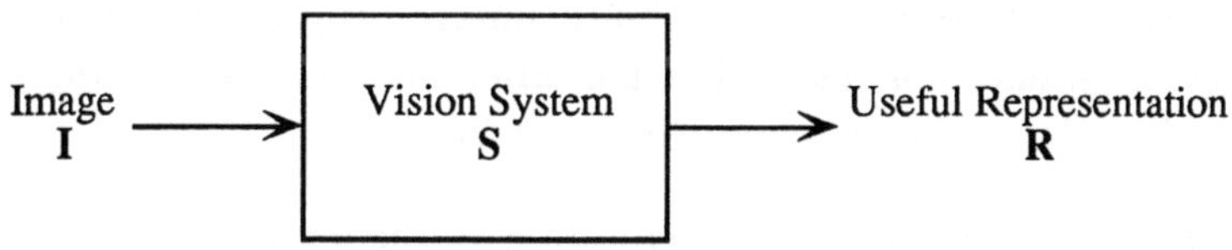

Figure 4.2.1 A vision system described as a black box **S**.

Design Support

The complexity of this design task makes it necessary to provide support by computerized tools. Developing such tools requires knowledge about which of the above activities can be either fully *automated* (i.e., done by a machine without human interaction), performed *interactively* (with a human partner), or are *not* supported at all.

Representations and Transformations

The minimum requirement for creating a *goal representation* **R** of a vision system is to know "what is important" in those images, i.e., what is the semantic frame of that application. Since the system lacks extensive "world knowledge", this information must be supplied *interactively*. Communicating this information to a (supposedly "intelligent") system requires to create links between the abstract (semantic) and the iconic representations, e.g., in the simplest case by simultaneously naming the relevant objects and pointing at them in a scene.

Once the goal representation has been found and related to the given images, a suitable network of intermediate representations and transformations must be developed. It is reasonable to assume that a wide class of vision applications can be implemented with a few *generic processing structures* that specify the key properties of the internal representations and transformations. The classic *hypothesize-and-test* structure for object recognition is a typical example, but other generic structures must be developed for different classes of applications. One could imagine tools to support the selection of existing processing structures, as well for developing new ones. In either case the process will be interactive and not fully automatic. Machine learning certainly is a key technology for creating and refining internal representations.

Reusability of solutions is one of the central issues in software engineering. The use of development tools for vision systems holds the promise that solutions developed for one application can be adapted for similar applications. Automatic self-documentation (in machine-readable form) is one first step towards this goal but is certainly not sufficient. With respect to reusability, the crucial problem appears to be how to discover the similarities (as well as the differences) between various applications. Again, this will most likely require human interaction.

Since there is usually no formal specification for the response of a vision system, *program verification* is not only difficult (as in most other programming tasks) but practically impossible, except for small components. Consequently, vision systems need extensive testing before they they can be deployed in a real operating environment. Often, the systems behavior will only be described in approximate or statistical terms, e.g., "... the system should not exceed a 5% error rate...". While a certain error rate can be tolerated in some applications (e.g., counting blood cells), other environments call for extremely reliable solutions (e.g., autonomous navigation). Consequently, the number of images (and the amount of time) required for testing may range from a few dozens to several millions, which may be beyond the life time of a project. While it is easy to *run* the tests automatically, *evaluating* the results and subsequently locating and correcting potential problems in the vision system requires additional intelligence and (most certainly) human interaction.

4.2.3 CAST Tools for Vision?

The aim of CAST is to support problem solving with systems-theoretical tools, following the original intent to make the theory practically accessible on the engineer's working desk. This is very much in contrast to machine vision, where no solid body of theory exists for solving a difficult engineering problem. So, can CAST be of any use for vision at all? Before answering this question, we look at some traditional links between Systems Theory and vision research.

Image processing operations, used for image filtering, image enhancement and restoration, spectral analysis, etc., have a strong formal background in Systems Theory. Another strong link comes from use of mathematical models for the behavior of biological vision systems, in particular the human visual system. Concepts from Linear Systems Theory, e.g., electrical circuit models, differential equations, spectral analysis, etc., are frequently used in modelling biological vision systems at the low-level. In history of machine vision, many close ties also exist to *cybernetics*, which is in turn one of the roots of modern Systems Theory and AI. Apart from the traditional "exact" approaches, there have been stimulating attempts to capture inexact behavior, such as *fuzzy logic* [Zadeh 66]. Similarly, research on artificial neural networks [Minsky, Papert 69; Rumelhart, McClelland 87; Lippmann 87] has been extremely influential, documented by their dramatic revival in these days.

While some areas in low-level vision are backed by strong formal concepts, many popular (and successful) methods are rather *ad hoc*, lacking a significant theoretical foundation. In addition, some issues that appear to be central in machine vision have probably not received adequate attention in Systems Theory and contemporary CAST research. E.g., the problems related to heuristics, combinatorial search, representation, learning, etc., which have been traditional areas of Artificial Intelligence

(AI), are vital to the design of any intelligent support tool that goes beyond the dumb toolbox-like systems that are abundant in engineering today.

The problem of developing sophisticated engineering tools is closely related to the concept of "method banks systems", which has been a traditional focus of CAST research (cf. Chapter 2 of this book or [Pichler, Schwärtzel 90]). Analogous to the situation in most other engineering disciplines, the current-generation design tools for vision applications are more similar to conventional CAD systems than intelligent method banks. While some of these tools support the construction and experimentation processes quite efficiently, practically all design decisions are still made by the developer.

Also, existing systems are often technology-oriented, i.e., oriented towards particular methods and implementations, and also driven by current computing technology, particularly by user interface techniques. However, there have been recent attempts to put CAST methods into a more universal, problem-oriented context. Of course, putting more intelligence into these systems is more than just a quantitative step. The goal should be to combine as much solid theory as possible with other problem solving techniques into a single tool.

Developing a useful problem-solving tool, it must be understood that different target domains may require radically different techniques and the difficulties involved may vary considerably. It may be quite straightforward to develop a CAST tool for a limited, formally specified, and "well behaved" domain but hopeless for another with the given technology. Machine vision may be one of the more unpleasant domains for CAST, lacking both a formal specification and a strong theoretical framework. On the other hand, vision is not the only domain of this kind, in fact, it seems that many other tasks in every-day problem solving exhibit a similar degree of complexity. Vision should therefore not be regarded as another exotic beast in the world of hard problems but it could serve as a pilot problem to develop a whole breed of more universal problem solving tools that have a much wider applicability than most of the systems we know today.

4.3 MODELLING OBJECTS FOR VISUAL RECOGNITION

Object recognition is of central importance in almost any practical application of machine vision. Traditionally, programming a vision system for a particular application has been the task of a human expert who applies his personal experience and knowledge about the application domain to come up with a suitable recognition strategy, and eventually a running recognition program. This process not only requires a significant amount of knowledge but also extensive experimentation in order to find optimal techniques and parameters. The resulting efforts and high development costs have probably kept machine vision out of many potential applications areas. Consequently, there exists an urgent need to support the development process in machine vision.

4.3.1 Modelling the Appearance of Real-Word Objects

Object recognition requires some internal description of the objects to be identified. Frequently, object descriptions are not explicit, but are hidden within the code of a special-purpose recognition program. Since these *imperative* models are difficult to manipulate, a system to support the construction of object models will preferably operate on explicit, *declarative* representations. Constructing suitable representations that support effective recognition of objects is therefore a central issue.

Several directions have been taken to support or even automate the task of building object models for machine vision, e.g., in statistical pattern recognition or CAD-based vision. In *CAD-based vision* [Bhanu 87], a three-dimensional CAD (computer-aided design) model of the object is given plus additional knowledge, e.g., knowledge about the optical properties of the material used. This is usually sufficient to predict the appearance of that object and allows the construction of a model for a vision machine which identifies the manufactured objects in real images. Typically, models for this class of objects are based on certain image structures, such as line segments, corners, and plane surfaces [Goad 83; Hansen, Henderson 1988; Ikeuchi, Kanade 1988; Lowe 1987]. Unfortunately, in many applications CAD models are not available (e.g., faces, fruits) or would be too numerous to be practicable (e.g. cars in a traffic scene).

In *statistical pattern recognition* [Hall 79], models usually consist of prototypical feature vectors or parametric descriptions of the feature space. Both types of models can be obtained in a training phase automatically from image examples. While the resulting models allow robust classification in some applications, the performance of statistical methods is insufficient in many other cases, particularly those with complex scenes. However, the idea of deriving object models from real images is a very powerful concept:

- *First*, the actual (runtime) vision engine itself is employed to construct the representation by *experimentation*. In other words, the very same mechanism is used for modelling as well as for the actual application. This, among other things, guarantees that the resulting representation is *operational* in the sense that every statement in this representation can be interpreted by the runtime vision engine. As a trivial example, knowing that an object is *red* would be useless (and thus make the description non-operational) if the runtime vision engine had no means of discerning colors.
- *Second*, all imaging effects such as lighting, distortions, blurring, sensor transfer function, etc., automatically go into the representation without the need for a separate (and probably complex) model of the imaging process.

Interactive Modelling

The goal of this work is to develop object representations from image examples in the context of *structural vision*. We assume that this cannot be accomplished in a fully automatic way, since the search space for suitable representations is usually too large. Instead, we propose an interactive scheme, in which a human instructor teaches the system "what" to look for in the sample images. The (complex) task of this system is to interpret the actions of its instructor, relate them to the current image and models, and build up a suitable internal representation (i.e., "how" to look) for the relevant objects and their mutual relations. We call this system a *Visual Modelling Tool* (VMT). The VMT is not intended ever to be completed as a complete system. Instead, its main purpose is to provide the motivation to design and implement some of its key components.

The Visual Modelling Tool

The central idea of the VMT is that a system should develop the details of its internal representations itself, using only hints and corrective inputs instead of direct manipulation[1] by the human instructor. The quality of an existing representation should be judged primarily upon its effectiveness in object recognition and localization when applied to actual images. Direct manipulation should only be necessary in exceptional situations, e.g., when the system's internal representation has become inconsistent or otherwise corrupted.

The operation of the VMT is split into the *modelling phase*, when internal representations are built or modified and the *runtime phase*, when actual object recognition is performed. Unlike in statistical pattern recognition, the two phases do not stand for strictly separate modes of operation. In fact, both phases alternate continually while the system is being trained. Figure 4.3.1 illustrates the use of the VMT. Given a *Primitive Vision Engine* which performs basic image operations efficiently, the final result is a running vision program for the particular task environment. The *Primitive Vision Engine* and the (continually changing) *Application-Specific Vision Program* combine to the *Runtime Vision Engine* (RVE).

Model building in VMT is an incremental process which starts with a crude model that is successively refined as more image examples are analyzed. A collection of object models for a particular application domain is called the *Application World Model*. In short, the process of model building proceeds as follows:

Initially, when the system knows nothing about the application domain, the human instructor must somehow indicate those objects that are relevant for the task at hand. This is done on the image itself using graphical tools. Given only this little information, the system then tries to *explore* the image using its own (predefined) repertoire of visual routines. The VMT then takes the (subjectively) most salient features it discovered and uses them to build an initial model. The key problem here is that the resulting model must be *general* enough to include the entire object class as well as specific enough to exclude all other objects within the application world model. This process is similar to what is called "concept acquisition" in machine learning [Michalski et al. 1986; Winston 1975]. As more image examples are analyzed, the models are continually modified. The task of the human instructor is to interactively

---

[1] In the sense of modifying program code or data structures manually.

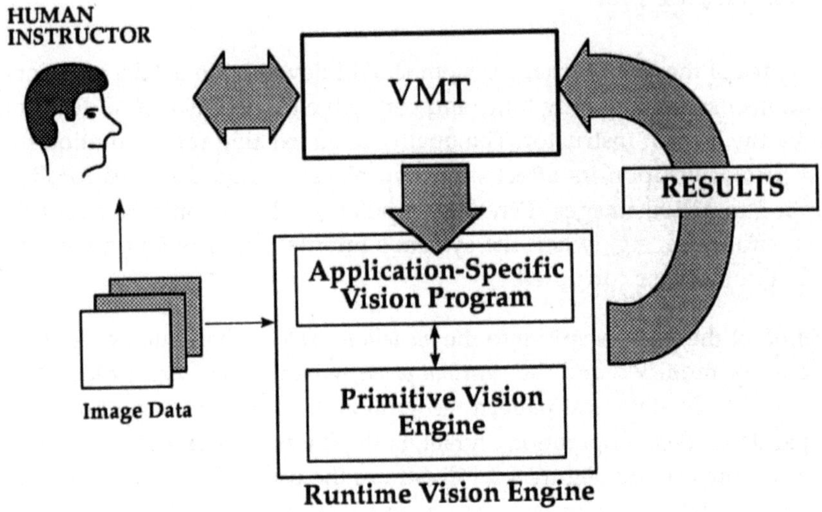

Figure 4.3.1 Application of the *Visual Modelling Tool* (VMT). The end product of the VMT is an *Application-Specific Vision Program* (AVP) which is developed interactively with a human instructor by analyzing real images. The AVP is derived from a declarative model internal to the VMT. The AVP combines with the given *Primitive Vision Engine* to the *Runtime Vision Engine* (RVE), which is exercised repeatedly during the modelling process.

correct the errors made by the system during this process. A textual interface is provided to communicate auxiliary information to the system, such as for naming objects and for expressing abstract properties and relations. During the entire modelling phase, the VMT keeps track of the physical images that were used to build each object representation. This is necessary to check the validity of modified object representations for images analyzed some time earlier.

In order to give an intuitive feeling of the ultimate goals of this study, the following section contains a sample dialogue between a (hypothetical) VMT and its human instructor. In Section 4.3.2, we discuss what we consider the basic prerequisites that would allow a system to carry on such a conversation and, even more important, to make useful inferences from it. The structure of the declarative model is explained in Section 4.3.3. Section 4.3.4 deals with the problem of transforming the declarative description into something executable, a process called model compilation.

4.3.2 Applying the VMT

Application Domains

The purpose of the VMT is to construct models for vision machines dedicated to particular, well defined tasks. This application task in conjunction with the operating environment are called the *Application Domain* of a vision system. Application domains are distinct and identified by unique names. For every application domain, the VMT maintains a representation of the relevant objects and their mutual relationships, referred to as the *Application World Model* (AWM). An AWM can be thought of as a closed world that describes a small portion of the global environment relevant for the given task. E.g., the AWM for a traffic control system would contain *streets, cars, trucks,* etc. as relevant objects, but would certainly not have to include thinks like *apples* or *telephone sets.*

Sessions

AWMs are built incrementally in a sequence of one or more *sessions* with the human instructor. The instructor is not required to be the same person in each session, but for now we assume that this is the case. One can imagine that different instructors in sessions for the same application domain could lead to more robust models but may also cause inconsistencies that need special care.

A Hypothetical Dialogue

Before looking into some of the details, we want to illustrate the purpose of the VMT by means of a hypothetical conversation between the system and its human instructor. This is (to hold back unrealistic expectations) not the transcript of a real man-machine interaction. Also, we have skipped over many details rather casually, although some of them are indeed respectable problems themselves. Consequently, the following sample dialogue is not to be taken too seriously. The main purpose of this example is twofold: *first,* we want to show the level of man-machine interaction that at which the VMT is supposed to interact and, *second,* we want to use it for pointing out some of the critical steps in the modelling process and provide hooks for developing specific components.

The application domain used for this example are *tools*, such as *hammers* and *screwdrivers*. We assume that, at the beginning of the session, the VMT already has some knowledge about tools, i.e., a corresponding AWM (named "TOOLS") exists. In the session shown below, the human instructor tries to communicate the new concept of a *screwdriver* to the VMT, using actual images with instances of screwdrivers. In the following dialogue, information between the system and the instructor is exchanged in both *textual* and *visual* form. The textual communication is shown on the right, the corresponding pictures are shown at the left side. The symbol ✎ in the text indicates that the instructor is requested to supply graphic input with the pointing device (i.e., the "mouse").

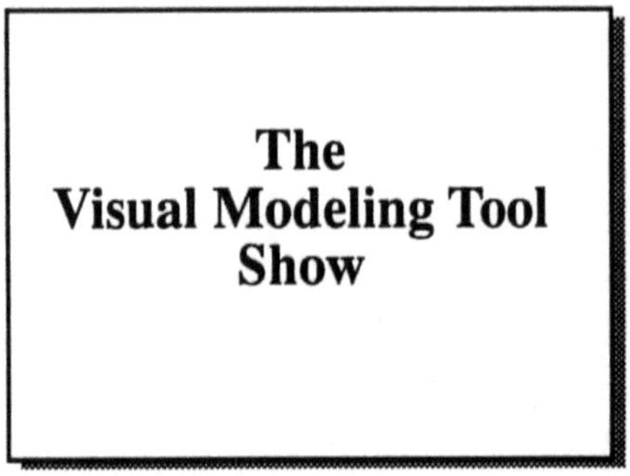

Starting a new learning session:
What is the application domain?....TOOLS•

In the TOOLS domain, I already know about
HAMMERS, PLIERS, and KNIVES.

Please select a training picture...

<selection>

OK. Let me have a quick look ...

I see
1 HAMMER and
1 KNIFE.

Is this correct?.....................NO•

Please show me what I did wrong........

Figure 4.3.3

So this is NOT a KNIFE.
Is it anything I should know?.........NO•

OK. I have to modify my current concept of KNIFE ...

Is anything else wrong?..............NO•

Did I miss any objects that I should know?........................NO•

Fine.

Does this picture show other relevant objects?...................YES•

Let´s begin with the first object:
How do you call it?......
SCREWDRIVER•

Please show me a SCREWDRIVER.........

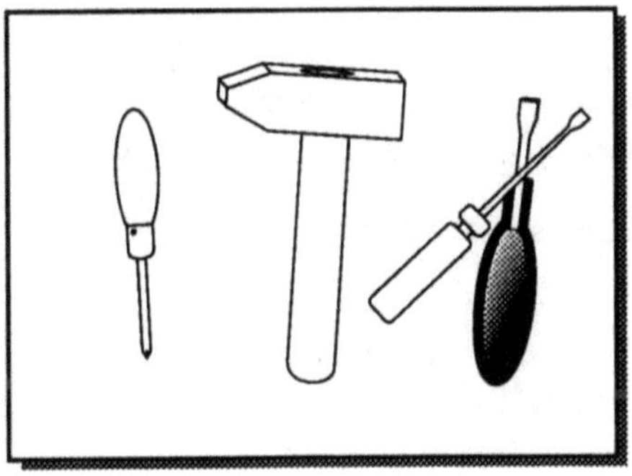

OK. Now let me try to outline this SCREWDRIVER ...Voila!

Did I include any void elements?......NO•

Figure 4.3.6

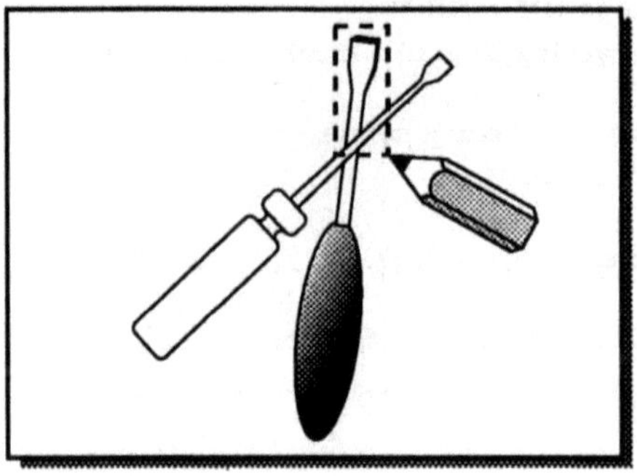

Did I miss anything?................YES•

Please point out the missing parts... .

OK. Let me try again...

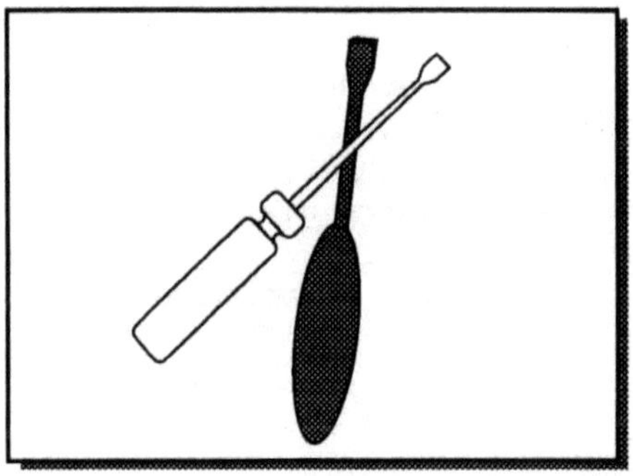

This SCREWDRIVER seems to be
occluded by something else.
Correct?...........................YES•

Is this SCREWDRIVER complete now?....YES•

Fine. I think now I know how a
SCREWDRIVER looks.

Do you want me to look for other
SCREWDRIVERS?.......................YES•

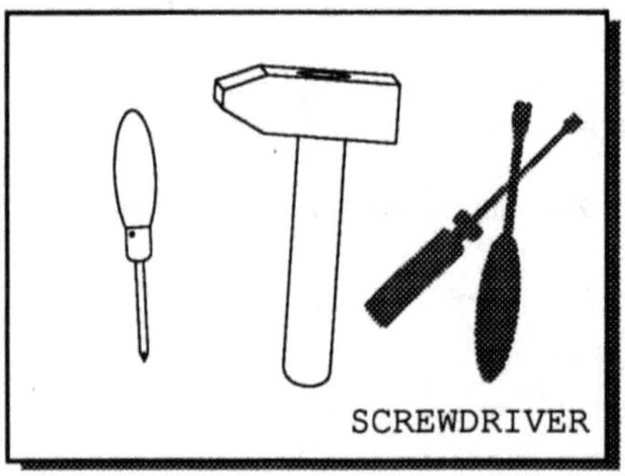

OK. I see 1 (one) more SCREWDRIVER.
Is the outline correct?.............YES•

Did I miss any SCREWDRIVERS?.........YES•

Please show me ....

Figure 4.3.9

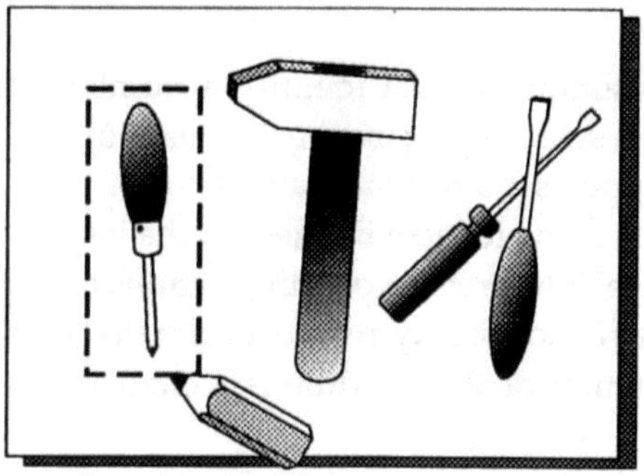

OK. Let me try again now ...

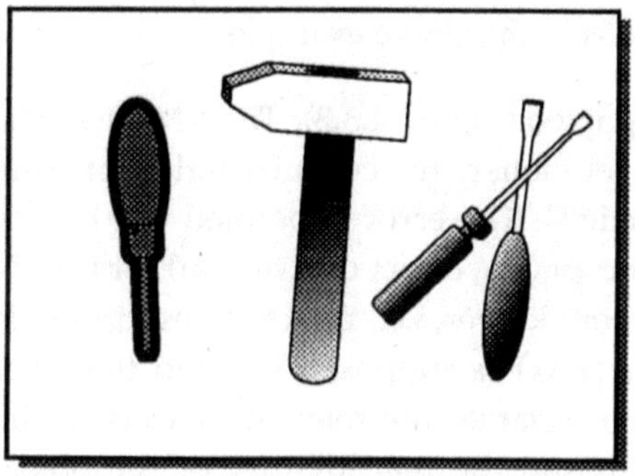

Is the outline correct?.............YES•

Any more SCREWDRIVERS?..............NO•

Any more relevant objects...........NO•

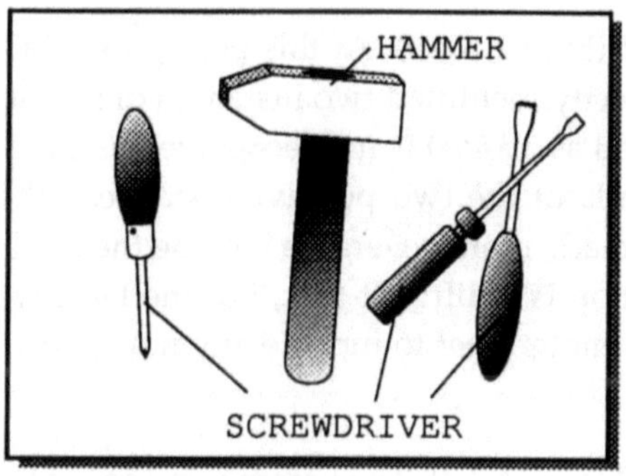

In summary, I see in this picture:
1 HAMMER and
3 SCREWDRIVERS.
Correct?..........................YES•

Would you like to continue with another picture from the TOOLS domain?........................

Figure 4.3.12

Discussion

By looking at this (seemingly) casual conversation, we can identify a number of essential prerequisites that are required by a system to perform such a task. In particular, we need a *Dialogue Facility* that supports both visual and textual man-machine interaction. For the visual part, it must allow to view images, let the instructor mark image elements with a pointing device, interpret the pointing actions of the instructor, correlate them with the image itself, and display results of the system's recognition engine. Although the implementation of this module is a challenging task, it is not the main focus here.

What we are mainly interested in is *(a)* how to use the information supplied by the instructor to update the *Application World Model,* and *(b)* how this model is actually modified. While the modelling process itself is discussed in Section 4.3.3, we want to point out some particularly interesting situations in the above example.

Situation 1: An unknown object is falsely recognized (Figure 4.3.3). The VMT has recognized the screwdriver on the left as a *knife* (obviously the current model contains knowledge about knives from earlier instruction). The error is pointed out by the instructor. The problem is that the existing concept (i.e., object description) for *knife* is not specific enough to separate screwdrivers from knives, i.e., the *knife*-description is too *general.* This situation is inevitable, as the VMT is supposed to build the most general object description that is sufficient to separate the relevant objects in the images analyzed in the past. Since, at this point, the real identity of the object thought to be a knife is unknown, the only possible fix to the problem is to make the description for knife more specific by searching for more discriminating image properties.

Situation 2: A known object is not recognized (Figure 4.3.9). At this point, the VMT has built a description of *screwdriver* and correctly identified two instances of it. The third screwdriver on the left (initially classified as a *knife*) is not recognized as such. For some reason (perhaps due to the flat ends of the two positive instances), the *screwdriver*-model is too specific and must be made more general to include the additional instance. Selecting the best generalization is a difficult task, but the least we can expect is that the model remains specific enough *not* to include any non-screwdrivers observed in the past.

Situation 3: An object is correctly recognized (Figure 4.3.3). The hammer in the center is correctly identified in this picture. There is no problem and the model does not need any updates. However, what the VMT can do is to use this instance to collect more knowledge about the appearance of hammers and add it to the existing description. This information may be useful for recognizing future instances or to rebuild the representation when necessary. This means that even without immediate assistance from the human instructor, the VMT may gather knowledge about the application domain. This raises the exciting perspective of a vision system that never stops learning at all, even during its application!

4.3.3 Building the Model

Declarative Model Structure

The declarative *Application World Model* consists of a set of *object* models which are in turn described by a collection of different *aspects* (Figure 4.3.13). Aspects are specified in terms of image features that are extracted from the image by the *Runtime Vision Engine* itself. Thus the declarative description and the operations performed by the vision engine are *congruent* in a sense, i.e., the description is implicitly *operational*. In

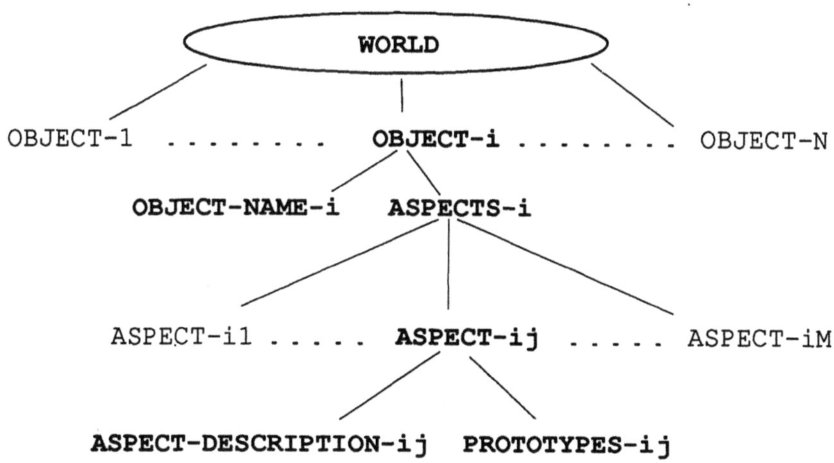

Figure 4.3.13 Declarative Model Structure

addition to this structural description, aspects may be linked to *prototypes,* which are selected instances of that object (i.e., images) encountered in the past. Prototypes are used to make sure that changes in the model are consistent with earlier classifications. Thus they form something like an *iconic memory* of the VMT.

Main Steps

The modelling process consists of the following steps that are iteratively executed by the *Model Builder* module:

(1) Receive instructions from the human instructor.

(2) Correlate these instructions with the current image.

(3) Determine the problem to be solved.

(4) Solve the problem by modifying the declarative model.

To resolve a particular problem situation that has been identified in step (3), the Model Builder has to decide how to modify the existing model. In particular, it must have means to *explore* the current image and to select image features (or feature groups) as candidates for augmenting the model. These features are obtained by actually applying certain parts of the recognition engine to the current image.

Applying the Runtime Vision Engine

In general, the search for a suitable model update involves the following (iterative) steps (Figure 4.3.14):

(a) Modify the declarative model.

(b) Transform the declarative model into a running vision program.

(c) Apply the vision program.

(d) Evaluate the results and modify the declarative model if necessary.

For this purpose, the *Model Builder* relies upon two additional components, the *Model Compiler* and the *Experimenter*. The *Model Compiler* turns the declarative model into an executable vision program, which is subsequently applied to the image by the *Experimenter*. In addition, the *Experimenter* allows the *Model Builder* to try specific visual routines in the image before actually changing the model.

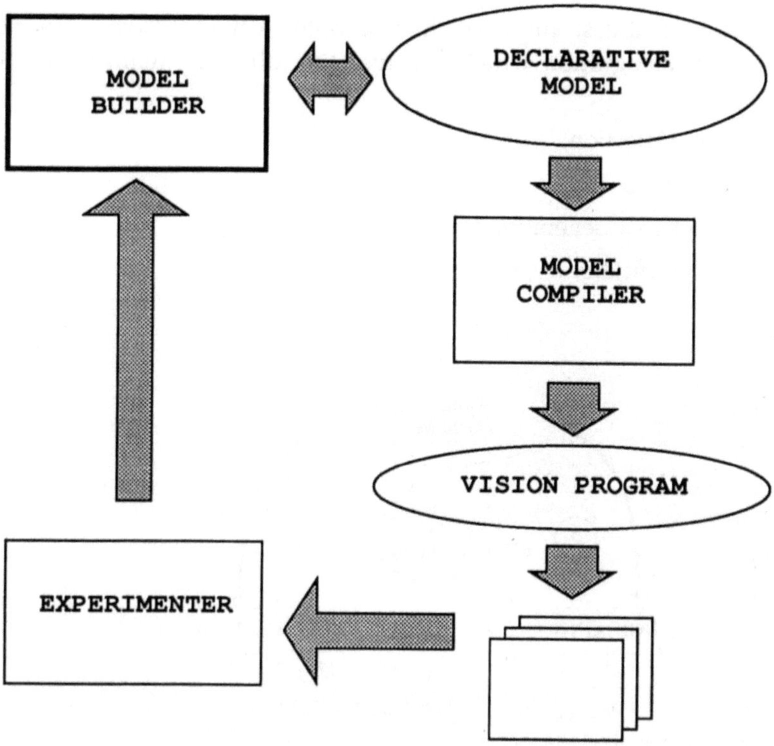

Figure 4.3.14 Updating the model. The *Model Compiler* transforms the declarative model into an executable vision program, which is applied to the image by the *Experimenter*. The results are evaluated by the *Model Builder*, who supervises this whole process.

4.3.4 Model Compilation

A Recognition Paradigm

While the declarative model can be kept largely independent of the recognition strategy, this is not true for the final recognition program. In object recognition, a tradeoff must be made between the effort spent for bottom-up *discovery* (filtering) and top-down, expectation-driven *search*. The vision program produced by the *Model*

Compiler must reflect this strategy. We have adopted a typical *hypothesize-and-test* scheme for object recognition which involves the following main steps (Figure 4.3.15):

(1) Basic Feature Detection,

(2) Feature Grouping,

(3) Object Hypotheses Generation (Indexing),

(4) Hypothesis Verification.

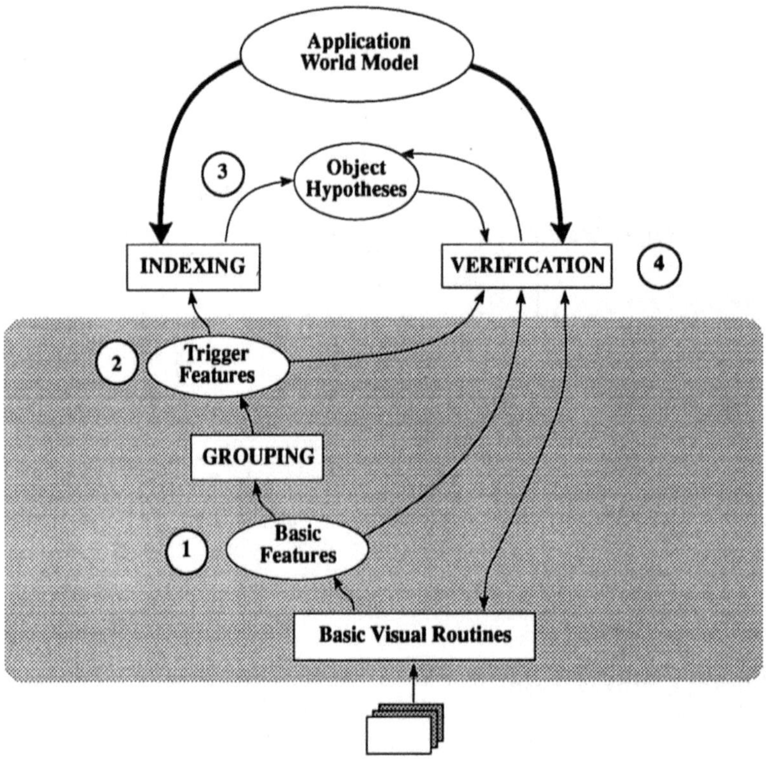

Figure 4.3.15 Main steps of the recognition process. *(1)* Basic Feature Detection; *(2)* Feature Grouping to create trigger features; *(3)* creation of object hypothesis (Indexing); *(4)* verification of object hypothesis.

Feature Detection and Grouping

Basic feature detection is accomplished by global application of primitive vision routines (e.g., line finders, contour detection, texture segmentation). These features are generally to unspecific to locate objects directly. *Feature grouping* assembles basic features too more complex, prominent features (e.g., straight lines, elongated structures, parallels). This is also a bottom-up process but the key features to be searched for may be defined by the application domain.

Indexing

Object hypotheses are created by matching key features and entries stored in the model. Each object hypothesis contains at least the following components:

(a) the *object* type,

(b) the matching *aspect*,

(c) the image location (i.e., position, orientation, and size), and

(d) associations between real image features and model features (matched features).

Result of the indexing step is a set of object hypotheses that are ranked according to their *saliency*, e.g., the number of matched features. At this point it is possible that one image feature is engaged in several object hypotheses and also that several object hypotheses refer to the same object.

Verification

The purpose of the verification step is to either confirm or reject existing object hypotheses by searching for additional evidence in the image. This process is goal-directed and driven by the features found in the stored object model. The steps involved are:

(1) Predict an image feature (from the information in the model).

(2) Try to confirm that feature.

(3) Modify the Object hypothesis.The modification in step *(3)* aims at completing the hypothesis (e.g., constraining the object's location) and also at computing a measure for its *plausibility*. In general, confirmed features increase the evi-

dence for a hypothesis and rejected features reduce it. Eventually, this can lead to the entire object hypothesis being rejected.

Recognition Strategy and Model Compilation

Indexing and hypothesis verification are characterized by inverse directions of search. Indexing leads from given image features to object hypotheses, while verification takes a hypothesis and looks for image features. Consequently, model compilation transforms the declarative description into two separate decision structures: the *Indexing Tree* and the *Verification Tree.*

An example for the effects of model compilation is given in Figures 4.3.16–17. We assume that the declarative model of a hammer contains two parallel lines (on its shaft) that are of approximately equal length (Figure 4.3.16). The corresponding aspect is linked to a prototype image of a hammer. During model compilation (Figure 4.3.17), features may be dedicated to indexing, if that feature is determined routinely in a bottom-up step, or verification, if that feature is better suited for top-down search. In

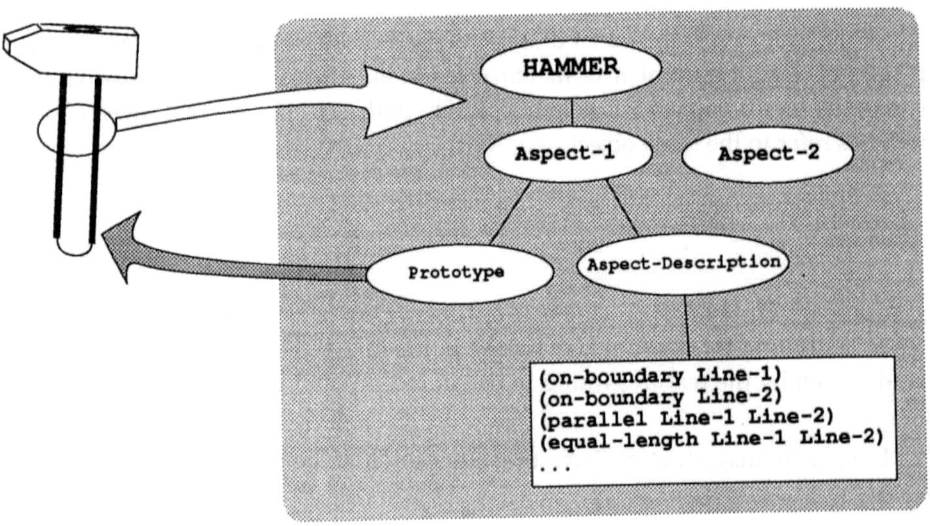

Figure 4.3.16 Example for a declarative object description. The two parallel lines of approximately equal length become part of one aspect of the HAMMER model. The aspect is linked to an actual instance (prototype) where this detail has been found.

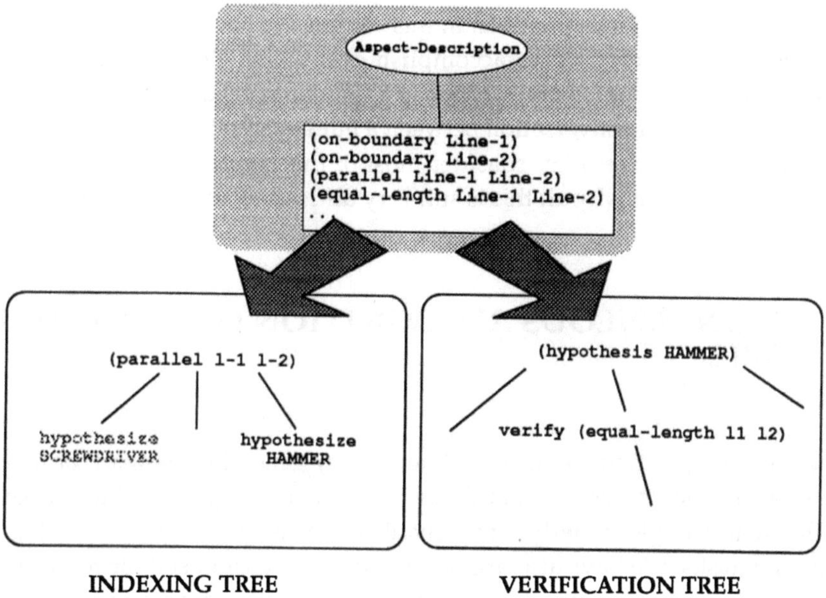

Figure 4.3.17 *Model compilation* transforms the declarative description into separate decision structures for the indexing and the verification process. Features are selected as trigger features for the indexing tree when they can be determined routinely in a bottom-up discovery process. In contrast, the verification tree is based on features that are better suited for top-down search.

our example, the parallel lines become the trigger feature for an indexing tree for the hammer. Since screwdrivers exhibit the same feature, they appear also in the same indexing tree. The verification tree for this aspect of a hammer now includes the feature of lines being of equal length, which is easy to verify once the two lines are given but would require expensive comparison of line pairs in a bottom-up step.

4.3.5 Conclusion

The main goal of the VMT is to streamline the construction of models for structural object recognition for a wide class of objects. Our approach is based on analyzing real images interactively with a human instructor and relies on textual as well as visual dialogue mechanisms. A central premise is to let the system determine the most suitable representations itself while hiding the details of this representations from the human instructor as far as possible. In this section, we have mainly tried to illustrate what the VMT is supposed to accomplish eventually. The simulated example in Section 4.3.2 should demonstrate, how easily fundamental and difficult problems can be hidden in a casual conversation between a human and the (non-existing) machine. In fact, a naive observer may consider most of this conversation trivial and hardly accept an explanation why all this is still very difficult to accomplish by a computer.

4.4 SPONTANEOUS RECOGNITION OF STRUCTURED OBJECTS

In the preceding section we have outlined a very general framework of research towards visual object recognition and interactive modelling. Central to this concept is the idea that a single underlying mechanism is used both for the training *and* the recognition tasks. We have also argued that the formalism used for representing objects and their relationships in the real world must depend upon the particular recognition strategy that is being used. E.g., if it were possible to recognize objects by iconic template matching, we could simply use prototype images to represent visual concepts. Recognition strategy and optimal representation are indeed two sides of the same coin. It is impossible to talk about one without considering the other.

4.4.1 Structural Object Recognition

The general paradigm we have adopted is *structural object recognition* which is based on the assumption that the relevant objects of a domain can be described by sets of characteristic structural tokens and the spatial relationships between these tokens. For the task of recognizing structured objects we assume that a set of object models already exists, in which each object is described in structural terms, without caring (for the moment) how these models were created.

The problem of structural object recognition can be stated as finding *similarities* between the detected image structures and the given model structures. Usually, such structural representations can be directly formalized with graphs, for which an extensive and solid body of theory is available. In the language of graph theory, this means to search for *isomorphic* graphs (or subgraphs, rather, since the subset of image components that belong to a particular object is not known in advance).

Straight-forward graph matching methods are troubled by two serious problems: *First*, all (known) exact methods for solving the structure matching problem are *NP*-hard, and, *second*, it is often not possible in vision to find exact matches between substructures at all. In addition, it is known that even heuristic approaches to structure matching (relaxation labeling, e.g.) are computationally expensive and often unreliable, even when only a small number of object models are considered.

Spontaneous Recognition

In remarkable contrast to this, the human visual system is able to classify objects from a vast number of alternatives in a highly spontaneous and robust fashion. It is clearly unconceivable that the human visual system could do anything like an exhaustive search over all possible structure mappings in such a short period of time. One could hypothesize that functions for recognizing specific arrangements are hard wired and can all be executed in parallel. However, such a model would raise numerous other questions, e.g., how new structures can be learned or how recognition is affected by cultural influences.

While the choice of a principal recognition strategy, i.e., "top-down" or "bottom-up", has been a traditional matter of debate, there appears to be wide consent today that only combinations of both hold a promise for success, presumably in some form of hypothesize-and-test approach. In object recognition, the bottom-up, hypothesize phase is frequently called "indexing", while the top-down part is referred to as "verification" (cf. Figure 4.3.15). Very generally speaking, the outcome of the indexing process is a set of object hypotheses (i.e., object names) that are to be verified subsequently.

We argue that the primary key to spontaneous object recognition is to accomplish *rapid indexing*. Consequently, our work towards an efficient recognition strategy is focused upon the indexing problem under the assumption of a relatively large model base of 100–1000 (or more) known objects. In this section, we layout the framework for a strongly sequential approach to structural object recognition which is motivated by several interesting observations.

Speech Recognition

Although speech recognition appears to be closely related to artificial vision, both areas have developed separately to a great extent, resulting in a large diversity of concepts. Structural approaches have a long tradition in both areas, however, some highly successful techniques from speech recognition have found no direct applications in vision yet. Of course, there are some fundamental differences between speech and pictures, a central one being the apparent lack of a natural ordering of image tokens, as opposed to the linear ordering of speech tokens over time. This brings us to a second observation. Humans do not seem to process pictures all at once but make sequences of fixations in order to scan the contents of the picture. We could hypothesize that these scan paths produce token linear sequences that are suitable for certain matching techniques known from speech recognition, e.g., probabilistic techniques.

Grouping and Context

Perceptual grouping phenomena have been widely studied in computer vision as a way to assemble simple tokens into more complex ones, potentially reducing the search space for recognition. Usually, the heuristics used for grouping are fixed for a given type of imagery, e.g., one may look for parallel lines in pictures of machined parts. Alternatively, one may treat the grouping problem as part of the structural recognition problem. Viewed this way, "indexing" could simply be the successive agglomeration of "salient" structures until the point where a successful classification can be made.

On the other side, *context* can play a significant role in the recognition process. The recognition of one object is often much easier and spontaneous when other objects in the scene are already known.

In the remaining parts of this section, we investigate the question whether it is possible to integrate the seemingly disparate aspects of structural object recognition, human scan paths, perceptual grouping, and context into a single computational framework. The following section gives a formal definition for the problem of recognizing structured objects, supplemented by a brief overview of current matching techniques in Section 4.4.3. In Section 4.4.4, we look at structural techniques that have been used successfully in speech recognition, followed by some sequential aspects of human vision in Section 4.4.5. A basic framework for structural object recognition is developed in Section 4.4.6, with a special focus on the problems of generating sequence of image tokens and the impact of global context in Sections 4.4.7, and 4.4.8 respectively.

4.4.2 Structural Object Representation

Describing Individual Objects

An assumption crucial to any structural object recognition approach is that the appearance of objects can be described as assemblies of certain structural primitives, called *tokens*. To be useful for recognition, tokens should *(a)* be detectable from images reliably in a bottom-up fashion, and *(b)* be *non-accidential* in the sense that they are unlikely to occur from arbitrary configurations in space. E.g., while a straight line segment in the image does not necessarily correspond to a straight line in 3-D, it is still very likely that this is indeed the case [Lowe 87]. Other types of structural image tokens are *line endings, corners, curve segments, blobs,* etc.

Specifically, we make the following assumptions:

- We have given a limited set of token classes and corresponding means to detect instances of each token class in real images.
- The appearance of each 2-D object of interest can be described by a set of structural image tokens and spatial relations between these tokens.
- Similarly, 3-D objects can be described by a finite set of distinct views, each of which can be treated as a separate 2-D object. Here we consider only 2-D (i.e., "flat") objects.

We define the structural representation of an object, denoted a "visual concept", as the triple $C = (V_C, P_C, R_C)$, where $V_C = \{v_{C1},\ldots,v_{Cl}\}$ is the set of tokens, $P_C = \{p(v_{C1}),\ldots,p(v_{Cl})\}$ are the token properties (i.e., unary relations on the set of nodes), and $R_C = \{r(v_{Ci},v_{Cj})\}$ is a set of binary relations between image tokens. Under the restriction of using only unary and binary relations[1], we can transform this representation into a directed graph $G = (V(G), A(C))$, where $V(G) = V_C$ denote the nodes and $A(G) = P_C \cup R_C$ are the arcs of the graph (Figure 4.4.1).

---

[1] Relations of higher order than two cannot be represented by simple graphs but with hypergraphs.

A World of Structured Objects

The set of visual concepts that an intelligent agent knows about comprises its visual world model. Thus, at least for the moment, the world model W is simply defined as a collection of structural object descriptions $W = \{C_1, C_2, \ldots, C_N\}$ (Figure 4.4.2). The number of objects (N) that a system needs to know about depends strongly upon the particular task environment. In much previous work in structural object recognition, only a few known objects have been assumed, often just a single concept. For many applications, e.g., in controlled industrial tasks, this may not be a severe limitation. However, intelligent machines operating in a natural environment need to cope with a much larger variety of different object types. We therefore assume that a realistic world model should contain at the order of 100–1000 individual objects (and perhaps even more).

While storage capacity is usually not the main concern with extensive world models, the problem is how to accomplish efficient recognition when the number of alternatives is relatively large.

Figure 4.4.1. Structural description of individual concepts. The assumption is that each object of interest can be adequately described as a collection of structural tokens and certain spatial relations between them.

Figure 4.4.2 A world model. We define the *world model* W of an intelligent agent as the set of known objects, given as structural descriptions $C_1, C_2, \ldots, C_N$.

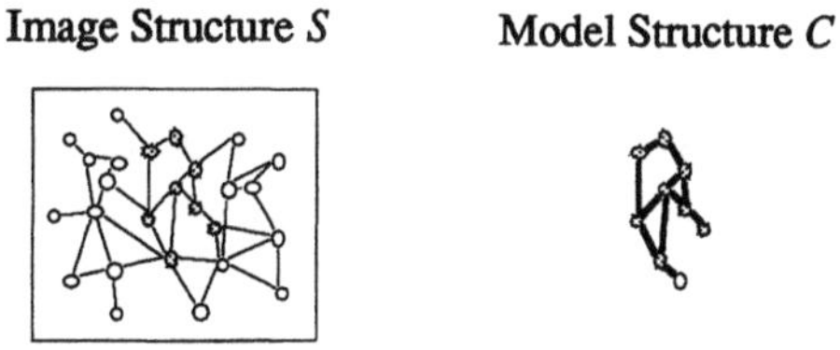

Figure 4.4.3 Structure Matching. Given a model structure C and an observed image structure S, the problem is to find an embedding of C in S.

4.4.3 Matching Structural Descriptions

For now we assume that the world consist of a single visual concept, given as a structural description C. Similarly, a set of tokens has been extracted from the image and spatial relations determined between its elements, thus producing a structural description of the image, denoted by S (Figure 4.4.3).

Graph Isomorphisms

The straightforward approach to recognize the object in that image would be to search for an embedding of the model structure C in the image structure S, i.e., a 1:1 and "onto" mapping between the nodes of C and S and the vertices of C and S. In other words, the problem is to find a subgraph of S' of S, such that S' is isomorphic to C. This is commonly referred to as the *subgraph isomorphism* problem.

In general, we cannot assume that the object C is entirely visible in the image, i.e., some parts of the object may occluded or missing. Consequently, we can only look for a *subgraph* C' of C that is congruent to a subgraph S' of S, a problem called *double subgraph isomorphism*. Although the latter problem is computationally not harder than the simple subgraph isomorphism, both problems are NP-hard[1], i.e., computationally expensive.

---

[1] This effectively means that no algorithm is known to solve the problem in polynomial time.

Structural Similarity

Crucially connected to the graph isomorphism problem is the notion of *equivalence* between two given graphs. In the strict sense, two graphs are considered equivalent (i.e., isomorphic) when a 1:1 mapping exists between their nodes and the labels attached to the corresponding arcs are identical. This form of equivalence is hardly ever encountered in structural recognition for various reasons. First, it requires a finite set of possible node labels. Node labels, however, represent properties and spatial relationships of tokens, including distances, angles, etc., from a potentially infinite range of values. Secondly, to be of practical use, matching must be tolerant against spatial variations to a certain degree.

Consequently, instead of searching for structural *equivalence*, matching must look for structural *similarity* that can be quantified by a suitable distance measure. A classic example for such a distance measure is the model proposed by Fischler and Elschlager [Fischler, Elschlager 73], where the spatial relations between tokens are abstracted as elastic springs. Their matching distance is a function of the spring loading, among other parameters.

The use of a matching *distance* instead of a matching *predicate* turns the original graph matching problem into an optimization problem with the goal to find the "best match" among a large number of possible ones. Practical graph matching techniques are often characterized by the use of heuristics to avoid exhaustive search. Examples are backtrack search algorithms [Ballard, Brown 82], association graph techniques [Ambler et al. 75], relaxation labeling [Bhanu, Faugeras 84], and various hashing schemes [Breuel 89, Lamdan 88], just to name a few. Considerable work has been done towards structural recognition in industrial environments, particularly under occlusion, i.e., when objects are only partially visible [Ayache, Faugeras 86; Bhanu, Ming 87; Bolles, Cain 82; Knoll, Jain 86]. All these methods have three things in common:

- They are computationally expensive (at least in the worst case sense).
- They work by enumerating mappings between tokens.
- They usually compute too much.

The Problem of "Early Assignment"

All combinatorial algorithms for structure matching are based on the same idea: (1) enumerate (in a possibly smart way) mappings between the observed structure elements and the model tokens, (2) evaluate the match, and (3) pick the best match. In each iteration, a hypothetical assignment between pairs of tokens is made and subsequently evaluated using a suitable distance measure. This "early assignment" of individual structure elements is a central problem of any combinatorial approach to structural object recognition.

In addition to the resulting complexity problem, the answers produced by combinatorial search are often much more detailed than is really needed. E.g., do we need to find a correspondence for each individual token in order to recognize an object? Is it necessary to know the exact position, orientation, and scale – and do we have to compute the viewpoint [Lowe 87] before we can recognize a 3-D object?

A solution to this problem appears to be possible only if we can find ways to determine similarities between structural descriptions without making preliminary assignments between tokens (i.e., by avoiding "early assignment") [Hayes-Roth 76]. In other words, can we find (and compute) properties of structural descriptions that can be used to determine similarity without explicit assignment of tokens (Figure 4.4.4)?

The intolerance against variations in the appearance of an object is another disadvantage of the pure graph-based methods. In fact, one could argue that there is almost no real-world object whose appearance can sufficiently be described by a graph of tokens. E.g., what would be the graph of the concept "House", considering the large variety of real houses? AND/OR graphs that have been applied in other areas of AI [Pearl 84] seem to offer only a partial solution to the problem of adequate object representation. Assuming that structural representations are useful at all, this raises the question which parts of a structure are actually required for recognition. In particular, it may not be necessary to create complete structural descriptions of a concept, but perhaps only of some essential parts. The parts could be assembled in a rather loose configuration (e.g., only constrained by adjacency) and still allow recognition. In this direction, some interesting results from visual experiments with pigeons can be found in [Cerella 86].

4.4.4 Structural Techniques in Speech Recognition

The problem of matching structural descriptions is, of course, not limited to object recognition or computer vision in general. In particular, the problem of inexact matching of sensory data to given structural models is of equal importance in the area of speech recognition, where structural approaches have been applied with good success. While most speech recognition systems involve considerable ingenuity

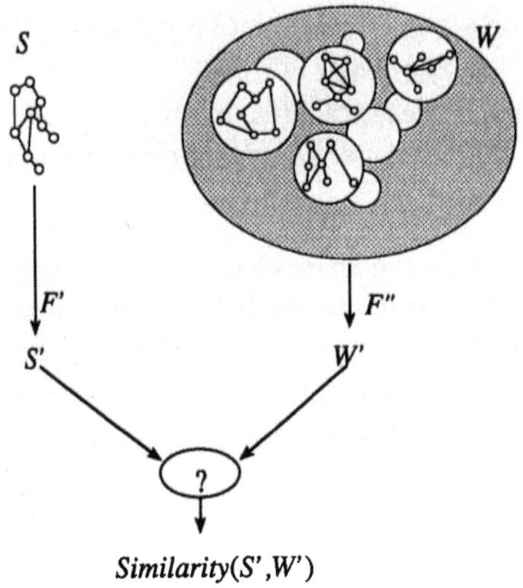

Figure 4.4.4 Matching without explicit token assignment? The problem is to find properties of structural descriptions that allow comparison without enumerating large numbers of possible mappings.

and highly specialized algorithms, it is surprising that most of these techniques have been widely ignored by the computer vision community. In the following, we take a brief look at probabilistic methods used in speech recognition as one way of avoiding the early assignment problem.

Probabilistic Modelling

The problem of matching structural descriptions appears at multiple levels of speech recognition [Levinson 1985]. For example, in the commonly used tri-level hierarchy, phonemes are extracted from a stream of acoustic features at the *acoustic level*. Phonemes are grouped to words at the *lexical level*, and words are combined into sentences at the *sentential level*.

At each level of this hierarchy, in principle the same problem has to be solved. Given is a sequence of measurements A (either directly derived from the acoustic input or as the result of the previous level),

$$A = a_1, a_2, a_3, \ldots a_l \qquad (4.4.1)$$

and a vocabulary Ω of structural tokens. The problem is to find a string

$$W = w_1, w_2, w_3, \ldots w_n \quad (w_i \in \Omega) \qquad (4.4.2)$$

such that W is the string most similar to A. In a maximum likelihood formulation, this means to select from all possible strings W' the string W that, given A was uttered, is most likely to correspond to A, i.e.,

$$P(W|A) = max_{W'}\, P(W'|A). \tag{4.4.3}$$

From Bayes' formula we get

$$P(W|A) = \frac{P(W)P(A|W)}{P(A)} \tag{4.4.4}$$

where $P(A)$ is constant for all W and the probability $P(A|W)$ can be trained from observations. Storing $P(A|W)$ in memory is usually impractical due to the large number of combinatorial possibilities for both A and W. Assuming that the observed patterns are independent of each other,

$$P(A|W) = \prod P(a_i|w_i) \tag{4.4.5}$$

where $P(a_i|w_i)$ is the probability[1] of observing the token a_i when the true token is w_i. For computing the missing *prior* probability function $P(W)$, we cannot assume independence among the elements w_i of W. Accounting for all possible interdependencies between string elements would require to calculate

$$\begin{aligned} P(W) &= P(w_1, w_2, \ldots w_n) \\ &= P(w_n|w_1, w_2, \ldots w_{n-1}) \cdot P(w_{n-1}|w_1, w_2, \ldots w_{n-2}) \cdot \ldots\, P(w_2|w_1)\,. \end{aligned} \tag{4.4.6}$$

Markov Models

Unfortunately, it is not possible in practice either to estimate or to store the values of the multi-dimensional probability function in (4.4.6). A common solution is to partition the set of possible strings into a set of *equivalence classes*, using a mapping M:

$$W = (w_1, w_2, \ldots w_n) \rightarrow M(W) \tag{4.4.7}$$

such that the probability of a new token w_{n+1}, given a string W, can be approximated by

---

[1] $P(a_i|w_i)$ is often called the "confusion probability".

$$P(w_{n+1}|W) \approx P(w_{n+1}|M(W)). \tag{4.4.8}$$

A particularly simple form of equivalence relation is obtained by considering only the last $k+1$ tokens in a string, i.e.,

$$M(W) = M(w_1,w_2,\ldots w_n) = w_{n-k},\ldots w_n \tag{4.4.9}$$

for $k \geq 0$. In other words, the string generating process is assumed to be a *Markov* source (of the order k) which allows to approximate the conditional probabilities in the form

$$P(w_{n+1}|w_1,w_2,\ldots w_n) \approx P(w_{n+1}|w_{n-k},\ldots w_n). \tag{4.4.10}$$

As a consequence, the resulting conditional probability functions (4.4.10) have small dimensions, so they can be stored and are easier to estimate from training examples. Moreover, efficient methods for sequential matching with probabilistic models have been developed, mainly using dynamic programming techniques. Examples are the algorithms by *Fano* [Fano 1963] and *Viterbi* [Viterbi 1967], and the *stack* algorithm by *Jelinek* [Jelinek 69].

These algorithms consider the string generator as a finite state machine and try to find the most likely paths in the graph of possible output strings (sometimes called a *trellis*). At each stage of the sequential matching process, they compute the set of "best" (i.e., most likely) paths with respect to the given input sequence. However, the algorithms differ by the number of alternative paths kept at each stage, the internal representation of partial strings, and by the appropriate size of the token vocabulary. A comparison of these algorithms can be found in [Blahut 85 (Ch. 12)].

The maximum likelihood technique, as it is used in speech recognition, is one way of avoiding the early assignment problem as depicted in Figure 4.4.4. Effectively, from a given structural description (of speech) a probabilistic model is derived that is then used for the recognition task (Figure 4.4.5).

Speech vs. Vision

Considering the good success of probabilistic modelling in speech recognition, it seems natural to investigate the use of similar techniques for visual recognition. After all, there are quite a few *similarities* between these two problem areas:

- Relevant entities are composed from primitive elements (i.e., tokens).
- Token measurements are neither robust nor invariant.

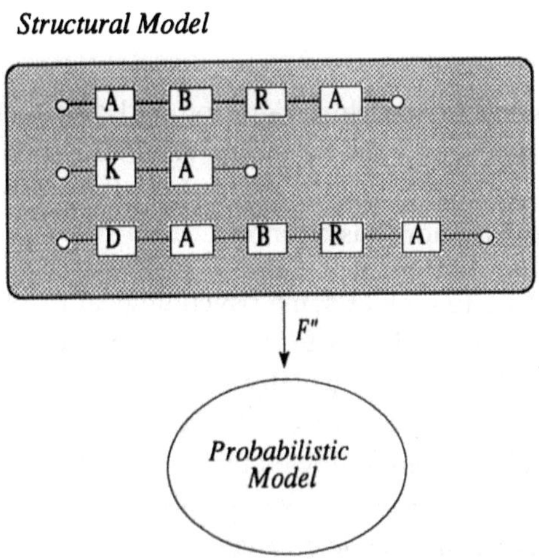

Figure 4.4.5 Probabilistic matching in speech recognition. The original (structural) description is compiled into a probabilistic model that is used for matching.

- Both are characterized by a hierarchical structure, i.e.,

  speech: phoneme → word → sentence

  vision: token → object → scene

Of course, there are also important points of *difference* between speech and vision that prohibit a direct application of certain specialized techniques:

- Unlike in speech, which is carried by a one-dimensional signal, there is no "natural order" of tokens in two-dimensional images.
- Similarly, pauses in speech serve as delimiters in token sequences that have no obvious counterpart in vision.
- The vocabulary of possible image tokens is much larger in vision.
- The (spatial) relationships between visual tokens can be complex.

Among the above aspects, the problem of generating token sequences from a two-dimensional image appears to be the most fundamental one. Any framework for structural object recognition based on sequential matching must include some means for producing "strings of tokens" in order to apply techniques similar to the ones described above.

4.4.5 Scan Paths in Human Vision

Our working assumption is that the problems of speech recognition and structural object recognition can be solved with similar, sequential methods. While the use of speech recognition methods for visual recognition is mainly a question of engineering, the applicability of sequential strategies is a more basic concern.

As a further motivation, it may be interesting to look at some psychological results regarding the way humans tend to analyze pictures. It is well known that humans do not "understand" pictures from a single glimpse, but usually scan over pictures actively and repeatedly. These so-called *scan paths* are series of "fixations", where attention is focused at a fixed point in the visual field, followed by a rapid transition to the next fixation. While some transitions provoke eye movements (called "saccades"), others are performed mentally within the field of view, without physically moving the eyes. The picture locations selected for fixations are usually "interesting" in some sense, e.g., regions of high information content or some salient feature grouping.

When they analyzed the visual scan paths of individual subjects, Noton and Stark [Noton, Stark 71] made an important observation. They found that a person seems to develop a "preferred scan path" for each known object, which is invoked almost identically every time that object is encountered again. Moreover, it is argued that human scan paths follow a global strategy in order to assemble what is called a "feature ring", as the internal representation of an object. In particular, they note that *"The feature ring establishes a fixed ordering of features and eye movements, corresponding to a scan path on the object."*

Not all the hypotheses stated in [Noton, Stark 71] seem to be plausible from a computational point of view. E.g., if preferred scan paths were indeed used to represent objects, how and when is the correct scan path selected during the recognition phase? In other words, if scan paths are the basis for structural object recognition, then what is the mechanism for *indexing* and at which point of the recognition process does it occur? At least, however, there seems to be sufficient evidence to believe that the human recognition engine works sequentially to a great extent. Moreover, the repetitive nature of the human scan paths could at least mean that linear sequences of structural tokens play a key role in the recognition task.

4.4.6 A Framework for Structural Recognition

Techniques from speech recognition have some nice properties that make them attractive for solving problems in visual recognition. Unfortunately, there does not seem to be a direct way to apply these inherently one-dimensional techniques to the analysis of two-dimensional images. On the other hand, psychological findings seem to indicate that sequences of tokens are probably used intensively by the human visual system for object recognition. The question is how to cast these complementary ideas into a computational vehicle for object recognition.

The general framework we propose consists of three layers of processes (Figure 4.4.6): *evidence accumulation, indexing,* and *identification*. Out of these, only the identification step operates on the original, structural world model, comprised by a set of individual object descriptions. Both evidence accumulation and indexing use a probabilistic model, that is derived from the structural model.

Evidence Accumulation

At the beginning of the recognition process, nothing is assumed to be known about the contents of the scene. Without invoking any of the known structural object models, evidence accumulation tries to find assemblies of image tokens that are salient with respect to the given world model. In terms of the probabilistic model, this means to find groups of tokens that are likely to belong to one of the relevant objects. To make the likelihood of a grouping computable, evidence accumulation does not search for arbitrary two-dimensional structures but rather for linear *sequences of tokens.*

Indexing

The purpose of the indexing step is to generate object hypotheses, i.e., indices into the base of known visual concepts. The token sequences produced by evidence accumulation are continually monitored to see if indexing is possible. Indexing will occur when sufficient evidence is available and the number of hypothesized objects is small. Successful indexing changes the recognition process from bottom-up to top-down mode by invoking the identification step.

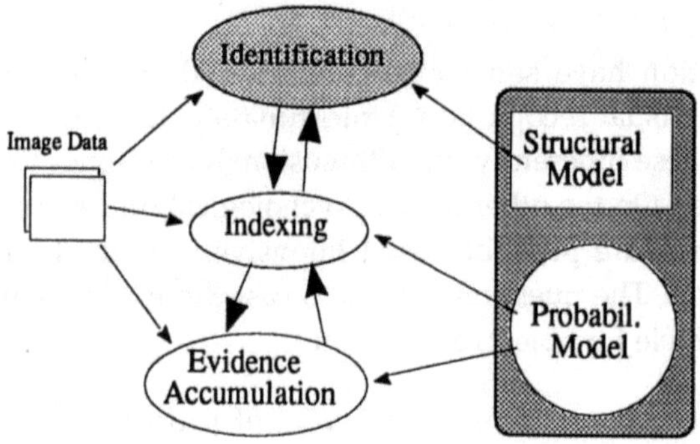

Figure 4.4.6 Framework for single object recognition. Three processes, *evidence accumulation, indexing,* and *identification* cooperate in a hypothesize-and-test fashion. Only the identification process makes direct use of the structural world model. The other two processes use a probabilistic model that is derived from the world model.

Identification

Given a set of object hypotheses, the task of identification is to reject false hypotheses by looking for more detail in the image data. Correct hypotheses must be verified using the structural information contained in the original model. Objects are delineated in the image, solved for viewpoint, etc.

In this work, we are primarily concerned with the tasks of evidence accumulation and indexing, i.e., we want to build a fast mechanism for producing the most specific object hypotheses. In the following, we describe two important aspects of the two lower layers of this concept: *(a)* the problem of finding salient token structures and *(b)* how contextual information can be used to control the recognition process.

4.4.7 Salient Strings of Tokens

The goal of evidence accumulation is to collect sets of structural tokens that are "interesting" with respect to the given world model. This involves solving two problems: *(a)* selecting tokens to be included into a token set and *(b)* computing the saliency of a token set.

Assembling Token Sets

Token sequences are created sequentially by successively adding image tokens to an existing set of tokens. To allow effective indexing, we require that there is a good chance of overlap between the generated token sequences and the token sets contained in the structural model. Assume now that we have a given set of image tokens grouped into a token string. The problem is how to *extend* this token string, i.e., which image token should be included next. There are several ways of doing this:

- *Random walk*: Select successive tokens by doing a random walk over image tokens, regardless of the current grouping. The chance of generating sequences that overlap (at least partially) with model sequences is minimal.
- *Biased walk:* Use some strong, fixed heuristics to bias the selection of successive tokens. Heuristics should be based upon the syntactic image contents only, i.e., similar images should create similar token sequences regardless of their semantic contents. Parameters could include local properties, such as *proximity* (i.e., prefer tokens that are close to the current set), *similarity* (i.e., prefer tokens that are of similar type), *good continuation* (i.e., try to continue along a locally smooth path), etc. Similar heuristics have been applied by Lowe [Lowe 87] for grouping straight line segments.
- *Context-dependent walk:* Let the selection of the next token depend upon the entire current token set. E.g., it may be desirable to have a fair strategy in the sense as to visit all image tokens within a certain region. Similarly, the search could be biased to follow a predefined global strategy, e.g., scanning along large circles in a predefined direction (Figure 4.4.7).

Coding Tokens and Spatial Relations

Token sequences, either extracted from the image or contained in the reference model, consist of structural elements and the spatial relations between those elements (Figure 4.4.8). A token sequence is thus a series of *structural events*, i.e., the entries in a "token string", for which we must find a suitable representation. For practical reasons, we want to consider only binary relations between tokens.

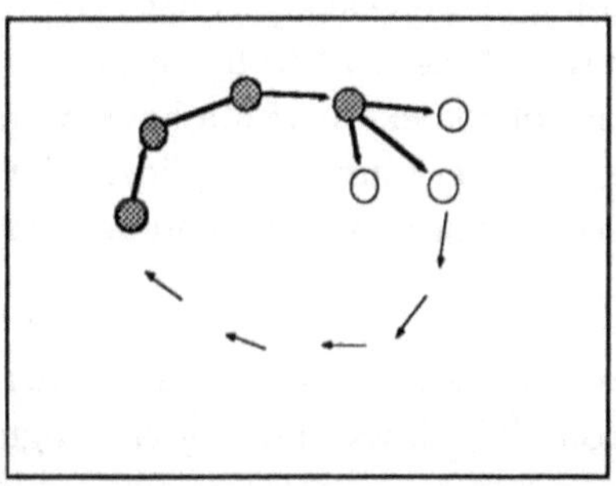

Figure 4.4.7 Applying a global strategy for selecting successive image tokens. New image tokens are added to the current token sequence e.g., by searching along circular paths in a predefined direction.

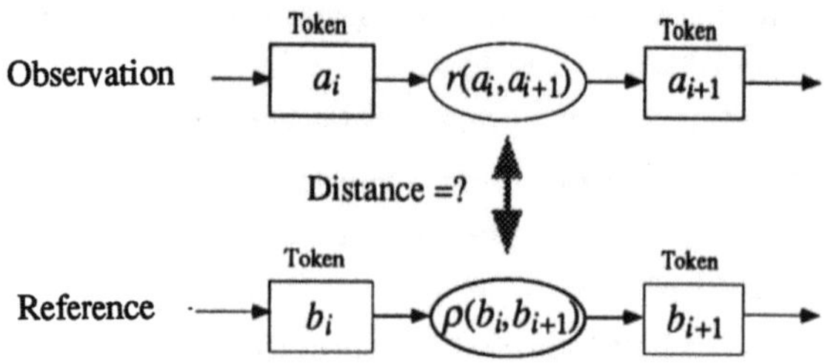

Figure 4.4.8 Token strings are alternating sequences of structural tokens and spatial relations between them. To compare the token strings extracted from the image and the token strings stored in the reference model, a suitable representation must be found.

A first approach could be to define a structural event E as the combination of a token b_i, and its relationship to the subsequent token b_{i+1}, i.e.,

$$E = [b_i, \rho(b_i,b_{i+1})] \in V_T \times V_R \tag{4.4.11}$$

Unfortunately, the size of the resulting vocabulary $V_T \times V_R$ of events would be very large and so would be the amount of storage for the conditional probability tables. Probably a better solution is to treat tokens and relations separately, i.e., the token string S is defined as an alternating sequence of token events E_T and relational events E_R:

$$S = (E_T^{(1)}, E_R^{(1)}, E_T^{(2)}, E_R^{(2)}, \ldots) \tag{4.4.12}$$

where $E_T^{(i)} = b_i \in V_T$ and $E_R^{(i)} = \rho(b_i,b_{i+1}) \in V_R$. The distance between two token strings would then be defined as a function of the distances between corresponding tokens $d_T[a_i,b_i]$ and the distances between corresponding spatial relationships $d_R[r(a_i,a_{i+1}),\rho(b_i,b_{i+1})]$. This should allow the use of parsing algorithms similar to the ones used at various levels of speech recognition [Levinson 85].

4.4.8 Global Context

The importance of context in object recognition and scene analysis has been recognized from the beginnings of computer vision [Baird, Kelly 74; Toussaint 78]. Unfortunately, successful examples for the actual use of context in vision have been rare (e.g., [Fischler, Strat 89]) and are often placed in a very limited application framework. The use of context in speech recognition appears to be limited as well (e.g. [Kuhn, De Mori 1990]).

Global context is different to the *local* context, as used for the context-dependent walk for assembling token sequences (cf. Section 4.4.7). If we view object recognition or scene analysis as a dynamic process that can be defined by a trajectory in some state space, the global context becomes part of the state descriptions and thus affects how the process continues. In particular, it should affect the classification of new objects.

Representing Scene Types

Recognizing an object means to remove the initial ambiguity by collecting additional information. However, if the type of scene is known, the classification of an object may be possible even when the remaining ambiguity is high, so that recognition is faster when the global context is known.

A scene is primarily characterized by the objects it contains. E.g., a house scene usually contains buildings, streets, trees, cars, etc., while an indoor picture will show completely different things. Unfortunately, the number of possible scenes is very large and possibly infinite, i.e., there is no obvious way of categorizing or naming scenes. Instead, we again try to use a probabilistic definition of a scene by expressing the probability that a given set of objects occur together in an image. Thus, we do not say something like

"A *house-scene* contains the objects *house, tree, car, street.*"

but ask for the probability that these four objects appear together, i.e.

$$P(house, tree, car, street).$$

Thus we do not classify scenes as such but we can express whether a certain collection of objects is familiar or not by the probability of their co-occurrence. This allows us to work with partial and new scenes.

The key question is how we can use global context for recognizing individual objects. We assume that objects are recognized sequentially, one by one. I.e., at some point in time t, we have a set of recognized objects $Q(t) = \{C_1', C_2', \ldots\}$, with $C_i' \in W$. We define $Q(t)$ as the *global context* (Figure 4.4.9). We also assume that for a new object to be recognized, we have an ordered set of candidate concepts $\{C_1'', C_2'', \ldots\}$ generated by *evidence accumulation* and *indexing* (cf. Section 4.4.6), but there is not enough information yet to make a final commitment towards one of the candidate concepts.

In order to apply the global context for recognition, we define the *bias* established by the current context $Q(t)$ towards the candidate concept C_j'' as the probability that concept C_j'' occurs, given the context set $Q(t)$:

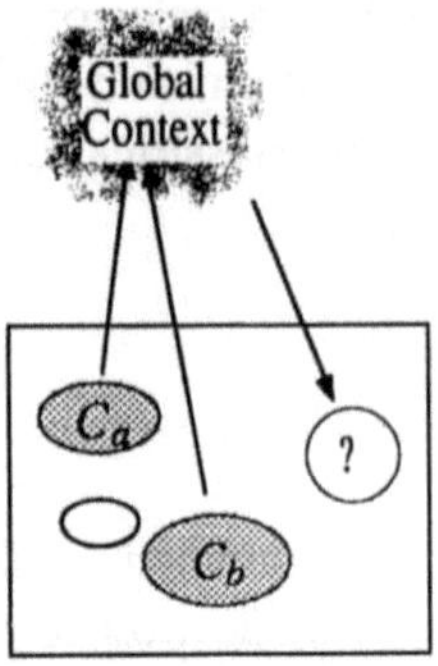

Figure 4.4.9 Global context is defined as the set of recognized objects. During the recognition process, the global context changes dynamically and the classification of a new object (?) is biased by the set of known objects $\{C_a, C_b\}$. The main problem is how to represent global context and how to use it effectively during recognition.

$$Bias(C_j''),t) = P\ (C_j'' \mid C_1',C_2',...) = P\ (C_j'' \mid Q(t))\ . \tag{4.4.13}$$

The task of *indexing* is to make a preliminary commitment to one or more model concepts based on the available evidence. Assume that at time t, evidence accumulation has collected a set of tokens $S(t)$ and we can compute for each model entry C_i a *compatibility value*

$$Comp(S(t),C_i) \in \mathbb{R} \tag{4.4.14}$$

which expresses the structural similarity of $S(t)$ with C_i. A simple way of introducing global context is to let the indexing decision depend upon both the compatibility value $Comp(S(t),C_i)$ *and* the bias for a certain object $Bias(C_j'',t)$, i.e.,

$$Ind(C_i,t) = f\,[Comp(S(t),C_i),Bias(C_j''t)] \tag{4.4.15}$$

where f is some useful combination function (to be developed yet). At each time t (i.e., whenever a new piece of evidence has been found), we can sort the set of object hypotheses $\{C_i\}$ according to the values of $Ind(C_i,t)$. Indexing will occur, when the values for the top ranking hypotheses exceed a certain threshold. The intention is that the global context does not only affect the ranking of object hypotheses but also the critical value for "triggering" *indexing*. I.e., indexing may occur earlier and with less evidence from structure matching when the global context creates a strong bias towards a certain object.

4.4.9 Summary

Many objects found in real-world images can be effectively represented by structural descriptions, consisting of structural tokens and the spatial relations between them. Unfortunately, recognition approaches based on structural descriptions are computationally expensive and usually intolerant against moderate variations of object appearance. In particular, the methods based on graph matching in the strict sense appear to be inadequate for this purpose.

In this paper, we have argued for a different direction for approaching the structural recognition problem, mainly motivated by the following three observations:

- The highly successful application of probabilistic techniques in speech recognition for solving the structure matching problem.
- The sequential and repetitive nature of the scan paths found in human vision.
- The aspects of perceptual grouping and global context for recognition.

We have tried to cast these confluent ideas into a framework for structural object recognition based on sequential matching, probabilistic representations, and the use of contextual information. The central idea is to assemble sequences (i.e., ordered clusters) of image tokens that are maximally salient with respect to the given model base. In order to allow efficient matching of these token sequences, we must make sure that similar images will produce similar token sequences (as it has been observed in human vision). The design of a suitable scanning mechanism is among our most important research goals.

The work described here is actually the first step of a long-term attempt to conceptualize an automated recognition engine that is capable of creating its own models and recognition strategies from examples, as described in [Burger 89]. Within this larger context, however, the development of a powerful and scalable recognition approach appears to be *the* crucial issue, that eventually defines all other components of the system.

There can be no doubt that numerous problems are still unanswered. Central to any probabilistic formulation is the problem of efficient training, which we have not addressed in this section at all. Evidence accumulation and indexing have only been coarsely defined and the proposed approach for incorporating global context still needs considerable refinement. With straight line segments as the basic primitives, a partial implementation of this framework is currently under way that will hopefully provide us with much more insight into the hidden details.

4.5 AN INTEGRATED DEVELOPMENT TOOL FOR MACHINE VISION

4.5.1 Introduction

The development of knowledge-based applications in computer vision creates completely new requirements in terms of hardware and software structures. Computer vision processes (see [Ballard, Brown 82], e.g.) are commonly separated into three layers, namely low-level, mediate-level, and high-level processes. Each of these three layers is characterized by *(a)* the kind of objects that are processed, *(b)* its major computational strategies, and *(c)* its relationship with other processing layers. Low-level processes do mainly pixel processing with potentially extensive parallelism; transition from pixel data to symbols occurs at the intermediate level, while the top level is concerned with the issues knowledge representation, inference, mission planning, etc.

Very generally speaking, as we traverse the processing layers starting from the lowest level, the style of computation changes from parallel to serial, while processing structures become less regular, and the raw amount of data is successively reduced. Interaction between processes at different layers may be intimate and complex, and may not be limited to simple bottom up data streams. For example, decisions made at the top level may be required to control low-level focussing mechanisms in an active vision system [Bajcsy 88]. Consequently, the designer of almost any multi-level vision system encounters a rich variety of different data types, control structures, and communication issues.

In traditional environments, the different tasks of low-level, intermediate-level, and high-level processing are often assigned to separate pieces of software or even hardware, mainly due to the limitations of the individual components. For example, such a system may consist of a dedicated image processor that cannot do much beyond rapid convolutions, followed by a universal computer (e.g., a VAX-type machine) for the more complex algorithms at the intermediate level, and finally a LISP machine for the AI part of the problem (see Figure 4.5.1). The *disadvantages* of this heterogeneous kind of architecture are obvious:

- The communication problems due to the logical or even physical separation of vision processes encourage simple forms of data flow, in particular linear pipeline structures (see Figure 4.5.2). More sophisticated forms of control and data flow, as that are required for knowledge-based vision, are very hard to implement on this kind of machinery.

- Efficient operation requires expertise on each of the machines involved. Usually each hardware platform comes with its own software (operating system, editors, programming languages, compilers, etc.) which means a high training effort for the user.
- The costs for installing and maintaining this complex kind of system may be prohibitive for many potential developers.

In this section, we report upon our work towards a software development tool for computer vision termed the *Vision Kernel System* (VKS) that can be run on a single general-purpose hardware platform. VKS aims at implementing all layers of vision processes in a *single*, uniform programming environment that is based on interactive, object-oriented programming concepts. It is not designed to produce final application systems that are subject to real-time constraints, because execution speed is limited by the underlying hardware. The main purpose of VKS is to provide a sufficient set of tools for developing the individual components of multi-level vision systems and for evaluating these various components in a complete system environment. Also, the dramatic progress of computer vision research makes it necessary to allow incorporating new techniques in an easy fashion and thus requires a completely open environment that is not biased towards any particular vision paradigm. Several similar developments are under way to create what has been referred to as "image understanding environments" [McConnell, Lawton 88], however, these systems are either not commercially available or follow a different design philosophy.

4.5.2 Developing Knowledge-Based Vision

The Design Process

The tasks involved in a full vision system are of extreme variety, ranging from "pixel crunching" at the lowest level, to sophisticated AI techniques at the top level. Interactions between individual processes may be complex. Consequently, developing successful vision systems requires careful design of its components and the possibility to evaluate them within a larger context. Usually, the design process requires extensive experimentation until a sufficient degree of performance is reached, often involving a number of cooperating specialists.

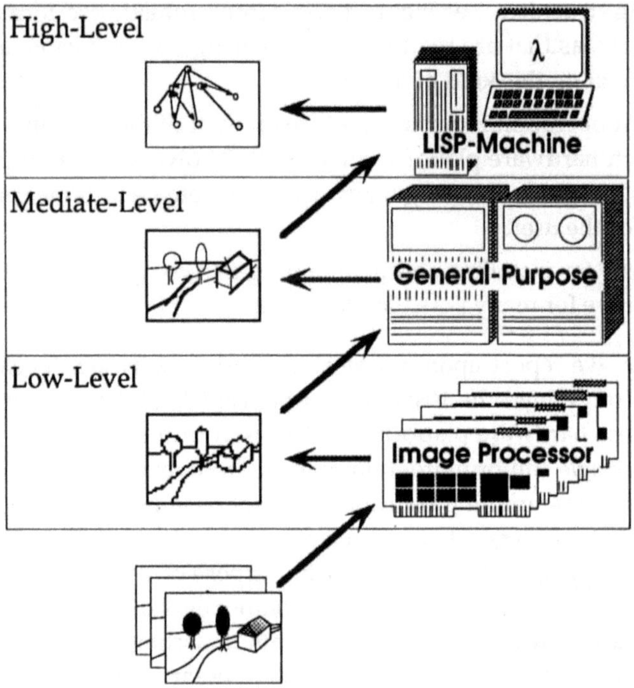

Figure 4.5.1 Commonly used classification of vision into low-level, mediate-level, and high-level processes. Frequently, the three layers are assigned to separate hardware platforms.

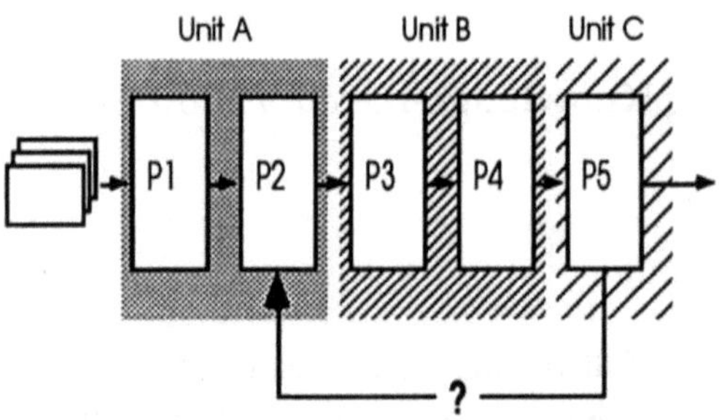

Figure 4.5.2 The communication problems inherent to a heterogeneous hardware and software architecture encourage the use of simple control flow mechanisms, linear pipelines in particular. More sophisticated control schemes necessary for advanced vision systems are hard to implement on this kind of machinery.

Most applications in computer vision have traditionally been extremely demanding in terms of computing power. Despite major technological advances, the popular notion saying something like "*...the recent dramatic increase in computing power has finally made computer vision practical ...*" has unfortunately remained a myth. Special-purpose hardware can be a solution for certain, well-defined applications. However, dedicated hardware often creates a strong bias towards specific computational techniques, thus discouraging experiments towards more innovative approaches.

We argue that, in principle, the development of architectures, methods, algorithms, and data structures should come *before* the specification of the final hardware platform. First the basic computational strategy should be devised and its viability demonstrated, then a suitable hardware should be designed as the second step. Consequently, processing speed is not a central issue during the initial development phase. In practice, of course, both steps will be intertwined to some extent, simply because hardware considerations cannot be ignored during any design phase.

Key Requirements

For a software tool capable of supporting the complex development process we have identified the following key requirements:

- Rapid prototyping capabilities;
- Representation and visualization facilities;
- Flexible control and data flow mechanisms;
- Availability of standard techniques;
- Easy extendibility and portability;
- Realization support.

Rapid Prototyping

In vision programming there is usually a large set of alternative strategies that should be explored, making mandatory short turnaround times during the experimentation process. There are several prerequisites that are essential for accelerating software design. *First,* the number of design languages used should be minimized, the languages should be adequate for the job, and they should be easy to learn. It should be possible to develop software *incrementally,* i.e., modifications are possible without reloading (or even recompiling) the remaining parts. The programming interface should be interactive to allow efficient debugging.

We have chosen *Common LISP* (CL) [Steele 84] as the main programming vehicle for VKS because it is a powerful and universal programming language, meets the above requirements, and has become a widely accepted standard. For the average VKS programmer, CL is the only language ever used, i.e., there is no separate command language that has to been learned. VKS uses an object-oriented extension of CL termed *Flavors*[1], which facilitates the design of well-structured, easy understandable software.

Representation and Visualization Facilities

Computer vision naturally deals with images and consequently *images* are a prominent class of objects in VKS. Visualizing 2-D images is generally straightforward and VKS provides extensive means for displaying, inspecting, and editing images in various modes. However, a multi-level vision system usually contains various other kinds of objects and structures that call for visualization to improve understanding, such as feature maps, structural tokens, relational structures (e.g., semantic nets), control structures, process states, data streams, processing histories, etc. In natural scene understanding, 3-D structures, e.g., terrain models, must be represented and possibly visualized. Typically, each research task in computer vision has its own representation and visualization requirements. The current state of VKS only reflects the most basic tools in this regard with many others to be added in the future.

Data Flow and Control

While simple bottom-up pipeline structures are adequate for feature based pattern recognition, advanced vision systems require more sophisticated modes of data flow and control. Some basic control structures are implicitly available in CL, others are easily added by appropriate extensions. In particular, we consider the following basic mechanism and combinations thereof:

- Conventional control structures (i.e., *if-then-else* and *loop* constructs): these are part of standard CL.
- Process-oriented control: Assigning different task to individual processes that execute "in parallel" is a convenient way of implementing distributed control structures, even on single-processor machines. Communication between these

[1] currently being converted to CLOS

processes is usually accomplished by (possibly bidirectional) data streams that queue date elements in a *first-in-first-out* fashion. Modern CL implementations can simulate multi-processing by so-called "lightweight" processes that share the same LISP environment, i.e., all processes have access to all global objects (e.g., functions and global variables). Multiple processes within LISP are easier to handle than most conventional schemes (e.g., *UNIX* processes) and do not disturb the simplicity and homogeneity of the software environment.

- Data-driven control: The most prominent representatives of data-driven control are pattern-directed inference systems, often referred to as "rule-based" or "blackboard systems" [Nii 86]. Most AI development tools (e.g., expert system shells) follow this paradigm in one or the other form. For good reasons, many advanced AI tools are themselves implemented in LISP, which permits direct integration of these systems with VKS. In fact, this has been a major argument for selecting *Common LISP* as the base language.

Standard Techniques

There exists a large body of standard methods, particularly in the area of low-level image processing. Numerous standard algorithms have been compiled into image processing libraries, using mostly FORTRAN or the C language [Everson 89, Havens 86]. Since most CL implementations allow to load "foreign functions" compiled from other languages, the integration of these programs into VKS is a simple task, provided that the algorithms can be invoked at the functional level and that the interfacing conventions are sufficiently documented.

At the mediate level, much fewer (if any) standard procedures are widely accepted. This may, in part, be due to the fact that the data structures and the processing goals are much more task dependent than they are at the pixel level. At this level, there is often a need for dynamic representations of varying size and structure, e.g., to represent sets of tokens extracted from an image. While a *list* is a natural representation for this kind of structure in LISP that can be printed and read by other LISP implementations, conventional languages do not directly provide this sort of comfort.

Similar arguments hold for the processes found in high-level vision. Standard AI techniques (e.g., heuristic search methods, structural matchers, natural language interfaces, etc.) are available as LISP code and therefore easily integrated with VKS. Because most commercial AI tools (e.g., ART[1] or KEE[2] can be used on top of VKS, a

---

[1] Automated Reasoning Tool (by *Inference*) [2] Knowledge Engineering Environment (by *Intellicorp*)

developer can select his favorite tool himself and there is consequently no bias towards any particular high-level paradigm.

Extendibility and Portability

VKS is a completely open system that can be extended *horizontally* (at the LISP level) as well as *vertically* by including foreign code or enclosing VKS within other environments. *Common LISP* itself is available on a large variety of different hardware platforms, workstations in particular. Visualization and the user interface are the crucial points as far as portability is concerned, which can only be warranted by adhering to standard interfacing techniques (as discussed below in more detail).

Realization Support

The main purpose of VKS is to develop the algorithmic components of a vision system and to demonstrate the performance of that system without any real-time constraints. To transfer the design from the algorithmic level to an actual runtime system may be an extremely difficult task depending upon the specific speed requirements, cost constraints, etc. While this task is outside the scope of VKS, we would like to have some rough estimate what performance can be expected under given technological (and financial) constraints, before someone actually attempts to build the system (Figure 4.5.3). What VKS can do to make the limitations visible is to supply runtime estimates for certain hardware solutions. For example, if an algorithmic solution turns out to be impossible to implement with current technology, a redesign is necessary in order to overcome these limitations (e.g., by making better use of inherent parallelism).

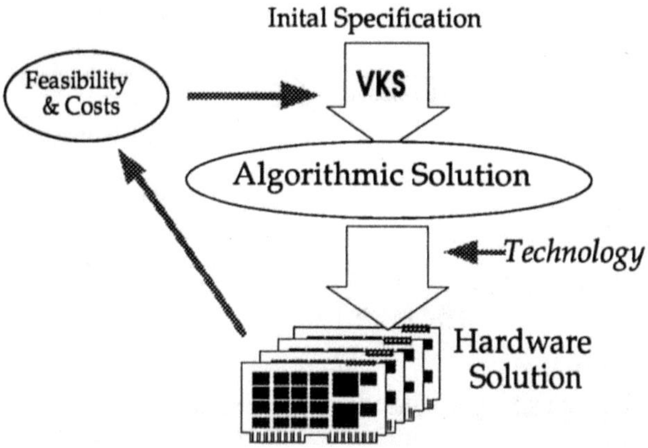

Figure 4.5.3 Computer vision design process. Ideally, the algorithmic components of the system should be specified before a suitable hardware platform is designed. Since hardware considerations cannot be ignored at any phase, we want to have some sense for the feasibility of the system under given technological constraints.

4.5.3 Implementation

Object-Oriented Preliminaries

The state of the VKS environment is defined by the set of objects that currently exist, the internal states of these objects, and the relationships between these objects. Each object is an instance of a certain class (or *flavor*) and classes are structured as a class hierarchy.

For example, VKS supports several classes of *raster images* that differ by the type of image elements they can hold, in particular the classes *binary-image, unsigned-8-image, signed-16-image, float-image, complex-image*. However, *raster-images* are in turn a specialization of a more general class called *image* which also includes other types of images, e.g., stroke (vector) images. The corresponding part of the VKS class hierarchy is shown in Figure 4.5.4.

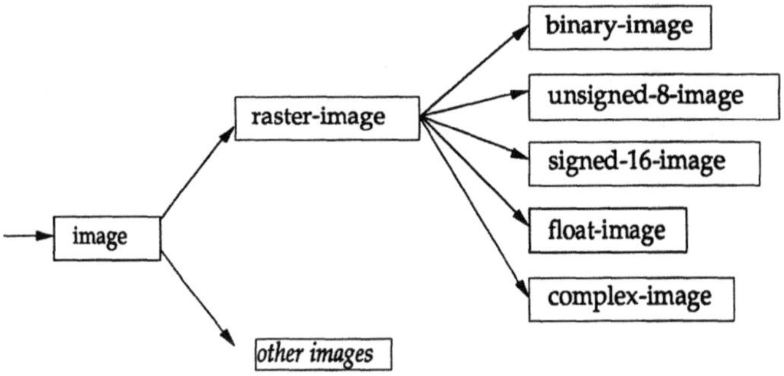

Figure 4.5.4 Part of the VKS class hierarchy for images.

Class definitions include a (possibly empty) set of named slots that each instance of that class will have, e.g.

```
(defflavor raster-image
  ((image-lines 256)      ;slot IMAGE-LINES
   (image-columns 256)    ;slot IMAGE-COLUMNS
   (image-array NIL)      ;slot IMAGE-ARRAY
   ...)                   ;other slots
  (image)                 ;superclass image
)
```

defines the class *raster-image* as a subclass of *image* that has three slots named *image-lines, image-columns* (with default values 256, 256, and NIL), and *image-array* which holds the image data. Similarly, the class *unsigned-8-image* is a subclass of the class *raster-image*. Since each class inherits the slots of its superclass(es), instances of the class *unsigned-8-image* have the same slots as instances of class *raster-image*. To create a new instance of *unsigned-8-image* we would simply write

```
> (setq im (make-instance 'unsigned-8-image
        :image-lines 100 :image-columns 200))
#<Instance UNSIGNED-8-IMAGE...>
```

which sets the value of the (global) variable *im* to the new object and initializes its slots *image-lines* and *image-columns*. The functionality of objects is specified by attaching *methods* to classes. E.g., the following piece of code defines a simple method (with no parameters) for computing the average value of an image:

```
(defmethod (unsigned-8-image :min-max-average) ()
  (let ((number-of-elements (* image-lines image-columns))
        (min-val 255) (max-val 0) (sum 0))
     (dotimes (lin image-lines)
        (dotimes (col image-columns)
           (let ((pixel (aref image-array lin col)))
              (setq min-val (min min-val pixel))
              (setq max-val (max (max-val pixel))
              (incf sum pixel))))
     (values              ;method returns 3 values
        min-val
        max-val
        (/ (float sum) number-of-elements))))
```

Notice that CL facilitates functions to return multiple values. The above method returns 3 values, the minimum and maximum pixel values, and the average pixel value of an image. Methods are invoked by sending a *message* to an object, e.g.:

```
> (send im :min-max-average)
7
253
150.7
```

Computational Model

As in the simple example above, a method in VKS is generally applied to an object to return one or more values without destroying its arguments, similar to ordinary functions. Most image processing methods are an exception in this regard because they transform images into other images. A purely functional model would therefore require the creation of many intermediate images which would be wasteful in terms of memory space. Consequently, instead of duplicating the original images, VKS does image processing operations mostly by *side effect*, i.e., the argument images are usually destroyed. For example, thresholding an image or applying an arithmetic operation destructively modifies its elements. Thus, after one of the messages

```
(send im :threshold :at 127)
(send im :add-constant 10)
(send im :mult-constant 1.5)
(send im :invert)
```

the variable *im* would still refer to the same object as before but the *contents* of that image may have changed. Of course it is always possible to save an intermediate image object by duplicating it. It is also straightforward to convert one type of image into another type, e.g., a *signed-16-image* to a *float-image* or vice versa. All image processing operations are performed upon image objects that reside within the LISP memory space. Consequently, the number of images that can exist concurrently is limited by the amount of virtual memory available. To create permanent copies, images (and other objects as well) can be given a name, stored on disk, and reloaded for subsequent processing.

Viewables and Viewports

A newly created image object exists logically but it is not automatically visible on the screen. Showing an image requires a *:create-viewport* message to be passed explicitly to the image object. This mechanism allows to display images selectively and it effectively decouples the issues of internal representation and display modalities. In addition to images, many other kinds of objects can be visualized, e.g., histograms, mapping functions, relational structures, processing histories, etc. In VKS, any object that can be visualized is a member of the class *viewable*. The number of viewports that can be attached to a *viewable* object (using *:create-viewport*) is not limited. Figure 4.5.5 shows the relationship between *viewables*, stored instances, and their viewports. When a *viewable* is discarded (by sending it a *:close* message) all its current viewports are removed as well.

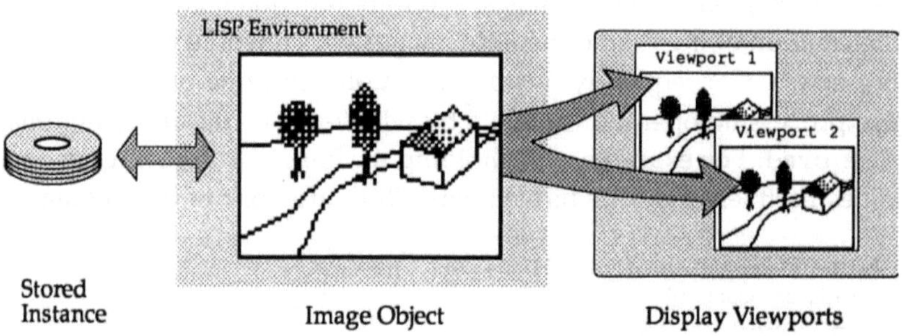

Figure 4.5.5 Image objects, stored instances, and viewports.

Software Structure

Figure 4.5.6 shows the main software and hardware components of VKS. The main programming language is *Common LISP*. Some time-critical parts, e.g., certain image processing operations, are implemented in C and linked as foreign code to the LISP environment. It is also possible to access existing function libraries written in other languages, e.g., *FORTRAN* or *Pascal*. For the high-level tasks, a knowledge-based executive (e.g., an expert system shell) may be installed as the top layer that has access to the full functionality of VKS. The user would then have the alternative to interact with the system at the LISP level (which is the regular mode of interaction) or at the executive level.

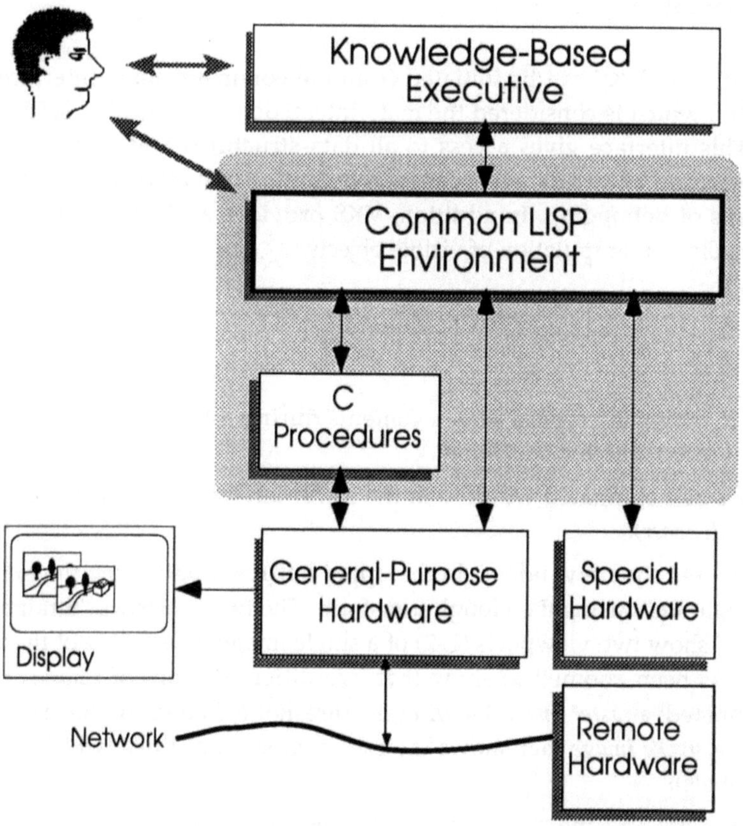

Figure 4.5.6 VKS software/hardware structure.

Since VKS has been designed to run on general-purpose workstation platforms, all processing is normally performed on the main CPU. However, provisions have been made to facilitate the use of special hardware within the same machine, e.g., add-in boards for doing fast convolution or other image processing tasks. Unfortunately, the use of dedicated hardware impairs the portability of the overall system, particularly when the additional components are tightly coupled to the system hardware (i.e., bus structure, interrupt handling, etc.). For this reason, VKS currently does not employ any special-purpose hardware at all. However, we are investigating the use of distributed special hardware that can be accessed in a high-speed network environment that allows remote execution (e.g., the OSF *Network Computing System*).

Interaction Facilities

Any operation in VKS can be initiated from the command lined interface (i.e., the LISP *listener*) which is considered the main interaction mode during software development. This interface gives access to all data structures that exist in the current environment, and allows data inspection, redefinition of methods and functions, and other means of debugging. In addition, VKS provides a flexible *menu* mechanism that allows direct manipulation of visible objects.

Example

Figure 4.5.7 shows the typical screen contents during a VKS session. Several details of this picture are worth mentioning:

- The small elongated window in the lower left-hand corner is the LISP interface (*listener*).
- The three square windows in the upper left-hand part show the extraction of straight lines using the Hough transform. The two windows "Straight Lines of PCB" show two viewports (1, 2) of a single image object. One of the viewports (1) has been zoomed to show that two different types of images have been combined: a *raster image* showing the thresholded edges of the original image, and a *stroke image* that shows the extracted straight lines (which are actually symbols).
- The "Girl" and the "Girl Landscape" windows are again two different viewports of the same image object. This time, the viewports are not only in a different state but are of different type (a *picture* vs. *landscape* viewport). The histogram on the left is yet another way of viewing an image.

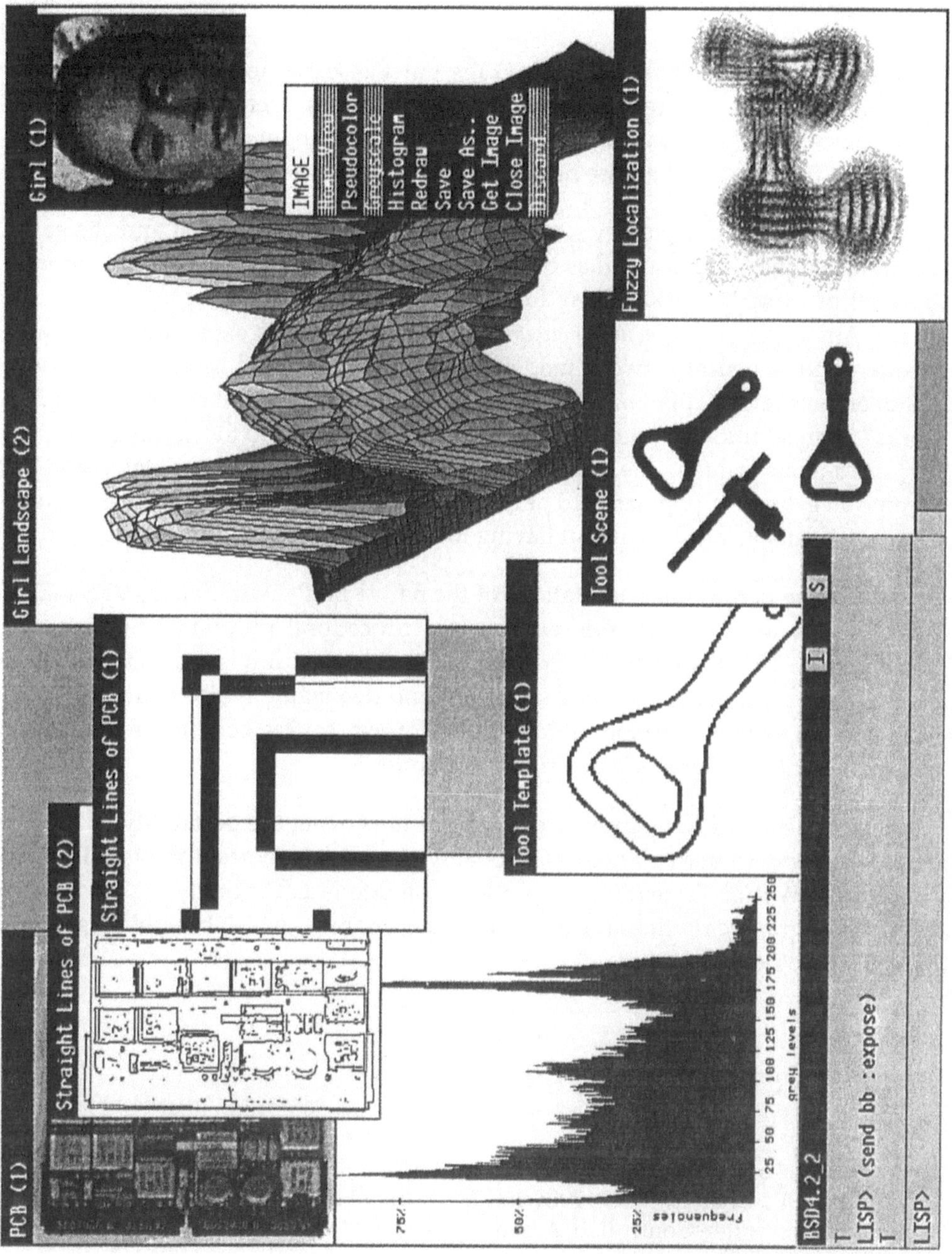

Figure 4.5.7 A typical screen image during a VKS session.

4.5.4 Summary

VKS is not a commercial software package but a research tool that is continually being enhanced as new requirements emerge. VKS has been used for several research projects at this department, typically shaving off about one man-year of work per project that would otherwise have been necessary to get started.

Advances made in computer vision have been dramatic in the past and the set of useful techniques in this area has become abundant. While it is impossible to integrate all of these into VKS, some fundamental tools will be adopted in the near future. Among these is a *neural network toolkit* that facilitates experimentation with several popular neural network models, such as feed-forward nets, Hopfield nets, Kohonen nets, etc. [Lippmann 87]. It will allow the use of "embedded neural networks" as an additional system component as an alternative to the use of neural nets in a stand-alone fashion. The application of neural networks in computer vision has become an extremely popular field of research. This tool will allow to make immediate use of available results without having to care about the low-level details.

Because of the performance limitations of the native LISP environment, VKS makes extensive use of "foreign" code written in a procedural programming language (mostly in C). Unfortunately, the creation of foreign modules is quite cumbersome (due to the typical *edit-compile-load* sequence) and debugging is also a difficult problem. A more elegant solution which allows foreign source code to be embedded directly into LISP programs is described in Chapter 5.

Another direction of our current work aims at improving the portability of VKS by employing widely used standards. Probably the most important move will be to adopt the *X-Window* system for the presentation interface. Currently, VKS uses vendor-specific graphic routines for the display environment which prohibits its immediate use on other platforms.

REFERENCES – CHAPTER 4

[Ambler et al. 75]
Ambler A.P., Barrow H.G., Brown C.M., Burstall R.M., Popplestone R.J., "A Versatile Computer-Controlled Assembly System," *Artificial Intelligence* **6** (2), pp. 129–156 (1975).

[Ayache, Faugeras 86]
Ayache N., Faugeras O.D., "HYPER: A New Approach for the Recognition and Positioning of Two-Dimensional Objects," *IEEE Transactions on Pattern Analysis and Machine Intelligence* **8** (1), pp. 44–54 (1986).

[Baird, Kelly 74]
Baird M.L., Kelly M.D., "A Paradigm for Semantic Picture Recognition," *Pattern Recognition* **6**, pp. 61–74 (1974).

[Bajcsy 88]
Bajcsy R., "Active Perception," *Proceedings of the IEEE* **76** (8), pp. 996–1005 (Aug. 1988).

[Ballard 82]
Ballard D.H., Brown C.M., *Computer Vision*, Prentice-Hall (1982).

[Beveridge et al. 89]
Beveridge J.R., Weiss R., Riseman E.M., "Optimization of 2-Dimensional Model Matching," Proc. DARPA Image Understanding Workshop, Palo Alto Calif., pp. 815–830 (1989).

[Bhanu 87]
Bhanu B. (ed.), "CAD-Based Robot Vision," *Computer* **20** (8), pp. 12–84 (1987).

[Bhanu, Faugeras 84]
Bhanu B., Faugeras O.D., "Shape Matching of Two-Dimensional Objects," *IEEE Transactions on Pattern Analysis and Machine Intelligence* **6** (2), pp. 137–155 (1984).

[Bhanu, Ming 87]
Bhanu B., Ming J.C., "Recognition of Occluded Objects: A Cluster-Structure Algorithm," *Pattern Recognition* **20** (2), pp. 199–211 (1987).

[Blahut 85]
Blahut R.E., *Fast Algorithms for Digital Signal Processing*, Addison-Wesley (1985).

[Bolles, Cain 82]
Bolles R.C., Cain R.A., "Recognizing and Locating Partially Visible Objects: The Local-Feature-Focus Method," *The International Journal of Robotics Research* **1** (3), pp. 57–82 (1982).

[Breuel 89]
Breuel T.M., "Adaptive Model Base Indexing," Proc. DARPA Image Understanding Workshop, Palo Alto Calif., pp. 805–814 (May 1989).

[Burger 89]
Burger W., "Towards Interactive Development of Object Models for Machine Vision," in *Wissensbasierte Mustererkennung*, A.Pinz (ed.), Verlag Oldenbourg, Wien, pp. 49–64 (1989).

[Burger, Grabner 89]
Burger W., Grabner S., "Eine homogene Softwareumgebung für Machine Vision," (in German) in *Wissensbasierte Mustererkennung*, A. Pinz (ed.), Verlag Oldenbourg, Wien, pp. 162–170 (1989).

[Cerella 86]
Cerella J., "Pigeons and Perceptrons," *Pattern Recognition* **19** (6), pp. 431–438 (1986).

[Everson 89]
Everson P.G., "ALV Public Domain Image Processing Toolkit for Sun Workstations," Technical Report, Computer Science Dept., Bristol University, UK (1989).

[Fano 63]
Fano R.M., "A Heuristic Discussion of Probabilistic Decoding, *IEEE Transactions on Information Theory* **9** (2), pp. 64–74 (1963).

[Fischler, Elschlager 73]
Fischler M.A., Elschlager R.A., "The Representation and Matching of Pictorial Structures," *IEEE Transactions on Computers* **22** (1), pp. 67–92 (Jan. 1973).

[Fischler, Strat 89]
Fischler M.A., Strat T.M., Recognizing Objects in a Natural Environment: A Contextual Vision System (CVS)," Proc. DARPA Image Understanding Workshop, Palo Alto Calif., pp. 774–796 (May 1989).

[Goad 83]
Goad C., "Special Purpose Programming for 3D Model-Based Vision," Proc. Darpa Image Understanding Workshop, Arlington, Virginia, pp. 94–104 (June 1983).

[Hall 79]
Hall E.L., *Computer Image Processing and Recognition*, Academic Press, New York (1979).

[Hansen, Henderson 88]
Hansen C., Henderson T., "Towards the Automatic Generation of Recognition Strategies," Proc. International Conference on Computer Vision ICCV'88, Tampa FL, IEEE Computer Society, pp. 275–279 (Dec. 1988).

[Havens 86]
Havens W., "PIPS: A Portable Image Processing System," Proc. IEEE Workshop on Visual Languages, Dallas, TX, pp. 125–135 (1986).

[Hayes-Roth 76]
Hayes-Roth F., "Representation of Structured Events and Efficient Procedures For Their Recognition," *Pattern Recognition* **8**, pp. 141–150 (1976).

[Ikeuchi, Kanade 88]
Ikeuchi K., Kanade T., "Automatic Generation of Object Recognition Programs," *Proceedings of the IEEE* **76** (8), pp. 1016–1035 (1988).

[Jelinek 69]
Jelinek F., "A Fast Sequential Decoding Algorithm Using a Stack," *IBM J. Res. Develop*. **13**, pp. 675–685 (1969).

[Knoll, Jain 86]
Knoll T.F., Jain R.C., "Recognizing Partially Visible Objects Using Feature Indexed Hypotheses," *IEEE Journal of Robotics and Automation*, **RA-2** (1), pp. 3–13 (1986).

[Kuhn, De Mori 90]
Kuhn R., De Mori R., " A Cache-Based Natural Language Model for Speech Recognition," *IEEE Transactions on Pattern Analysis and Machine Intelligence* **12** (6), pp. 570–583 (June 1990).

[Lamdan, Wolfson 88]
Lamdan Y., Wolfson H.J., "Geometric Hashing: A General and Efficient Model-Based Recognition Scheme," Proc. Int'l Conf. on Computer Vision ICCV'88, Tampa, FL, pp. 238–249 (1988).

[Levinson 85]
Levinson S.E., "Structural Methods in Automatic Speech Recognition," *Proceedings of the IEEE* **73** (11), pp. 1625–1650 (Nov. 1985).

[Lippmann 87]
Lippmann R.P., "An Introduction to Computing with Neural Nets," *IEEE ASSP Magazine* **4** (2), pp. 4–22 (April 1987).

[Lowe 87]
Lowe D.G., "Three-Dimensional Object Recognition from Single Two-Dimensional Images," *Artificial Intelligence* **31**, pp. 355–395 (1987).

[Marr 82]
Marr D., *Vision*, W.H.Freeman, San Francisco (1982).

[McConnell, Lawton 86]
McConnell C.C., Lawton D.T., "Image Understanding Environments," *Proceedings of the IEEE* **76** (8), pp. 1036–1050 (1988).

[Michalski et al. 86]
Michalski R.S., Carbonell J.G., Mitchell T.M., *Machine Learning – An Artificial Intelligence Approach*, Vol. II, Morgan Kaufmann Publishers, Los Altos CA (1986).

[Minsky, Papert 69]
Minsky M., Papert S., *Perceptrons*, MIT Press, Cambridge, MA (1969).

[Nii 86]
Nii H.P., "Blackboard Systems," *AI Magazine* **7** (2–3), pp. 38–53, 82–106 (1986).

[Noton, Stark 71]
Noton D., Stark L., "Scanpaths in Eye Movements During Pattern Perception," *Science* **171** (3968), pp. 308–311 (Jan. 1971).

[Pearl 84]
Pearl J., *Heuristics: Intelligent Search Strategies for Computer Problem Solving*, Addison-Wesley, Reading, MA (1984).

[Pichler, Schwärtzel 90]
Pichler F., Schwärtzel H., *CAST Computerunterstützte Systemtheorie*, Springer-Verlag, Berlin (1990).

[Steele 84]
Steele G.L., *Common LISP: The Language*, Digital Press, Bedford, Mass. (1984).

[Toussaint 78]
Toussaint G.T., "The Use of Context in Pattern Recognition," *Pattern Recognition* **10**, pp. 189–204 (1978).

[Viterbi 67]
Viterbi A.J., "Error Bounds for Convolutional Codes and an Asymptotically Optimum Decoding Algorithm, *IEEE Transactions on Information Theory* **13** (2), pp. 260–269 (1967).

[Winston 75]
Winston P.H., "Learning Structural Descriptions from Examples," in *The Psychology of Computer Vision*, P.H. Winston (ed.), pp. 157–209, McGraw-Hill, New York (1975).

[Zadeh 66]
Zadeh L.A., "Fuzzy Sets," *Information and Control* **8**, pp. 338–363 (1966).

5 IMPLEMENTATION ISSUES

5.1 SOFTWARE DEVELOPMENT FOR CAST

For the realization of CAST systems, the classical top-down software development cycle has proved to be inadequate. An inherent feature of all our CAST projects has been that no full requirement specification ever existed. Instead, specifications were developed incrementally while projects were in progress. In many cases, the developer only has a (vague) idea what the system is meant to do and how a solution could look like. Frequently, only an implementation can show the feasibility of the original idea and the viability of the proposed solution. Many problems, e.g., complexity issues, only will ever be tackled when a running prototype is available. Moreover, the actual implementation brings up new ideas for improvements and extensions.

Consequently, we do not rely upon classical software engineering methods for CAST developments but instead prefer *explorative software engineering* (also called *evolutionary prototyping*) [Stritzinger 91; Budde et al. 84; Connel, Shafer 89]. By explorative software engineering we mean an incremental, iterating development strategy. To come from an initial idea to a satisfying final implementation, we iterate many times through the phases of *problem analysis, specification, design, implementation, test,* and *solution evaluation* [Stritzinger 91] (see Figure 5.1.1).

Programming languages and software environments for CAST system development must satisfy several requirements not found in most conventional programming systems. In particular, such a programming environment should

- have enough expressive power to allow the formulation of a problem solution on a high level of abstraction without getting involved into any computational details (e.g. like storage management or representation of numbers);

- support quick and convenient debugging and testing;
- be a contiguous language [Meyer 88] in the sense that small changes in the specification only require small changes in the implementation; thus our programming language has to support software maintainability and extendibility;
- facilitate the realization of modern user interfaces to allow convenient user interaction and visualization of program output;
- facilitate software reusability, i.e., the realization and use of predefined software components; application frameworks [Stritzinger 91; Bobrow, Stefik 83; Weinand et al. 88] should be available to provide reusable building blocks for quick user interface implementation;
- execute efficiently.

We have chosen LISP as our main programming vehicle, because it fulfills most or all of the above requirements. While we used the *InterLISP-D/LOOPS* system [Xerox 83, Bobrow, Stefik 83] for our first implementations, we have adopted *Common LISP (CL)* [Steele 84] and its object-oriented extension *Common LISP Object System (CLOS)* [Bobrow et al. 88, Keene 89] for our recent CAST developments. Beyond this, we plan to incorporate the upcoming standard for building user interfaces, the *Common LISP Interface Manager (CLIM)*, as well as object-oriented database management concepts [Atkinson et al. 89; Kim 90; Kim et al. 90; Itasca 90] into our future developments.

The CL language and programming environment fulfills the above stated requirements through the following features:

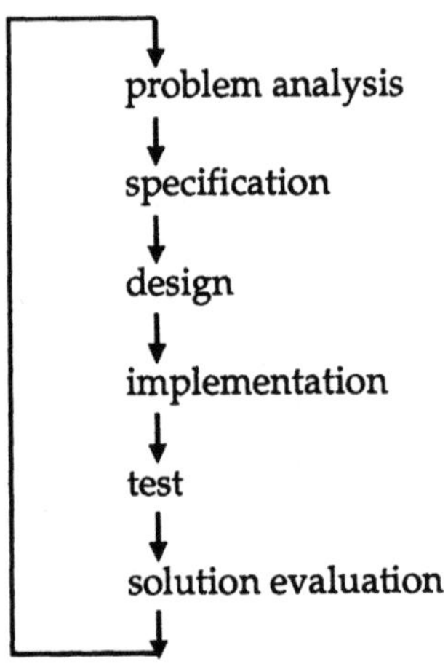

Figure 5.1.1 Explorative software engineering.

- LISP has been designed as a language for symbolic computation and not so much as a language for numeric computation. Thus it allows the efficient coding of high-level problem solving methods that are typical for CAST systems.
- LISP languages use lists as principal data structure which provide a lot of flexibility. Through lists and data abstraction, dynamic and arbitrarily complex data structures can be implemented.
- CL frees the user from as much detail from the computer as possible. E.g., CL provides virtual memory and automatic memory management to free the programmer from physical memory considerations. Also, it provides arbitrary precision numbers such as integers with unlimited length and rational numbers to make the programmer independent of internal number representation schemes.
- In CL, functions (= programs) are first-class objects, facilitating to manipulate programs like ordinary data.
- By the use of functions and especially macros, it is possible to extend the CL language and tailor the syntax to special needs. In such a way it is easy to build a new language which resides upon the CL kernel (CLOS was originally implemented that way).
- CL uses dynamic typing in contrast to static typing. Although static typing is regarded to be a great advantage of imperative languages in conventional software engineering, dynamic typing is less restrictive and allows quicker software development. The advantages of static typing (type conflicts can be detected at compile time, compilers can generate more efficient code) are compensated by CL´s support of interactive, explorative software development and testing (see below).
- CL does not require the typical *edit–compile–link–run* cycle. Instead, most CL implementations provide an interpreter (called a *listener*) and allow incremental compilation and arbitrary mixing of compiled and interpreted code. These are the main prerequisites for interactive, explorative software development and testing in LISP.
- The CL debugger efficiently supports the detection and elimination of programming errors. Any data structure can be examined and altered from the debugger.
- CL programs can be compiled and then run significantly faster. CL also allows incremental compilation, i.e., to use compiled and interpreted functions together.

- The CL environment is a *persistent* environment, i.e., one works in a permanent main program. All definitions have indefinite *extent* in the sense that they exist until they become obsolete, i.e., until there is no way to access them. There is no principal difference between system-supplied functions and user-defined functions. There is no principal difference between programming, testing, and running an application. CL environments provide means for long-term persistency, i.e., to store the entire working memory on disk and resume from that copy in a later session.
- LISP languages have the reputation to be slow. However, this is not true for modern implementations on special LISP machines as well as Unix workstations. The difference in speed between optimally compiled LISP and C code is often less than a factor of 2, even in numerical applications. Beyond that, most modern CL implementations allow to call procedures written in other languages (e.g., C or FORTRAN) for implementing critical sections of a program. In some cases, e.g. processing large arrays of numbers, the gain in speed may be considerable. Another application of the foreign language interface is, of course, the integration of already existing programs written in other languages or to access certain system libraries.

CL is a standardized language (ANSI) that is widely accepted by the Artificial Intelligence community. It is available on several different hardware platforms and, as it is standardized, software written in (pure) CL is portable. CL is the most frequently used language in the AI community, thus making our implementations compatible with many other AI programs.

Object-oriented programming is regarded to be the most promising approach to tackle the central issues in software engineering, viz. *software reusability, extendibility,* and *maintainability* [Meyer 88]. Object-oriented programming is programming with abstract data types and has characteristics that are essential for explorative software engineering:

- Object-oriented programming implies the straight forward mapping of real world objects into program objects. Therefore, changes in one real world object will only affect the corresponding and some closely related program objects.
- Object-oriented programming implies encapsulation and information hiding, i.e., there is a clear distinction between the interface specification and the implementation of objects. The users of an object (other objects) only see the interface, while the internal structure of objects is hidden.
- The concept of inheritance in object-oriented languages facilitates to set up hierarchies of object types (classes), where more special types are derived

from more general ones. Features of the more general types are inherited by the more special ones. In that way, new object types can be specified by reusing and extending existing ones.

- Object-oriented languages use generic functions (message passing and methods) and usually dynamic binding.

Most LISP environments incorporate an object-oriented extension. *LOOPS* [Bobrow, Stefik 83] is the object-oriented superset of *InterLISP-D*. *Flavors* [Apollo 90b] is an early object-oriented extension of *Common LISP*, originally introduced by *Symbolics*. The *Common LISP Object System (CLOS)* [Bobrow et al. 88, Keene 89] has been defined as the *standard* extension of CL for object-oriented programming. The strong interest in *CLOS* promises that it is indeed going to be accepted as a standard. This highly flexible and very powerful object-oriented system is developed in CL but, on the other side, it integrates CL itself by including the type system of CL in its own class hierarchy. CLOS provides all features known from other object-oriented languages but, in many respects, goes beyond existing object-oriented languages. In particular, it provides:

- multi-methods,
- several different method combination types,
- multiple inheritance,
- a meta-object protocol to implement new object-oriented schemes.

All these features make CLOS the most powerful object-oriented language in existence today.

The *Common LISP Interface Manager (CLIM)* is the emerging standard for user interfaces for Common LISP. CLIM provides the capabilities to build complex graphical, interactive user interfaces quickly and easily. As another feature, CLIM is designed to be portable across a variety of windowing systems and platforms (like *X-Windows*, *Genera*, *SunView*, *Macintosh*, etc.).

Although CL environments allow long-term persistency by providing means to store the current LISP image on file, often it is required to have *selective* persistency, i.e., not to store the entire LISP image on file but only selected, user defined parts of it. *Object-oriented database management systems (OODBMS)* [Atkinson et al. 89; Kim 90; Kim et al. 90; Itasca 90] provide an elegant solution to this problem. OODBMS combine the advantages of object-oriented programming with the advantages of database systems, viz. long-term data persistency, secondary storage management, concurrent users, and recovery from hardware and software errors. Currently the first commercial OODBMS are emerging, including implementations that are strongly linked to LISP. For future CAST systems, the use of an OODBMS as a vehicle for long-term, selective, distributed data persistency should be considered.

In the following sections, we discuss several problems and approaches arising in connection with CAST software development. Section 5.2 deals with problems and solutions to represent the knowledge of various system types and system specification formalisms, employing object-oriented class hierarchies. Two approaches are presented, the *first* proposing an algorithm for setting up optimal inheritance hierarchies, and the *second* introducing the concept of *dynamic classes*. Section 5.3 discusses the concept and relevance of browsers for CAST implementations. Browsers in object-oriented systems are used to visualize complex object structures and serve as a command unit to interactively initiate operations on the objects. Two extensions of standard browsers in LOOPS - the *system entity structure browsers* (see also Section 3.5) and the *block diagram browser* - are presented. Section 5.4 outlines a simple approach to implement long-term, selective data persistency without the use of an OODBMS. Section 5.5 describes an interface for using procedural languages in conjunction with LISP.

5.2 BUILDING CLASS HIERARCHIES

The task of Systems Theory is to provide general problem solving methods under a model-based strategy. Therefore, it has to provide techniques for model building and operations to solve a given problem employing models, viz. model-application (see Chapter 1). To accomplish the requirements of different models for different applications, Systems Theory provides a number of different system types. A complex taxonomy of system types has been established. Specific system transformations can be used to convert a system of one type into a system of another type. Analysis operations can be used to compute properties of systems. It must be emphasized that the applicability of an operation depends mainly on the *type* of a system. There are operations which can be applied to most types of systems and there are operations which are applicable only to systems of a very specific type.

A software basis supporting the development of CAST method banks – a CAST *kernel system* – therefore has to provide these systems-theoretical concepts for model construction, the different system types, and the basic operations for systems-theoretical problem solving. The experiences with CAST implementations proved that there are very important system types and less important ones. Thus we will select these most important system types first to investigate and design a kernel system.

The most important system types include all systems with a local dynamic behavior (general system type *generator*), different types of functions (*black box*) and different types of *networks*. It has also turned out to be useful to distinguish between *systems-theoretical* objects and *mathematical* objects. Each of them demands special treatment. Systems-theoretical objects always consist of a number of mathematical objects (e.g. an automaton consists of *sets* and *functions*). A special case in this context are *functions*. Functions can be regarded as mathematical objects or as system-theoretical objects. In the following, we call mathematical functions just "Functions" and systems-theoretical functions "Function Systems."

For different reasons, the most appealing software environment for CAST implementations has shown to be the object-oriented programming paradigm [Müller-W. 86; Praehofer 86, 88; Mittelmann 88; Zeigler 87]. This new software engineering approach is characterized by the idea that the program is organized around objects which contain object specific data as well as "methods" to specify their functionality. Similar objects, i.e., objects which own the same kind of data and the same methods, are combined to a "class". The class then defines which variables and which methods the instances of that class own. Another important feature of the object-oriented programming paradigm is inheritance. Classes are organized as a hierarchy, where variable and method definitions are inherited from the super- to the subclasses. Thus the main work in the implementation of an object-oriented software system is the design of an appropriate class system hierarchy which can take the form of a tree or, when using multiple inheritance, the form of a directed, non-cyclic graph.

All these features make object-oriented programming useful for CAST implementations. System types can be readily implemented by classes. Particular system specifications correspond to the instances of these classes. Variables define the parts of a system specification and methods represent the operations which can be applied.

The main goal of the implementation of a kernel system is to build a very compact and consistent software basis. Code should not be doubled and a very high degree of reusability, maintainability, and especially extendibility should be achieved. We recognized two main problems, the first how to design the data structures implementing the systems, the second how to represent the different system types and mathematical objects types, especially how to represent common and special features of these different types. It has turned out that both problems can be solved by an object-oriented implementation.

5.2.1 Data Structures

The problem of implementing the different system types is easily solved by pursuing a pure object-oriented implementation. System specifications always consist of mathematical objects. The problem to choose data structures for the system specifications can be transferred to choosing data structures for the parts of the system. If the parts are implemented by abstract data types and hence by class definitions, the system itself just knows how to use the parts but the inner representation is hidden. Thus it does not matter any longer if the transition function of a system is implemented by a table or by an expression or something else. One just has to ensure that the names and interfaces of the exported methods are equal. Then different types of functions can be treated uniformly and full flexibility is achieved. In the system definition phase the user just has to specify of which generic types the parts of his system are. For instance, a finite state machine consists of the arbitrary finite sets A, B and Q and of arbitrary functions δ and λ.

A further advantage of a pure object-oriented approach is that it leads to a very compact user interface. By once establishing a class system which provides important types of sets and functions and implementing interactive, user-friendly programs for defining these mathematical objects, they can be used for the definition of all kinds of systems. It is also very easy to come up with new types of functions and sets and they can be easily integrated into the whole environment.

5.2.2 Algorithm to Compute an Optimal Inheritance Hierarchy

As Systems Theory provides a very complex framework of system types, the computer representation of this framework is very important and a hard problem. But it can be solved by using the object-oriented programming paradigm.

It is straight forward to map system types to class definitions, system specifications to instances of these classes. Left to solve is how to design the inheritance and specialization hierarchy of the resulting class system to gain a compact and consistent implementation and to avoid doubling of code. For that problem we present an algorithmic solution in the following.

A class implementing a system type is defined by a set of variables defining the parts of the system specifications by constraints on these variables (types of the variables), and by a set of methods which can be applied to the systems. We call these the *features* of a system specification. For instance, a sequential machine (class SeqMachine) may be defined by variables *A, B, Q, δ,* and *λ* (with the constraints that A, B, and Q are arbitrary finite sets and δ and λ are arbitrary functions) and by the methods *DefineSeqMachine, ComputeLattice, Output?, NextState* and *Reduction*. In a CAST system we have many such system type definitions. Further examples are:

FunctionSystem
: consists of *A, B* of type *set,* and *l* of type *function*. Methods are *DefineFunc, Output?*.

BooleanFunctionSystem
: consists of *A, B* of type *Boolean set,* and *l* of type *Boolean function*. Methods are *DefineFunc, Output?*.

SeqCircuit
: consists of *A, B, Q* of type *Boolean set,* and *d* and *l* of type *Boolean function*. Methods are *DefineSeqCircuit, Output?, ComputeLattice, Reduction, GateDiagram*.

LinearSeqMachine
: consists of *A, B, Q* of type *finite field,* and *d* and *l* of type *linear function*. Methods are *DefineLinSeqMach, Output?, ComputeLattice, LinReduction, D-Transformation*.

LinearSeqCircuit
: consists of *A, B, Q* of type *Boolean set,* and *d* and *l* of type *Boolean function*. Methods are *DefineLinSeqMach, Output?, ComputeLattice, LinReduction, D-Transformation*.

DiscreteEventSystem
: consist of *A, B, Q* of type *set,* and *d* , *l* and *ta* of type *function*. Methods are *DefineDEVS, Output?, NextState, TimeAdvance?*

Network
: consists of *A, B* of type *set,* components of type *set of systems*. The coupling methods are *DefineNetwork, Output?,* and *NextState*.

As one can see, these types share a lot of common features (every system specification has sets *A, B,* and a method *Output?*) and some features are very specific. We will present an algorithm to compute a class system with an optimal (no unique feature is defined twice) inheritance hierarchy. It should be emphasized that this algorithm is not only applicable to the organization of the class system of system types but is more general and can be applied wherever the organization of an inheritance hierarchy in object-oriented programming is the problem.

Algorithm for an optimal inheritance hierarchy:

Definitions:

We call classes real classes, if it makes sense to create instances of these classes.

We call classes abstract classes, if incarnation of the class is not possible or not sensible.

We call methods, variables and constraints features of an object.

The predicate feature (f, X) is true iff f is a feature of class X.

Algorithm:

1 Determine the set rC of real classes.

2 Determine the set F of all features f1, f2, ... for each real class. The following rules have to be obeyed: Features in different classes which have the same name but different meanings and therefore different implementations have to be renamed to differing names in F. Features with different names but equal semantics must be contained only once in F under a common name. (The latter case should not arise in the object-oriented paradigm anyway.)

3 Compute the set K of subsets of rC (K is a subset of *P* (rC)) in the following way:

$$K := \{X \in P(rC) \mid [\exists f \in F : ((\forall x \in X : \text{feature } (f, x)) \wedge (\forall x \notin X : \neg \text{feature } (f, x)))] \vee \text{card } X = 1\}$$

In other words: K is the union of the set of all equivalence classes of real classes modulo each feature in F, and the set of those sets containing exactly one real class.

4 Form the set

$$Kf \subset K \times P(F)$$
$$Kf := \{ (X, Y) \mid X \in K \wedge \forall f \in Y : [\forall x \in X : \text{feature}(f,x) \wedge \forall x \notin X : \neg\text{feature}(f,x)] \}$$

5 Each element of Kf has to be implemented by one class. The first component determines the inheritance hierarchy, the second the features to be implemented in that class.

6 The inheritance hierarchy is fixed by the subset relation of the first component of the elements of Kf as follows: Let (X_1, Y_1) and (X_2, Y_2) be elements of Kf. (X_2, Y_2) inherits from (X_1, Y_1), if X_2 is a subset of X_1. (X_1, Y_1) is a direct superclass of (X_2, Y_2) if there is no (X_3, Y_3), element of Kf, with X_2 is subset of X_3 and X_3 is subset of X_1.

When we apply this algorithm to our previous example of system types we obtain the set K and the optimal inheritance hierarchy shown in Figure 5.2.1. The nodes in italic font are abstract classes, the other nodes are real classes. This approach to system type representation is better but still similar to the representation in CAST.FSM.

The main advantage of this method is the pure algorithmic formulation of steps 3–6, which suggests an implementation of this algorithm in an advanced object-oriented software development system.

The need for totally recomputing the class system (steps 3–6) in the case of a later change, such as adding a new method or variable is one major drawback, in particular if the class system becomes larger. In our case, when we want to implement many different system types we have to come up with many classes and a very complex inheritance hierarchy is the result. But as the recomputation could be done automatically by the development system, this drawback does not weigh too much.

5.2.3 System Types Definition Employing Dynamic Classes

We now propose a different approach employing so-called dynamic classes. The idea is not to implement the system types itself by class definitions but to represent special properties of a system by class definitions.

Let us take an example of sequential machines. Sequential machines may or may not have the property to be linear, to be a circuit, to have a shift register structure, to be of type *Mealy* or *Moore* or to be autonomous. Non-linear machines may have the property of being deterministic or partial. These properties should be implemented by classes. The classes define features which only machines with the special property own. Figure 5.2.2 exposes such a class system. The nodes in this tree shown in italics are abstract classes representing the choice between exclusive properties (most times a property and the opposite). A sequential machine owning a set of properties must be an incarnation of a class of which the classes defining the properties are superclasses. For instance a linear sequential circuit of type *Moore* and no shift register property and no autonomous property has to be an incarnation of a class having the classes CircuitMach, LinearMach, MooreMach,NonShiftRegMach and NonAutMach

K = { { FunctionSystem, BooleanFuncSystem, DiscreteEventSystem, SeqMachine, SeqCircuit, LinearSeqMachine, LinSeqCircuit, Network } =

{ FunctionSystem, BooleanFuncSystem, DiscreteEventSystem, SeqMachine, LinearSeqMachine, LinSeqCircuit } =

{ Network } =

{ DiscreteEventSystem, SeqMachine, SeqCircuit, LinearSeqMachine, LinSeqCircuit }

{ SeqMachine, SeqCircuit, LinearSeqMachine, LinSeqCircuit } = Automata,

{ SeqCircuit, LinSeqCircui = CircuitAutomata,

{ SeqMachine } = SeqMachine,

{ SeqCircuit } = SeqCircuit,

{ LinearSeqMachine } = LinearSeqMachine,

{ LinSeqCircuit } = LinSeqCircuit,

{ DiscreteEventSystem } = DiscreteEventSystem,

{ FunctionSystem } = FunctionSystem,

{ BooleanFuncSystem } = BooleanFuncSystem }

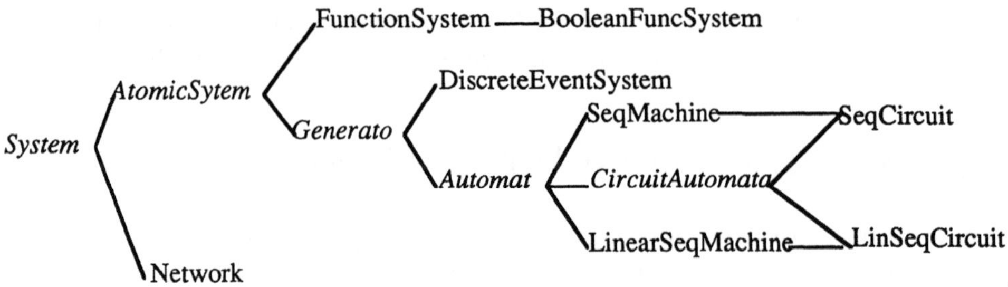

Figure 5.2.1 Optimal inheritance lattice.

as its superclasses. The selection of the properties can be constrained by regarding the inheritance hierarchy as a special graph, a so-called *AND / 1 out of N graph* . For every abstract class representing a choice one has to select 1 out of N properties.

The concept of dynamic classes which is provided by some object-oriented programming systems (e.g.: LOOPS) is of great help to implement these ideas. When a system with some properties should be created, first a class has to be created which has the property classes as its superclasses. Dynamic classes facilitate the creation of such classes at the time of incarnation when they are needed. In LOOPS [Bobrow, Stefik 83] one just has to specify the superclasses. For example, by using the message

(SEND (super1 super2 ....) New),

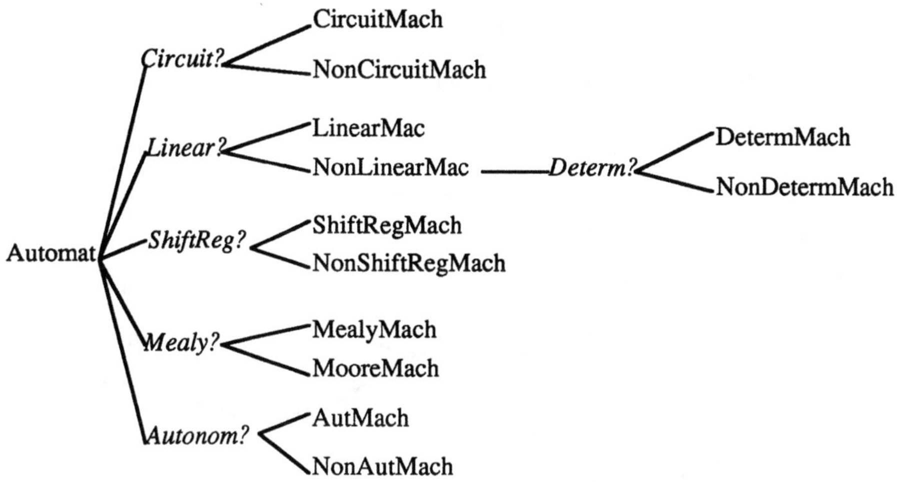

Figure 5.2.2 Property class system of automata.

an instance of a class is created which has super1, super 2, ... as its superclasses. The class itself remains unnamed but is available for reusing when another instance of the same type should be created.

Figure 5.2.3 shows a possible inheritance hierarchy representing properties of mathematical functions. Figure 5.2.4 shows the inheritance hierarchy representing properties of sets. A function may or may not be Boolean, linear, the value may be scalar or a vector. Linear functions may be table functions or defined by an algebraic expression or by a matrix. Non-linear functions may be partial or total and defined by a table or by an expression. Sets may be defined by enumeration or by a predicate, an order relation may or may not be defined and it may be a cross product. New properties can easily be introduced.

In property classes, constraints (types) on variables are defined. These constraints are used to make a preselection of property superclasses of the classes implementing the parts of the system. The union of the constraints of all selected property classes of a system defines the minimal set of property classes the class of the part has to have as its supers. In our example, in the property class CircuitMach of sequential machines, it would be stated that δ has to be a Boolean and vector valued function In the property class LinearMach, it would be specified that δ has to be a linear function. Hence, when a linear sequential circuit is to be defined, the property classes BooleanFunc, LinearFunc and VectorFunc are fixed and the other choices are left to the user.

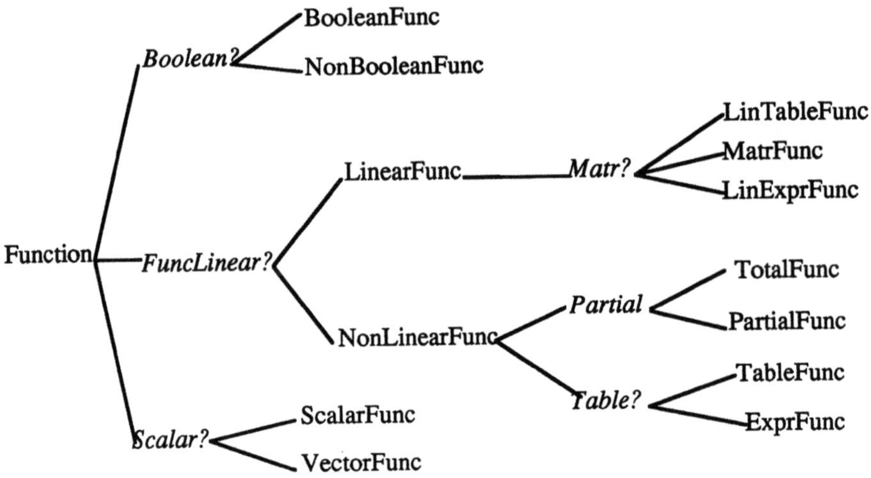

Figure 5.2.3 Property class system for mathematical functions.

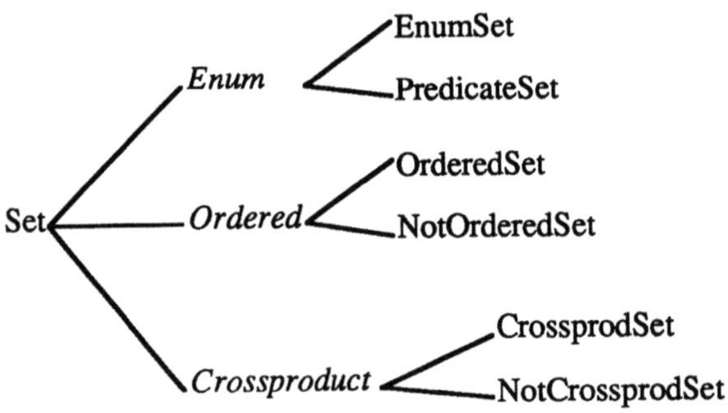

Figure 5.2.4 Property class system for mathematical sets.

5.3 BROWSERS FOR CAST TOOLS

5.3.1 Standard LOOPS Browsers

In most cases, object-oriented programming environments come with window systems [Scheifler 87; Xerox 83; Bobrow, Stefik 83; Goldberg 84] which, together with a mouse, facilitate the development of highly interactive user interfaces. The most important interactive tool for object-oriented systems are so-called *browsers* that are built on top of the basic window system. Browsers can be found in (and are inherently connected with) many object-oriented programming systems but appear in quite different forms. E.g., in *Smalltalk class browsers* [Goldberg 84], the classes appear as a simple list and the inheritance hierarchy is indicated by indenting the subclasses relative to the superclass. Common to all browsers is that they are used to show objects or classes together with their interrelations and, most important, that browsers are used to initiate operations on these objects.

The LISP programming environment InterLISP-D [Xerox 83] and its object-oriented extension LOOPS [Bobrow, Stefik 83] has been used as software basis for different CAST implementation at the authors' department [Müller-W. 86; Mittelmann 88; Pichler, Praehofer 88; Praehofer, Spalt 88; Praehofer 88]. LOOPS provides easy to use browsers as built-in tools. There is the LOOPS *class browser* used to represent the inheritance hierarchy of classes and to initiate operations concerning class implementations. Then there is the LOOPS *instance browser* to show user-defined relations of arbitrary objects and to initiate user defined operations on these objects. These standard LOOPS browsers use a graphical representation and operations are initiated by selecting the object with the mouse and choosing an operation from the appearing pop-up menu (Figure 5.3.1).

LOOPS browsers themselves are implemented using the object-oriented paradigm. Figure 5.3.1 shows the class system implementing standard LOOPS browsers. The task of a browser is two-fold, namely graphical representation and method invocation. Also, its implementation can be partitioned along this line. For the representation part, the GRAPHER package is used. This package is based on InterLISP-D's window system, graphical primitive functions, and records and facilitates the representation of acyclic non-directed graphs. The *LatticeBrowser's* method *Show* takes a

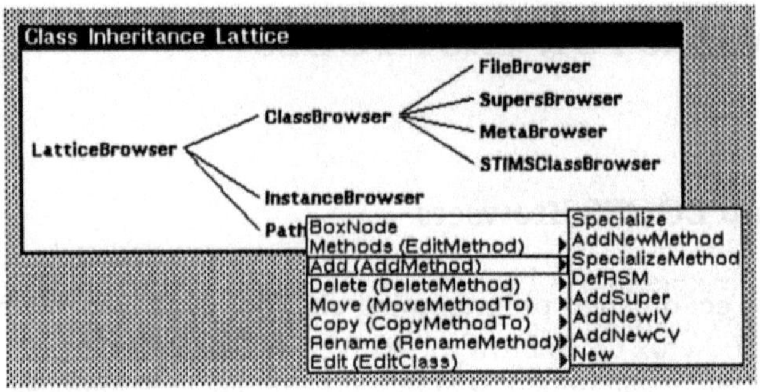

Figure 5.3.1 Class Browser with pop-up menu.

list of objects and the specification of their interrelations and generates a list of node records. The GRAPHER then undertakes the layouting and the displaying. The specification of the interrelations of the objects is incorporated in the objects themselves and also depends on the browser. LOOPS class browsers always take superclass - subclass or class - metaclass relations while the instance browser variable *subIV* is employed to specify the name of the instance variable of the object which contains the list of the related objects.

The entry point to a method invocation cycle is the method *ButtonEventFn* which is invoked whenever a mouse click in the browser is recognized by the InterLISP-D mouse process. From this point on, the following activities are performed: find the selected object, determine the mouse buttons pressed, generate the menu, wait for a menu item to be selected, regard the item selected as a method selector and send this method selector either to the object or, when not understood by the object, send it with arguments *object* and *object-name* to the browser itself.

A software engineer who wants to use a LOOPS browser with this standard functionality in his own application program can build his individual browser without doing any programming but by specifying it in a declarative way. First he has to choose the right class of browser, create an instance of this class and eventually set the variable *subIV* to specify the kind of interrelation to present. The items of the pop-up menus generated by object selection are determined by the browser instance variables *LeftButtonItems* and *MiddleButtonItems* and the software engineer has full control over the menu items by setting these two variables. Then, when sending the method *Browse* with an argument holding the list of root objects to the browser, the root objects and all their subobjects will be displayed and methods specified by the menus can be invoked.

5.3.2 LOOPS Browser Specializations

The object-oriented implementation make LOOPS browsers easy to modify and to extend. This has been exploited in implementing special browsers which modify the functionality of standard LOOPS browsers. One of the first extensions was done to facilitate (in contrast to standard instance browsers where menus are browser-specific) object-specific menus [Müller-W. 86]. Hence, in this kind of browsers, one gets different menus when selecting different types of objects. The main idea in this extension was to retrieve the menu from a class variable of the selected object instead from a variable of the browser.

The goal of another extension was to implement a browser for *system entity structure (SES)* representation and editing [Murauer 89]. The system entity structure [Zeigler 84, 87; Praehofer 89] is a knowledge representation scheme related to Minsky's frame concept. It is a labelled tree representing knowledge about taxonomy, decompositions, and coupling constraints of a real system. Taxonomic and decomposition knowledge is represented by different arcs in the tree and hence a browser which uses a graph with different arcs was required. The *SESBrowser* accomplishes this task by adapting the underlying GRAPHER package for graphical representations. A new browser (Figure 5.3.2) has been implemented. From a set of objects representing the system entity structure, its new method *Show* generates a GRAPHER data structure and does the layouting. For the actual display, the standard GRAPHER functions can be used.

For CAST implementations, especially for modelling and simulation environments based on Systems Theory concepts [Zeigler 87; Praehofer 88; Praehofer and Spalt 88], coupled systems are of great importance. Hence, a graphical interactive tool for coupled system representation and specification is being implemented. This tool will represent coupled systems in the form of block diagrams and it will have the functionality of browsers, i.e., it will react to mouse events in the same way as other browsers do (Figure 5.3.3). Hence the basic mechanism of method invocations will be adopted from standard browsers, while this *BlockDiagrBrowser* will use an entirely different package for the graphical representation.

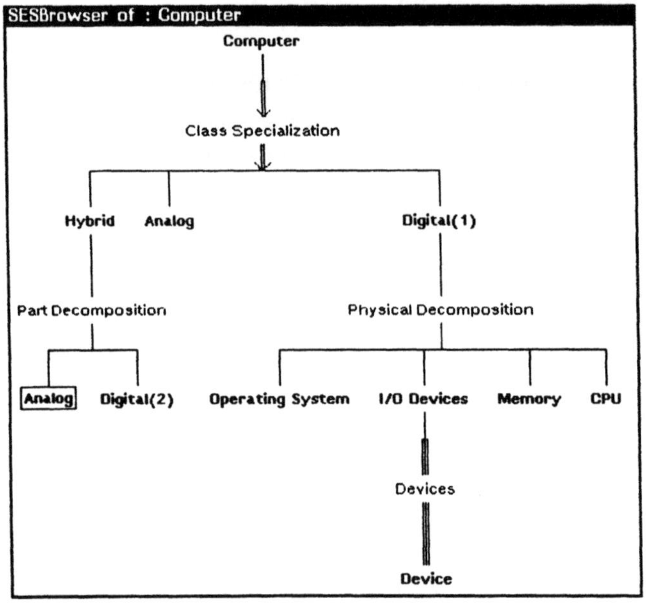

Figure 5.3.2 SES Browser of a computer system entity structure.

5.3.3 Implementing Browsers in Common LISP, CLOS, CLX and CLUE

In the course of porting CAST implementations from the InterLISP-D/LOOPS environment to the more powerful and standardized Common LISP/CLOS system [Steele 84; Bobrow et al. 88; Keene 89] and X-Windows [Scheifler et al. 88; Kimbrough et al. 89], an interesting challenge will be the implementation of browsers. They should be implemented in such a general way that they can use different schemes of graphical representations and different schemes of method invocations.

The X-Windows system is a general window system of the client-server model type. The machine-dependant *server* part of X runs on workstations and graphical terminals. It presents program output in a graphical way and accepts user input from a pointing device such as the mouse and the keyboard. The application or *client* part of X may run on the same workstation, or any other computer connected via a powerful network. The application is seen as a collection of passive programs, each reacting to

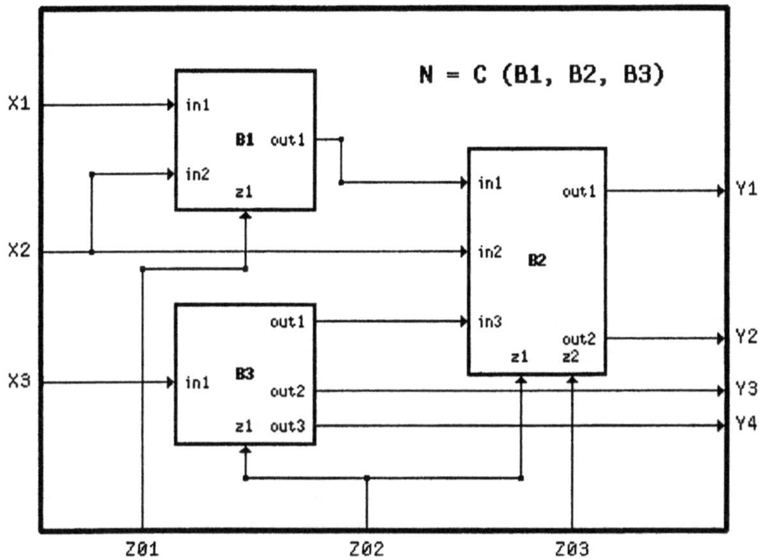

Figure 5.3.3 BlockDiagrBrowser representing coupled system.

a certain kind of user input event like clicking or dragging the mouse in a distinct region of the screen, or typing.

The key feature of the X-Windows system is the machine-independent definition of the protocol for all communication between the server and the clients. The LISP implementation of the X protocol is CLX, and is available for all major Common LISP implementations. As the X system is intentionally totally policy-free to allow for the realization of any kind of user interface, it is also very low level compared to the InterLISP-LOOPS environment. On top of CLX there exists the toolbox CLUE, which allows for an object-oriented user interface programming. CLUE uses the Common LISP Object System (CLOS). But still, a simple pop-up menu application, presenting the user with some choices in a menu and returning the item chosen by the user, prints as 16 pages of source code, in contrast to InterLISP-D, where the same functionality would be achieved with one function definition of twenty lines of code. Therefor, a higher-level toolbox for usage in CAST implementations has to be constructed. It has to include various forms of ready-to-use graphical interface devices, such as menus, fill-up tables, and browsers.

Basically, browsers represent objects and relations between these objects. CLUE defines the class of contact objects, which may receive input via the pointing device or the keyboard, and have some kind of graphical representation on the screen defined. Obviously, the objects in the browser are to be displayed as incarnations of specialized contacts. In the class browser example, the contact would be displayed as the printed class name, whereas in the block diagram browser, the contact could have a rectangle picture representation. In any case, the contact would accept selection type events and respond in a way suitable for the represented object, like calling the application for popping up a command menu.

The browser itself would also be a kind of contact. CLUE has a general class for hierarchical composed contacts, named *composite*. A composite is responsible for the geometric layout of its children contacts. In the class browser case, the browser composite has to place the node contacts according to their superclass relation, and to draw the arcs. A SES browser had to draw different kinds of arcs for the four relations between the objects, and should also use visually different representations for the different types of node classes. In the block diagram example, the blocks itself are composites, allowing to open a block, resulting in a subbrowser showing the inside of the block as a part of the next lower level of the block diagram.

Essential for the overall design of this (and any other) software is the separation of input/output related parts and the essential functions of the application. As an example, we look at the selection of a node in a class browser. When the user clicks the mouse button while the cursor points to an object label in the browser window, the contact object receives a "mouse button pressed" event. The behavior and the visual feedback for the user is defined in the contact class: When clicked, the contact changes its graphical picture (the label with the class name), e.g. to a different color or video inverse, and calls the browser application via an appropriate callback function, e.g. named *object-selected*. What to do with a selected object is determined by the application, each kind of browser will have its own methods. The class browser will present the user with a set of command choices, depending on the object selected and on the mouse button pressed. This is accomplished by the browser by creating and opening a menu contact. The menu contact waits for the user to choose a command item. The menu contact's exact behavior is again separate from the browser application. The menu contact fires a callback to the browser containing the chosen item. Then the browser may decide to process the command by itself, or sending the command to the browsed object.

In this way, a clearly structured browser package can be constructed. It will be used in the implementations of several CAST systems. The object-oriented programming style leads to reusable software tools, were the distinction between user interface components and application dependent functionality helps keeping the software systems maintainable and adaptable to different user interface policies. Changes in the user interface style will require changes in the contact classes, but not in the application specific programs.

5.4 IMPLEMENTING DATA PERSISTENCY IN LISP

5.4.1 Data Persistency

Implementations of Common LISP (CL) usually provide means to store an image of the current LISP environment, i.e., a copy of the working memory, onto disk and to resume from this image at any time later. Since this operation (e.g. `disksave` in *Lucid* CL [Apollo 90]) copies the contents of the entire LISP environment, any unwanted changes that had been made to the objects therein are also memorized. The current LISP environment may contain obsolete function definitions, symbols, and other garbage that one may not want to carry along. On the other hand, data structures may have been created during a session that one wants to use in a future session. Thus there is a need for a mechanism which permits to save only selected parts of the environment, to subsequently start a new session from a "clean" LISP image, and eventually to load the previously saved data again.

A major motivation for implementing data persistency is the idea of a *knowledge base* (KB) as a collection of task-specific data. Internally, i.e., within the working memory of the application program, a KB may be represented as a structured collection of LISP objects. In many applications, the KB is not static, but is successively modified during a session. One alternative would be to keep the knowledge base *external* all the time, i.e., as a data base on disk together with a set of access functions. The KB would then be saved by simply closing the data base which is then kept until it is reopened in another session. The disadvantages of using an off-the-shelf data base are the following:

(a) We are forced to use a predefined data base model (e.g. a relational model), thus giving up the full structural flexibility of LISP.

(b) We would be faced with yet another mechanism for class definition, inheritance, object instantiation, etc., as is already available in object-oriented LISP (e.g., CLOS[1] [Keene 89], *Flavors*[2] [Apollo 90b]).

---

[1] CommonLISP Object System. [2] Originally introduced by *Symbolics Inc.*

Of course, this would also mean a performance penalty, since every database query, search, or modification would require disk access operations. However, if the contents of the KB can be mapped easily onto a given data base management system (DBMS), this may be the preferred way to go. Other advantages of using a DBMS are its explicit data encapsulation, independence of programming language, and the possibility to share data between multiple users.

In our application, we have no clear model for the structure of the required knowledge bases yet. Presumably, there will be several KBs which are again not homogeneous. In particular, we want the entities within the KBs to be virtually any kind of object and we want to manipulate these objects inside our LISP environment.

The approach we are taking is simple. The creation and manipulation of data structures is done within the LISP working memory, thus preserving the flexibility, generality, and efficiency of the native environment. However, we provide means to save selected parts of the LISP working memory onto disk and reinstall the stored data into another session. The key prerequisite for this purpose are functions to store and load arbitrary LISP expressions, as they are provided in *InterLISP-D* (Xerox), *ZetaLISP* (Symbolics), and probably other LISP dialects as well. In particular, we have defined two functions, **dump-form** and **load-form**:

dump-form *form filename* [*Function*]

load-form *filename* [*Function*]

The function **dump-form** stores an arbitrary LISP expression *form* on the file *filename*. **load-form** reads the external representation stored in file *filename* and reconstructs an equivalent internal LISP form, which is returned. The use of these functions should become clear from the following example, where the list `'(1 2 (3 4) (5 6))` is dumped to a file named "`dumpfile`" and subsequently loaded again:

```
> (dump-form '(1 2 (3 4) (5 6)) "dumpfile")
#P"dumpfile"

> (setq a (load-form "dumpfile"))
(1 2 (3 4) (5 6))

> a
(1 2 (3 4) (5 6))
```

5.4.2 Internal LISP Objects

The above example could have been handled just as well using the regular print and load functions in Common LISP. However, for some LISP objects the printed representation (as produced by the print function) is not sufficient to reconstruct equivalent objects. Among the objects not handled sufficiently by the print function are specialized arrays, cyclic structures, and instances of classes defined in *Flavors* or *CLOS*.

Arrays

Arrays in Common LISP are either general (i.e., each element can hold any type of LISP object) or have elements of a specified type. While there is no problem with general arrays, the output from the standard print function is not sufficient to reconstruct a specialized array, as shown in the following example:

```
> (setq a (make-array 5 :element-type '(unsigned-byte 8)
                        :initial-contents '(7 8 6 5 4)))
#<Simple-Vector (UNSIGNED-BYTE 8) 5 69196B>
> (print a)
#<Simple-Vector (UNSIGNED-BYTE 8) 5 69196B>
nil

> (let ((*print-array* t)) (print a))
#(7 8 6 5 4)
nil
```

The regular printed representation of the array, #<Simple-Vector (UNSIGNED-BYTE 8) 5 69196B>, is non-readable. With the value *print-array* being t, the printed representation of the array #(7 8 6 5 4) is sufficient for the read and load functions to create a new array with the same contents. However, this would be a *general* array, i.e., the specific type information (unsigned-byte 8) is lost. The new array created by **load-form** would therefore *not* be equivalent to the original one.

Cyclic Structures

Talking about cyclic structures, we are mainly concerned with cyclic *lists* but similar things apply to cyclic references in arrays and other structured objects. A *cyclic list* is a list that contains itself as a *sublist*. The following example shows how a cyclic list may be constructed:

```
> (setq a '(1 2 3))
(1 2 3)

> (setf (cdddr a) a)
(1 2 3 1 2 3 1 2 3 ...)
```

The resulting list is *infinite* and cannot be printed by the `print` function, since a regular iteration over the list elements (e.g., with `do-list`) would not terminate. Also, `(length a)` is undefined. **`dump-form`** cannot handle this kind of list.

A somewhat different situation occurs when a list contains itself as an *element* (not as a sublist):

```
> (setq b '(1 2 3))
(1 2 3)

> (setf (third b) b)
(1 2 (1 2 (1 2 (1 2 ...))))
```

Again, `print` cannot produce a suitable representation of the modified list. However, `do-list` would iterate over the 3 elements of the list and terminate properly and `(length b)` would return the value `3`. **`dump-form`** *does* handle this kind of list as well as recursive references in arrays and objects (*CLOS* or *Flavors*).

Instances of User-Defined Classes

Instances of classes defined with *CLOS* or *Flavors* are characterized by their class name and a set of named slots together with the current slot values. Slots may again contain arbitrary LISP objects, including other instances. **`dump-form`** does not save the class definitions. Therefore, when **`load-form`** is executed, any class of which an instance is to be reconstructed must be defined in the current LISP environment.

The following refers mainly to instances in *Flavors* but is mostly valid for *CLOS* objects as well. When instances are constructed by **`load-form`** the following sequence of events is executed:

(1) A new instance of the specified class is made but not initialized, i.e., the usually automatic `:init`[1] method is suppressed.

(2) The slots of the new instance are set to the slot values of the original instance.

(3) Finally, the `:init` method is applied to the new instance.

This sequence must be kept in mind when designing specialized `:init` messages for object classes. Particular care must be taken that slot-values set by step *(2)* of **`load-form`** are not deliberately destroyed by the `:init` method.

Structures

Common LISP *structures* are currently not supported by **`dump-form`**. This is not a severe drawback, since *structures* can easily be replaced by using *Flavors* and *CLOS* objects instead.

Multiple References

Multiple references to the same object within *form* are a more general issue. Consider the following example:

```
> (setq L1 '(a b c))
(a b c)

> (dump-form (list L1 L1) "dumpfile")
#P"dumpfile"
```

In this case, we want **`load-form`** to reconstruct the list `(a b c)` only once, such that first and the second element of the reconstructed list refer to one and the same object again, i.e., they are equal with respect to the predicate `eq` :

```
> (setq L2 (load-form "dumpfile"))
((a b c) (a b c))

> (eq (first L2) (second L2))
t
```

The function **`dump-form`** resolves multiple internal references to *non-atomic* objects and replaces them by unique external references. Any non-atomic object is repre-

---

[1] `initialize-instance` in CLOS

sented only once in the external environment. This is covered in more detail in the following.

5.4.3 External Representation

The external representation produced by **`dump-form`** is a readable (ASCII) character file. It contains 4 sections that are written (and read) in the following order:

1. class templates,
2. object type definitions,
3. object contents definitions,
4. the returned object.

External References

Each non-atomic object is given a unique external reference (UER) by **`dump-form`**, which is simply a positive integer. In the external representation, a reference to a non-atomic object is denoted by `#!`*n* , where *n* is the particular UER. References may occur only *after* the corresponding object has been defined, i.e., there are no forward references.

Class Templates

A class template is provided for every class (*CLOS* or *Flavors*) of which objects are contained in the dumped form. Class template definitions have the general format

```
#[C class-name {slot-name}*].
```

Each template indicates unambiguously, in which order the slot-values of that class are stored, such that the slot-names do not need to appear later in every instance definition. For example, the line

```
#[C my-class slot-1 slot-2 slot-3]
```

indicates that the slot-values for all instances of class `my-class` are stored in the order `slot-1 slot-2 slot-3`.

Object Type Definition

Each object type definition specifies the type of an object referenced by a particular UER. Separate formats are provided for lists, arrays, and instances:

```
#[D n L no-of-list-elements]
#[D n A array-dimensions array-specifier]
#[D n I class-name]
```

Lists: the integer `no-of-list-elements` is equal to `(length list)`, which is the number of cons-cells used to represent that list. (Notice that `length` of a *dotted list* is one less than the apparent number of elements.)

Arrays: the list `array-dimensions` corresponds to the `dimensions` argument of the Common LISP function `make-array`. The list `array-specifier` contains additional keyword/value pairs suitable for `make-array`, in particular the values for `:element-type`, `:adjustable`, and `:fill-pointer`.

Instances: `class-name` must be a valid class name and a class template must be available.

Object Contents Definition

The general format of a content definition (*fill*) is

```
#[F n {elem}*] ,
```

where *n* is the UER of the referenced object and `{elem}*` is a sequence of elements. This sequence is interpreted as either the successive elements of a list, the elements of an array (in row-major order), or the slot-values of an instance (in the order given by the appropriate class template), depending upon the type of the referenced object. Notice that a dotted list is distinguished only by the fact that its contents definition contains one element more than the value of `no-of-list-elements` given in the corresponding object type definition.

Returned Object

The first entry in the file which is not a *class template, object type definition,* or *object contents definition* is returned as the final object by the function **`load-form`**. Usually this entry is a reference to one of the external objects defined earlier in that file (denoted by `#!n`) but it may also be an atomic object acceptable to the LISP `read` function.

Example

Assume we defined a flavor `window-size` with the two slots named `lines` and `columns`:

```
> (defflavor window-size
     ((lines 512) (columns 512)) ())
WINDOW-SIZE
```

Then we evaluate the following top-level forms:

```
> (setq a '(1 2 3))
(1 2 3)

> (setq b (make-array '(2 2) :element-type '(signed-byte 32)
                              :initial-contents '((5 6) (7 8))))
#<Simple-Array (SIGNED-BYTE 32) (2 2) 6D3753>

> (setq c (make-instance 'window-size))
#<Instance WINDOW-SIZE 6EBBFB>

> (dump-form (list a b c) "abc")
#P"abc"
```

This would create a file `abc` with the following contents (comments have been included to improve clarity):

```
;-------------------------- FILE abc -------------------------------
#[C WINDOW-SIZE LINES COLUMNS ]                        ;a class template
#[D 4 I WINDOW-SIZE ]                                  ;an instance definition
#[D 3 A (2 2) (:ELEMENT-TYPE (SIGNED-BYTE 32)          ;an array definition
            :ADJUSTABLE NIL :FILL-POINTER NIL) ]
#[D 2 L 3 ]                                            ;a list definition
#[D 1 L 3 ]                                            ;a list definition
#[F 4 512 512 ]                                        ;slot values
#[F 3 5 6 7 8 ]                                        ;array contents
#[F 2 1 2 3 ]                                          ;list contents
#[F 1 #!2 #!3 #!4 ]                                    ;list contents
#!1                                                    ;the returned object
;-------------------------------------------------------------------
```

Loading this file again (into the same LISP environment) would reconstruct a new form that is equivalent to the originally dumped form (with respect to the predicate `equalp`):

```
> (setq d (load-form "abc"))
((1 2 3)
 #<Simple-Array (SIGNED-BYTE 32) (2 2) 703A0B>
 #<Instance WINDOW-SIZE 70302B>)

> (equalp (list a b c) d)
t
```

5.4.4 Implementation

Dumping Forms

Dumping a form is accomplished in 4 steps:

(1) First the form supplied as the first argument to **dump-form** is traversed recursively and all non-atomic objects are collected in a map[1] which relates internal to external references (UERs). Also the class of the instances in the form are collected.

(2) All necessary *class templates* are written to the file.

(3) The *object type definitions* are written to the file.

(4) All *object contents definitions* are written to the file.

(5) The *final form* or an external reference to the final form is written to the file.

Package Considerations

Common LISP provides the *package* concept to support modularity. Each package carries its own symbol table and there is a simple mechanism for inheriting symbols between different packages. Thus, in general, each symbol in the LISP environment is identified by its name *and* the package it resides in. When a form is dumped, we must make sure that the symbols referenced (e.g., class identifiers) are the same when the form is reconstructed. Therefore, **dump-form** operates from within a special

---

[1] This is similar to the *mark phase* in garbage collection schemes.

package that we call the *null* package. This package does not inherit symbols from any other existing package. As a consequence, symbol names in the external representation are always *fully* qualified, such that symbols can always be referenced unambiguously when a form is loaded.

Loading Forms

For loading a form from a file, two special *reader* functions are used. The functions `DEFINITION-READER` and `REFERENCE-READER` respond to the dispatching macro sequences `#[` and `#!` respectively[1]. When either of these character sequences is encountered by the standard **`read`** function, the corresponding reader function is invoked. The special reader then reads successive characters from the input stream (i.e., the file) up to the delimiting character `]`, parses the character sequence, and eventually returns control to the standard reader. The standard read function terminates after the first object (i.e., the final form) has been read from the stream. Therefore the `DEFINITION-READER` returns zero values.

Loading a form is accomplished in 4 steps, reading the file only once:

(1) The *class templates* are are read and memorized.

(2) The *object type definitions* are read and *empty* objects are created. A map is created that relates external to internal references.

(4) The *object contents definitions* are read and the corresponding objects are *filled* appropriately.

(5) The *final form* or a reference to the final form is read and the corresponding internal object is returned.

The *Flavors* environment has been augmented to support the dump-form mechanism. In particular, methods have been defined for `VANILLA` flavor (a superclass to any other *Flavors* class) for

`:class-of` (returns the most specific flavor name of the instance),

`:slot-names` (returns the list of slot names),

`:slot-values` (returns the list of slot values),

`:accept-slots` (sets given slots to specified values).

---

1 CommonLISP explicitly reserves these dispatch macro sequences for the user.

In *CLOS*, similar methods (for the class `standard-object`) either exist or have been defined.

5.4.5 Open Problems

Sometimes the slots of instances may themselves be quite large or complex, e.g., window bitmaps or very large arrays (images). Often we may not want to dump these objects, because they should be replaced when the form is loaded again, such as screen bitmaps. Sometimes we want to keep large objects in separate files because this allows a more compact but non-readable way of storage. This decision must be made by the programmer, since only he knows about these details.

Of course, the programmer must also take care to properly install the correct slot-values again when the instance is re-initialized. This could be done either in a specialized `:init` method or with a separate method like `:init-special-slots`. At the moment we do not have a final strategy for this issue. *CLOS* is certainly more flexible in this respect than *Flavors*.

Efficiency is another area of concern. Since the dump/load-form mechanism uses an external ASCII representation and the built-in LISP `read` function for loading forms, speed is not too impressive. Also, the recursive traversal of the original form during **`dump-form`** could probably be more efficient.

5.4.6 Summary

The way the dump/load-form mechanism works is independent of the particular LISP implementation and thus portable. In fact, a dumped form could even be loaded and used by an appropriate C program, provided that it knows how to handle lists, general arrays, and classes, of course. Thus what we have is a portable way for transporting arbitrarily complex data structures between different LISP sessions and possibly different programs as well. Details can be found in [Burger 89].

Of course, this is far less than a full-blown object-oriented DBMS can offer. However, persistent programming in LISP (and other languages) using OODBMS is just in its beginnings, so we may have to live with poor man's tools like the one described for another while. Currently, the dump/load-form mechanism is used in several applications, e.g., for storing graphic structures in the VKS system (see Section 4.5).

5.5 DIRECT EMBEDDING OF PROCEDURAL CODE IN LISP

5.5.1 Introduction

While programming in LISP offers many advantages, it is sometimes appropriate or necessary to revert to other languages, in particular procedural languages such as C or *FORTRAN*. In this section, we look at the problem of using LISP and *procedural* (*"foreign"*) code together, without the need to sacrifice the key benefits of LISP. Among the reasons for using foreign code are:

(1) Use of existing code without recoding in LISP (e.g. certain FORTRAN subroutine libraries).

(2) Superior speed. Procedural languages still have an advantage over compiled LISP in terms of execution speed.

(3) Access to systems resources that are not directly available from LISP (e.g., Input/Output libraries, graphic routines, window management, networking).

Assuming LISP as our main programming environment, there are, in principle, two different ways to employ code written in other languages. *First,* one could write complete *programs* in the foreign language and execute them (in separate processes) from within the LISP environment. Data exchange with these programs could be accomplished through files or inter-process communication. The *second* way to interface LISP with foreign code is at the *procedure* (or subroutine) level, i.e., a foreign procedure is called from LISP similar to a real LISP function. In this case, parameters and return values are used to transport data. Modern LISP implementations (e.g., *Lucid* Common LISP [Apollo 90a]) support *both* methods for using foreign code.

5.5.2 Conventional Use of Procedural Code

In the following, we focus upon the use of foreign code at the procedure level, i.e., by calling *foreign procedures* from within the LISP environment. The usual sequence of steps for creating a foreign function (or a set of foreign functions collected in a file) and the corresponding LISP interface are the following (see Figure 5.5.1):

1. Define the LISP interface description (referred to as the "LISP socket") for each foreign procedure, including the list of parameters, the parameter types, return values, and the name of the procedure. This is done with the LISP-form `def-foreign-function`[1].
2. Write the foreign procedure by editing the foreign source file. This includes specifying the parameters again in the foreign language.
3. Compile the foreign source file with the corresponding language compiler, thus creating the foreign object file.
4. Dynamically link the foreign object file to the current LISP environment (using the form `load-foreign-file`).
5. Use the foreign procedure like any ordinary LISP-function.

The details of this process shall be illustrated in the following, simple example. The task is to create a C-function with a single parameter (a floating point number) which the procedure shall print:

1. Define the LISP interface socket for the foreign function:

```
LISP> (def-foreign-function put_it_out (lpar :signed-16bit))
PUT_IT_OUT
```

2. Create a file with the C-function named "output.c"

```
UNIX> edit output.c
```

```
/* File: output.c     Language: C */
#include <stdio.h>
void put_it_out (short cpar);
    { printf ("%d\n",cpar); }
```

3. Compile the File "output.c" to the foreign object file "output.bin"

```
UNIX> /bin/cc output -O output.bin
No errors, no warnings, 1 info msg, C compiler 68K Rev 6.8(168)
```

4. Link the procedural code to the LISP-environment:

```
LISP> (load-foreign-file "output.bin")
#P"output.bin"
```

[1] The current Common LISP standard does not contain a specification of a foreign language interface, i.e., any implementation must be in the form of *extensions* to Common LISP. In the following, any specific interface syntax refers to the *Lucid* Common LISP implementation (Version 4.0).

Now the foreign function is completely defined.

5. Call the foreign function

```
LISP> (put_it_out 10)
10
```

While the mechanism itself is quite flexible and powerful, this kind of software development is also extremely inconvenient and error prone. In addition to the complexity of this process, its disadvantages are the following:

- The parameters of the foreign function have to be specified twice. *First* in the declaration part of the foreign function (Step 1) and a *second* time when the interface socket is defined in LISP (Step 2). This means double work and provokes all sorts of programming errors.
- The programmer has to take care of compiling the foreign source file manually (Step 3). In general he needs a separate shell to develop the foreign source code (edit, compile etc.).
- The programmer has to take care of linking the foreign object file after editing and compiling (Step 4). Otherwise, changes to the foreign code will not be put into effect.
- The need to edit at least two files (the LISP file and the foreign source file) that semantically belong together.

Frequent errors are:

- The type mapping of two parameters is not correct because of different C- and LISP-interface definitions.
- Working with old code by simply forgetting to compile and/or link the new procedural code.
- The procedural code cannot be called because of a wrong and/or multiple procedure name.

5.5.3 A Foreign Language Toolkit (FLT)

Motivated by the deficiencies of the standard procedure for using foreign code described in the previous sections, we have developed a Foreign Language Toolkit (FLT) as an extension to the foreign language interface provided in Lucid Common LISP [Burger, Mayer 92]. The main improvements for the user (i.e., the programmer) are:

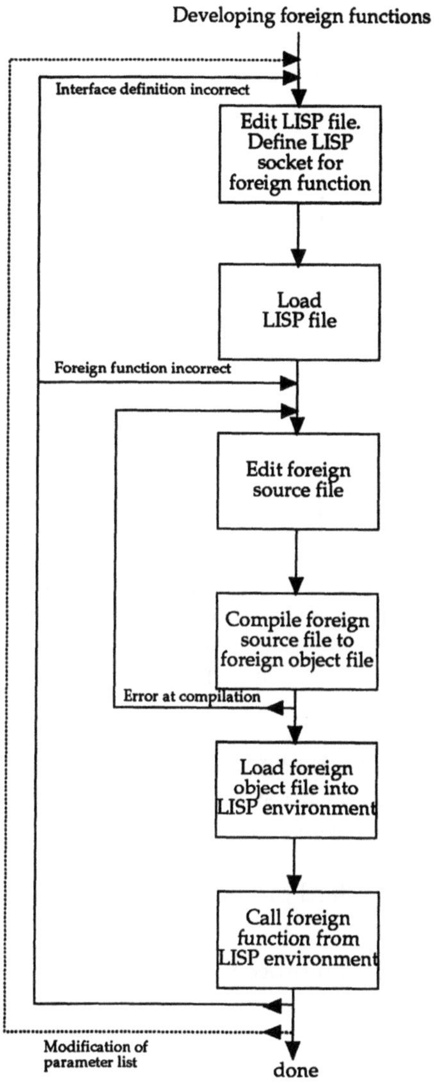

Figure 5.5.1 Usual sequence for creating foreign procedures that can be called from within LISP like ordinary LISP functions.

- It is possible to define the interface socket only once in a language-independent syntax. The language-dependent definitions are generated from there.
- Code that belongs together (LISP and procedural) can be stored and processed together in a single file. The foreign code is included directly in the LISP file.
- The compilation and linking of the foreign code has been automated.

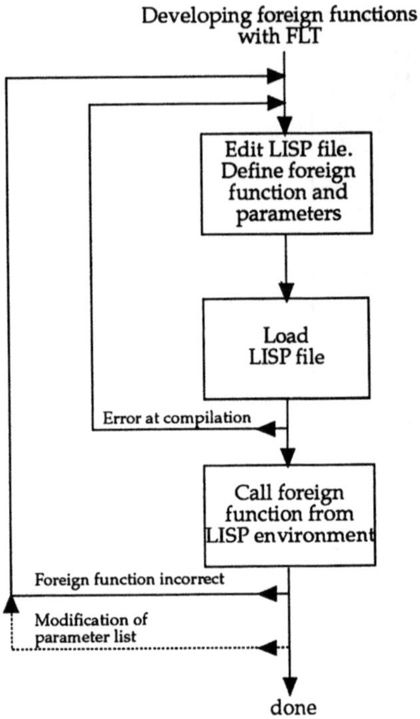

Figure 5.5.2 Sequence for creating foreign procedures with the Foreign Language Toolkit (FLT).

The sequence of steps necessary to define a foreign procedure (or a set of foreign procedures) has been greatly simplified, as illustrated in Figure 5.5.2. Thereby, several potential sources of errors have been removed.

To compare the new interface with the standard procedure for defining foreign procedures, we revert to our earlier example:

1. Create a LISP-file containing the foreign procedure definition `def-f-function` named "output.LISP":

```
;;; File: output.LISP     Language: LISP
(def-f-function (put_it_out (:language :C)
     ("par" :signed-16bit)
     (:include "<stdio.h>"))
  #{ printf ("%d\n",par); }#      ; this is the C source
  )
```

2. Load the file "output.LISP" into the LISP-environment:

```
LISP> (load "output")
;;; Loading source file "output.LISP"
```

```
;;; DEF-F-FUNCTION: Checking output.lbin
;;; DEF-F-FUNCTION: Creating output.lc
;;; DEF-F-FUNCTION: Compiling output.lc to output.cbin
No errors, no warnings, 1 info msg, C compiler 68K Rev 6.8(168)
//hardy/user_data/mm/test/output.LISP loaded. Contains foreign code
#P//hardy/user_data/mm/test/output.LISP
```

3. Call the foreign function:

```
LISP> (put_it_out 10)
10
```

Direct Embedding of Foreign Source Code

The main vehicle of the FLT is the macro `def-f-function`, which facilitates the both the LISP and the foreign components of a foreign procedure. The foreign code itself (a string) is handed to the macro as the last argument. The LISP reader macros `#{` and `}#` have been provided as string delimiters to allow unchanged embedding of foreign code that contains strings itself. Code between these two strings is converted by a special reader into a string (see examples in Section 5.5.6). There is no need to write a procedure head in the foreign language any longer, because it is created from the specifications given to `def-f-function`. Other arguments to the `def-f-function` macro specify

- the language of the procedural code,
- local INCLUDE-files, which are only valid for that foreign function,
- the name of the foreign function.

Generation of Source-Files

The FLT constructs the head of the foreign procedure from the arguments of `def-f-function`. The LISP data types are converted to foreign data types according to a user-definable mapping rule. Requests to include other foreign source files are translated to corresponding preprocessor instructions.

Compilation Using Foreign Compilers

The generated source text of the foreign procedure head and the pre-processed procedural code itself is then written to a separate file. Then the foreign compiler is invoked from LISP (as a separate program) depending upon which foreign language (and thus which compiler) is given by the `:language` argument to `def-f-function`. Since the interface operates on the procedural level, the foreign compiler must handle code sections without a "main" program.

Linking Compiled Files

After compilation, the generated object file is linked to the LISP-Environment with the LISP form `load-foreign-file`. The foreign object file must be relocatable in order to be linked dynamically to the LISP environment.

Generation of the Interface Socket and Type Mapping

The interface socket is the interface between the LISP environment and the foreign procedure. It is usually specified with the `def-foreign-function` form available in the underlying LISP implementation (see Section 5.5.2). The syntax of the new FLT macro `def-f-function` is very similar to the original `def-foreign-function` in order to allow an almost direct mapping between the two.

Similarly, the type mapping mechanism relies upon existing concepts. In the original LISP implementation, foreign type definitions are necessary to create a correct parameter vector for each foreign call. Primarily, this requires information about the size and the alignment assumed by the foreign procedure, and how conversions can take place between LISP types and foreign types.

The type mapping mechanism in the FLT, mainly visible through the macro `def-c-type`, adds the capability to infer a foreign language type definition from the corresponding LISP type. E.g., for the LISP type `:unsigned-8bit` would correspond to the type definition "`unsigned char`" in the C language. In addition to these primitive types, the mapping mechanism also handles complex type specifications, e.g. the complex LISP type "`(:pointer :character)`" would be translated to "`char *`" for a C procedure.

Inline Expansion of LISP-Expressions

The FLT allows to insert arbitrary LISP expression into the foreign source code. Each form following a @-character is taken as LISP form and evaluated by the LISP-Reader before the procedural code is compiled. The string *@expression* is replaced by the result of evaluating *expression* in the current LISP environment. This is similar to the preprocessor instruction `#define` in C (see examples in Section 5.5.6).

Dynamics

There are four different situation when LISP has to handle procedural code:

1. A foreign procedure is defined by evaluating `def-f-function` through the LISP interface (the so-called LISP *listener*). Each time `def-f-function` is evaluated, the generated procedural code is written to a temporary source file, compiled to a temporary object file, which is subsequently linked.
2. A LISP file containing procedural code is loaded. In this case, all the foreign functions included in one file are gathered into a single foreign source file, which has the same name as the LISP-file but with another extension (e.g., "foo.lc" for the C code contained in the LISP file "foo.lisp"). The foreign files are preserved after compilation, so that the generated source-code can be inspected. Additionally, the corresponding LISP socket is included as a comment with each foreign function for easier debugging. When the same LISP-file is loaded again, the FLT checks if the corresponding foreign object-file is younger than the LISP-file. In this case, only the LISP-sockets are redefined and the existing foreign object file is linked.
3. A LISP-file containing procedural code is compiled by the LISP compiler. LISP-files which include foreign functions can be compiled as usual with `compile-file`. The contained procedural code does not interfere with the LISP compiler and is fully contained in the compiled file.
4. A compiled LISP-file (FASL-file) containing procedural code is loaded. When loading such file, the FLT reacts in the same way as when loading a LISP source file with foreign code.

5.5.4 FLT Reference

def-c-type *LISP-type c-type* [*Macro*]

The macro `def-c-type` associates a LISP-type with the equivalent C-type. *LISP-type* is a symbol, *c-type* a symbol or a string which contains a valid C-type or a combination of LISP-types and a C-type. It returns the list (*LISP-type . c-type*) if successful, *nil* otherwise. Here are two examples for the use of `def-c-type`:

```
LISP> (def-c-type test "struct c_test {char stru_test;}")
(TEST . "struct c_test {char stru_test;}")
LISP> (def-c-type :string (:pointer "char"))
(:STRING :POINTER "char")
```

def-f-function *name-and-options* [*Macro*]
*{arg-description}**
procedural-code-string

The macro def-f-function creates a foreign procedure and makes it accessible to LISP. *name-and-options* is a symbol or a list. If it is a symbol, it specifies only the function-name. The list contains the function-name and several options like local include-files etc. *arg-description* is a list containing the parameter-declaration(s). *procedural-code-string* is a string which contains the procedural code. Usually, the *procedural-code-string* is enclosed within the dispatching macro sequences #{ and }#. def-f-function returns the function symbol if the definition was successful, *nil* otherwise.

Example:

```
LISP> (def-f-function (add-one (:name "add_one_proc")
                        (:language :C)
                        (:include "<stdio.h>")
                        (:return-type :signed-16bit))
              ("par1" :signed-16bit))   ;a single parameter
   "/* this is C-Code now */ return (1+par1);")

No errors, no warnings, 1 info msg, C compiler 68K Rev 6.8
ADD-ONE
```

def-aux-code *procedural-code-string* [*Macro*]

The function def-aux-code defines auxiliary procedural code to be included in the corresponding foreign source file at the specified position. The *procedural-code-string* is a string which contains the foreign code. This is useful for defining data types, global variables, shared procedures (that are not accessible from LISP), etc., in the foreign language. The *procedural-code-string* may also contain LISP expressions marked with @. Example:

```
LISP> (def-aux-code "struct header {unsigned short rows; unsigned short cols;};")
```

default-c-include {*include-file*}* [*Function*]

The function `default-c-include` defines one or more files to be included into the corresponding C-file. *include-file* is a string which contains a valid argument for a `#include` statement in C. The result is a list of the filenames. `default-c-include` has to appear before the first `def-f-function` in a LISP-file. The `#include` statement(s) will be inserted at the top of the C-file. The purpose is to avoid multiple declarations of often used "includes", e.g.

```
LISP> (default-c-include "<stdio.h>" "<float.h>")
("<stdio.h>" "<float.h>")
```

5.5.5 Examples

Example 1

In the simplest case, the foreign function definition is directly fed into the LISP-evaluator. A temporary foreign source file is created and deleted after compilation:

```
LISP> (def-f-function (abc) #{printf ("hello world\n");}# )

;;; DEF-F-FUNCTION: Creating /tmp/temp.lc
;;; DEF-F-FUNCTION: Compiling /tmp/temp.lc to /tmp/temp.cbin
No errors, no warnings, no info msgs, C compiler 68K Rev 6.8(168)
ABC
```

Now call the new function:

```
LISP> (abc)
hello world
0
```

Example 2

Now we load a LISP-File which contains more than one `def-f-function` definition. The file "example2.lisp" has the following contents:

```
;;; File: example2.lisp

(in-package 'user)
(defvar *za* 10)

(default-c-include "<stdio.h>")

(def-f-function (c_fun (:name "the_c_fun")
                       (:include "<float.h>")
                       (:return-type :signed-16bit))
        ("par1" :signed-16bit)
        ("arr" (:array :single-float (3 5)))
        (par2 :signed-16bit)
#{int i,j;
  for (i=0; i<3; i++)
    for (j=0; j<5; j++) { printf("%5.2f\n",arr[i][j]); }
  return (par1 + PAR2);
  }# )

(def-aux-code
#{
void abc()
{printf ("hello world\n");}
}# )

(def-f-function (inp_test (:include "<limits.h>"))
#{int x,y;
  printf("Enter an integer -- ");
  scanf("%d", &x);
  y = x * @(* (/ *za* 10) 2) ;
  printf("\n%d ist twice %d\n", y, x);
  abc();
  }# )
```

When the LISP file is loaded "example2.lisp, a new C source file named "example2.lc" is created:

```
LISP> (load "example2.lisp")

;;; Loading source file "example2.lisp"
;;; DEF-F-FUNCTION: Checking example2.lbin
;;; DEF-F-FUNCTION: Creating example2.lc
;;; DEF-F-FUNCTION: Compiling example2.lc to example2.cbin
```

```
No errors, no warnings, 1 info msg, C compiler 68K Rev 6.8(168)
//hardy/user_data/mm/test/example2.lisp loaded. Contains foreign code.
#P//hardy/user_data/mm/test/example2.lisp
```

Now we test the new foreign functions. Please note that the function abc can only be called from other function in the file but not from LISP.

```
LISP> (setq arr (make-array '(3 5)
        :element-type 'single-float
        :initial-contents '((1.2 2.3 3.4 4.5 5.6)
                            (1.0 2.0 3.0 4.0 5.0)
                            (9.9 8.8 7.7 6.6 5.5))))
#<Simple-Array SINGLE-FLOAT (3 5) 9EA8A3>
LISP> (c_fun 1 arr 2)
 1.20
 2.30
  ...
 5.50
3
LISP> (inp_test)
Enter an integer -- 12

24 ist twice 12
hello world
0
LISP>
```

The second time the file "example2.lisp" is loaded, the FLT recognizes that the corresponding file "example2.cbin" is newer than "example2.lisp" and only reloads the compiled foreign code.The printing of info-messages can be suppressed by setting the global variable **verbose-foreign-load** to *nil.*

```
LISP> (load "example")
;;; Loading source file "example2.lisp"
;;; DEF-F-FUNCTION: Checking example2.bin
;;; DEF-F-FUNCTION: C-File newer than LISP-File. Reloading foreign file
;;; DEF-F-FUNCTION: Loading example2.cbin
;;; Warning: Reloading foreign file "example2.cbin"
//hardy/user_data/mm/test/example2.lisp loaded. Contains foreign files
#P//hardy/user_data/mm/test/example2.lisp
```

Figure 5.5.3 shows the contents of the generated C-File "example2.lc". Notice that the expression following the @-character in the file "example2.lisp" (procedure INP_TEST) was evaluated and substituted during the generation of the file "example2.lc".

```
/* File: example2.lc   Language: C  */

/* Name of LISP-Function: C_FUN */
/* LISP-Definition:
 (DEF-FOREIGN-FUNCTION (C_FUN (:NAME "the_c_fun") (:RETURN-TYPE :SIGNED-16BIT)) (PAR1
:SIGNED-16BIT) (ARR (:ARRAY :SINGLE-FLOAT (3 5))) (PAR2 :SIGNED-16BIT)) */
#include <stdio.h>
#include <float.h>
short the_c_fun (short par1, float arr [3][5], short PAR2)
{
int i,j;
for (i=0; i<3; i++)
  for (j=0; j<5; j++) { printf("%5.2f\n",arr[i][j]); }
return (par1 + PAR2);
}

void abc ()
{printf ("hello world\n");}

/* Name of LISP-Function: INP_TEST */
/* LISP-Definition:
 (DEF-FOREIGN-FUNCTION (INP_TEST (:NAME "INP_TEST"))) */
#include <limits.h>
void INP_TEST (void)
{
int x,y;
printf("Enter an integer -- ");
scanf("%d", &x);
y = x * 2 ;
printf("\n%d ist twice %d\n", y, x);
abc();
}
```

Figure 5.5.3 Contents of the file "example2.lc" created by processing the LISP file "example2.lisp".

5.5.6 Summary

The *Foreign Language Toolkit* has been designed to ease the use of foreign, procedural code segments in conjunction with Common LISP. In summary, its features are the following:

- It is possible to integrate procedural source code directly into the LISP source. Therefore it is no longer necessary to manipulate multiple files (the LISP-file and the procedural code-file).
- The parameters of foreign procedures have to be specified only once in LISP. Both the effective LISP interface and the foreign procedure head are derived from a single specification.
- The automatic generation of the procedure-head of a foreign function and the automatic definition of the interface socket reduce the redundancy and the chance of errors when foreign functions are used. Changes are made easier too.
- It is possible to define new data-types for the parameters in an easy way.
- Procedural code can be defined via the LISP evaluator or contained in LISP source files.
- Compilation of LISP files that contain foreign code is completely transparent. The LISP functions `load` and `compile-file` operate in the usual manner.
- Arbitrary LISP expressions can be inserted into the procedural code.

The FLT has been used extensively within the Vision Kernel System (VKS, see Chapter 4) and other CAST software, primarily for performance reasons and for procedures at the systems level. Also some existing image processing software [Khoros 1990; Everson 89] has been made accessible in VKS through the use of the FLT.

REFERENCES – CHAPTER 5

[Apollo 90a]
Domain/CommonLISP Advanced Users Guide, Hewlett Packard, Order No. 015738-A01, 1990.

[Apollo 90b]
Domain/CommonLISP Tool Reference (Volume 2), Hewlett Packard, Order No. 017146-A00, 1990.

[Atkinson et al. 89]
M. Atkinson, F. Bancilhon, D. DeWitt, K. Dittrich, D. Maier, S. Zdonik, "The Object-Oriented Database System Manifesto," *Proceedings of the First DOOD Conference*, Kyoto, pp. 40–56, Japan, 1989.

[Bobrow, Stefik 83]
D. G. Bobrow, M. Stefik 1983, *The LOOPS Manual*, Xerox PARC, Palo Alto, CA, 1983.

[Bobrow et al. 88]
D. G. Bobrow, L. G. DeMichiel, R. P. Gabriel, S. E. Keene, G. Kiczales, R. P. Moon, "The Common LISP Object System Specification (X3J13-88-002)", American National Standard Institute, 1988.

[Budde et al. 84]
R. Budde, K. Kuhlenkamp, I. Mathiassen, *Approaches to Prototyping*, Springer Verlag, Berlin, 1984.

[Burger 89]
W. Burger, "A Proposal for Creating External Environments in Common LISP," Technical Report TR89-1, Johannes Kepler University, Dept. of Systems Science, Linz, Austria, April 1989.

[Burger, Mayer 92]
W. Burger. M. Mayer, "Enhancements to the Interface between LISP and Procedural Languages," Technical Report TR92-1, Johannes Kepler University, Dept. of Systems Science, Linz, Austria, Jan. 1992.

[Connel, Shafer 89]
J. L. Connel, L. B. Shafer, *Structured Rapid Prototyping*, Yourdon Press, Prentice Hall, 1989.

[Everson 89]
ALV Public Domain Image Processing Toolkit for Sun Workstations, Technical Report, Computer Science Dept., Bristol University, UK, 1989.

[Goldberg 84]
A. Goldberg, *Smalltalk-8 – The interactive programming environment*, Addison-Wesley, 1984.

[Itasca 90]
Itasca Systems Inc., *Itasca Distributed Object-oriented Database Management Systems*, Technical summary, Itasca Systems Inc., Minneapolis, MN, 1990.

[Keene 89]
S. E. Keene, *Object-Oriented Programming in Common LISP*, Addison Wesley, 1989.

[Khoros 90]
Khoros Manuals, The Vision Lab, Dept. of EECE, Univ. of New Mexico, Albuquerque, NM, 1990.

[Kim 90]
W. Kim, "Object-Oriented Databases: Definition and Research Directions," *IEEE Transaction on Knowledge and Data Engineering*, Vol. 2, No. 3, pp. 327–341, 1990.

[Kim et al. 90]
W. Kim, J. F. Garza, N. Balou, D. Woelk, "Architecture of ORION Next-Generation Database System," *IEEE Transaction on Knowledge and Data Engineering*, Vol. 2, No. 1, pp. 109–124, 1990.

[Kimbrough et al. 89]
K. Kimbrough, O. LaMott, "Common LISP User Interface Environment," Texas Instrument Incorporated, 1989.

[Meyer 88]
B. Meyer, *Object Oriented Software Construction*, Prentice Hall, New York, 1988.

[Mittelmann 88]
R. Mittelmann, "Object-oriented Implementation of Petri Nets Concepts," *Cybernetics and Systems Research Conference '88* (R. Trappl ed.) Kluwer Academic Publishing, pp. 759–766, 1988.

[Murauer 89]
J. Murauer, "The InterLISP-D/LOOPS Implementation of the System Entity Structure," (in German), Internal Memo, University of Linz, Austria, 1989.

[Müller-W. 86]
T. Müller-Wipperfürth, *LISAS – A LOOPS-Implementation of a Systolic Array Simulator*, Master Thesis (in German), Univ. Linz, Dept. of Systems Science, 1986.

[Pichler 88]
F. Pichler, "CAST–Computer Aided Systems Theory : A Framework for Interactive Method Banks," in: *Cybernetics and Systems '88* (ed. Trappl, R.), Kluwer Academic Publishers, pp. 731–736, 1988.

[Pichler, Praehofer 88]
F. Pichler, H. Praehofer, "CAST.FSM – Computer Aided Systems Theory: Finite State Machines," in: *Cybernetics and Systems '88* (R.Trappl ed.), Kluwer Academic Publishers, pp. 737–742, 1988.

[Praehofer, Spalt 88]
H. Praehofer and A. Spalt "An Interactive Simulation Environment Based on Systems Theory Concepts and Object Oriented Programming," *Proc. of the SCS European Simulation Multiconference '88*, SCS Press, Nice, France, pp. 199–203, 1988.

[Praehofer 88]
H. Praehofer, "STIDEVS – Systems Theory Discrete Event Simulation," in: *Cybernetics and Systems '88* (R.Trappl ed.), Kluwer Academic Publishers, 1988.

[Praehofer 89]
H. Praehofer, "A Kernel System for Modelling and Simulation," *Proc. of the 1989 SCS Eastern Simulation Multiconference*, Tampa, Florida, 1989.

[Scheifler et al. 88]
Scheifler, Robert W. et al. , "CLX – Common LISP Language X Interface," Version 11, Rel. 3, Texas Instrument Incorporated, 1988.

[Steele 84]
Steele, Guy, *Common LISP: The Language*, Digital Press, Burlington, 1984.

[Stritzinger 91]
A. Stritzinger, *Reusable Software Components and Application Frameworks*, PhD Thesis, University of Linz, Austria, 1991.

[Weinand et al. 88]
A. Weinand, E. Gamma, R. Marty, "ET++ – An Object-Oriented Application Framework in C++," *OOPSLA '88 Proceedings*, San Diego, CA, 1988.

[Xerox 83]
Xerox, *InterLISP Reference Manual*, Xerox Corporation, Palo Alto, CA, October 1983.

[Zeigler 84]
B. P. Zeigler, *Multifacetted Modelling and Discrete Event Simulation*, Academic Press, London, 1984.

[Zeigler 87]
B. P. Zeigler, "Hierarchical, Modular Discrete Event Simulation in an Object Oriented Environment," *Simulation Journal*, Vol. 49:5, pp. 219–230, 1987.

APPENDIX: CAST.FSM USER'S GUIDE

A.1 CAST.FSM BACKGROUND

CAST.FSM deals with finite state machines, Boolean functions, switching circuits, linear machines, Petri nets, and related topics. Application areas of CAST.FSM are the design of digital circuits, design for testability, cryptological analysis, Systems Theory education, and others [Praehofer 86]. The principal idea of CAST implementations is to provide the user with a sort of systems-theoretic calculator for easy and explorative manipulation of mathematical systems.

A systems problem (Figure A.1), stated as a set of initial systems, a goal system, and a set of constraints, is solved in a CAST session in the following way:

The user of the system, e.g. an engineer designing a digital circuit, is provided with the graphical representation of the implemented system types (Figure A.2).

The user selects a system type suitable for modelling his problem, creates an incarnation and fills in the values. Then he chooses an appropriate system transformation, analyzes the result, transforms it further, and so on, until the constructed realization tree shows an acceptable solution of the problem. In CAST.FSM, all this is as easy as moving the mouse and clicking in the graphic representations of the various objects. The keyboard of the workstation is used only for typing names of created objects or naming elements of sets.

A simple example would be the construction of a gate level circuit realization of a given finite state machine with some constraints. First the user selects "Create Instance" from the pop-up menu of the class Deterministic Sequential Machine (DetSeqMach) in the class browser (Figure A.2) by clicking on it with the pointing device (mouse). In the appearing window the input, state and output sets have to be defined by typing in the elements or by giving a LISP expression yielding a set. Then the drawn tables (Figure A.3) of the state transition function and the output function can be filled using the mouse to select from pop-up value menus.

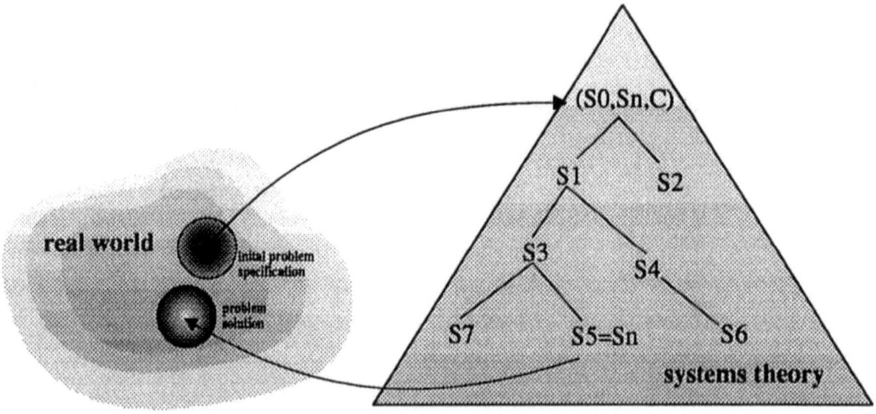

Figure A.1 Real world and systems problem solving.

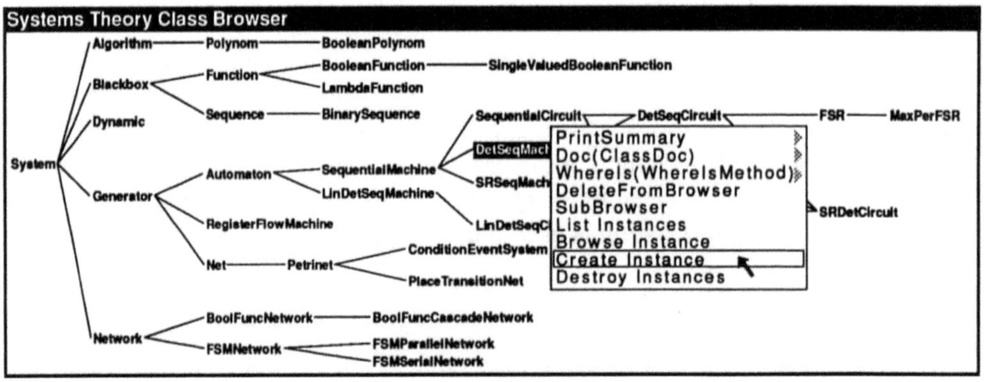

Figure A.2 The CAST class browser.

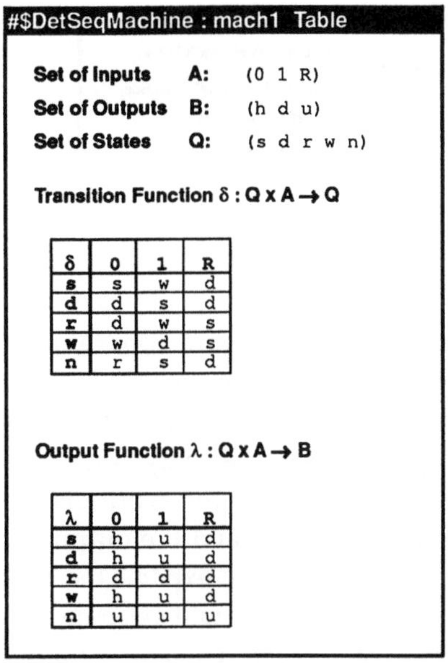

δ	0	1	R
s	s	w	d
d	d	s	d
r	d	w	s
w	w	d	s
n	r	s	d

λ	0	1	R
s	h	u	d
d	h	u	d
r	d	d	d
w	h	u	d
n	u	u	u

Figure A.3 Table definition.

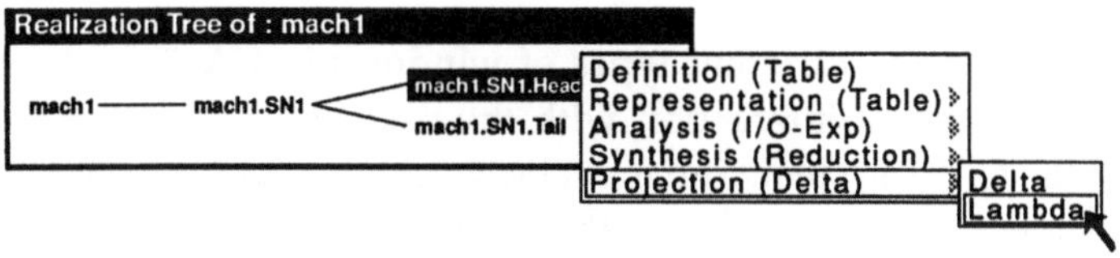

Figure A.4 CAST instance browser.

When the definition is complete, an instance browser, called *realization tree* in this context, appears on the screen. It contains the name of the created machine object. Clicking on this node shows a menu containing five groups of applicable methods (Figure A.4).

The first group contains a redefine command, the second methods for representation (graphical, table oriented, block diagram, ...), the third contains analysis methods, and the fourth group contains synthesis methods for transforming the system into another type. The last group consists of projections on subobjects.

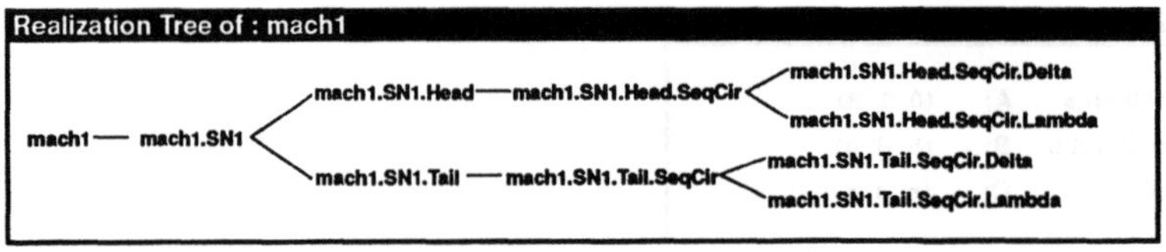

Figure A.5 Realization tree.

The submenu in each group shows only those operations which are applicable to the clicked system object. In the case of a DetSeqMach instance there would be (among others) table and state diagram representation methods, methods for computation of losslessness, the lattice of congruences, testability features in the analysis group, and state reduction, decomposition, circuit and linear realizations in the synthesis group. There would be projections on the state transition function δ and the output function λ. Application of a synthesis operation or a projection will expand the realization tree shown in the instance browser. The new realization will become a child node with a standard name, consisting of the original name and an extension which is a mnemonic for the applied realization method. This new object has to be refined into more specific objects in further steps until the leave nodes of the generated realization tree fulfill the problem requirements (Figure A.5).

Although there are many different types of windows in CAST.FSM, the browsers with their unique kind of interaction with the user build the main user interface component of the CAST system. The main browser types used in CAST.FSM are the *class browser*, shown in Figure A.2, and the *instance browser* or *realization tree*, shown in Figure A.5.

A.2 SYSTEM REQUIREMENTS

A.2.1 Hardware

CAST.FSM runs on Dandelion workstations. Dandelions from XEROX Corp. are named 1108, those from SIEMENS AG EMS 5815. CAST.FSM should run on the successor machines (XEROX 1186 or SIEMENS 5822) also. CAST.FSM needs at least 1.5 MByte main memory and a 42 MByte hard disk. An Ethernet connection to a NS file service is strongly recommended for storing font files, *sysouts* (LISP images) and source files.

A.2.2 Software

CAST.FSM needs InterLISP-D and LOOPS. The latest InterLISP-D version used for development is "Intermezzo", the LOOPS version is "Buttress". CAST.FSM will not run smoothly with different InterLISP or LOOPS versions, because of several upwards and downwards incompatibilities between older and newer versions of the XEROX software environment. One exception is the InterLISP-D "Koto" release. CAST.FSM is expected to be easily adaptable to work with this release by the CAST user.

It should be emphasized that CAST.FSM is not a product. It is a prototype only and has to be used with patience and care. The user has to live with some flaws and errors (also see Section A.8).

A.3 CAST.FSM INSTALLATION

To install CAST.FSM on a Dandelion, one has first to install and boot a fresh InterLISP/LOOPS sysout (refer to the Dandelion User's guide). CAST comes on floppies, and one floppy contains a TEdit document named "CASTReadMe.TEdit" describing the further installation procedure in detail. The TEdit document may be opened by selecting "TEdit" from the right button background menu, and then filling in the filename in the "Get" Option in the TEdit command menu.

A.4 STARTING A CAST SESSION AND BROWSING

Once CAST is installed and saved as a CAST-sysout preferably on a file server, a CAST session can be started by right-button clicking in the CAST icon. If the kernel software has to be loaded, the opening of the class browser takes some time. When the browser is finally opened, the user has access to the full LOOPS class browser functionality.

A.4.1 The CAST Class Browser

Each (class) browser offers the usual InterLISP-D window commands in the right button menu anywhere in the window. The general browsing methods are located in the title menu. One has to click (left or middle) in the browser's title bar for these commands. Finally, on each node shown in the browser, there is a left button menu and a middle button menu. The left button menu contains commands for displaying information about the class in various forms. The CAST browser offers methods dealing with incarnations of the class in this menu as well. The middle button menu consists of all methods needed to modify the class system, i.e. editing commands for system developers.

A.4.2 The CAST Instance Browser

In a usual CAST session, the user will select a system type (left button) and pick the "Create Instance" command from the menu. After typing a new name for the system in the browser's own prompt window, the user is asked to place an instance browser on the screen. The instance browser is called "realization tree" in the CAST context. It offers similar right button window commands and general browsing title menu items like the class browser.

The *left button menu* for a node in an instance browser consists of several groups of commands which are specific to the selected object (i.e. specific to the objects system type or class). For the most important systems there are the following groups:

Definition Methods for defining or redefining the system's components in one or more editing styles, such as table editing, graphical editing, and others.

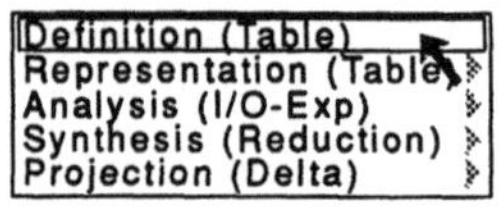

Representation The selected system may be displayed in several ways on the screen. The representation windows may also be printed.

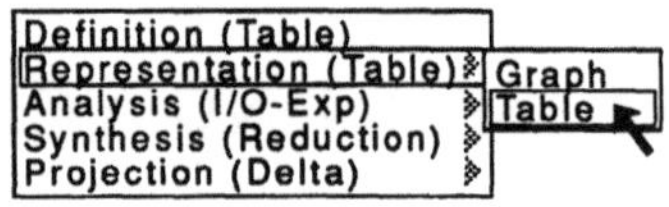

Analysis In this group of commands there are methods for computing some mathematical properties as well as experimental re-

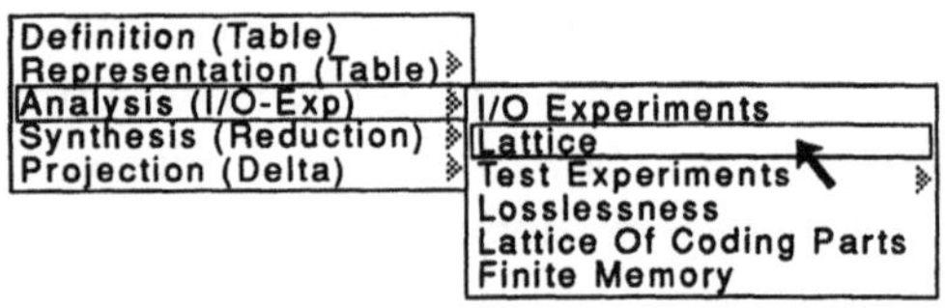

sults of the selected system. In most cases the results will be displayed in special windows.

Synthesis Those system transformations which yield a new realization of the selected system are collected in the synthesis group. One example would be the state reduction of a finite state machine. When the computation of the reduced machine is finished, it will be named by extending the name of the original system with an appropriate extension and a serial number. In the browser, the new system is shown as a child node of the original selected system. If in turn this new node is expanded by selecting a synthesis operation for it, the graph grows further. In the general case the resulting graph has a tree structure, hence the name realization tree. The leaves of the tree represent the most detailed description of the realized system.

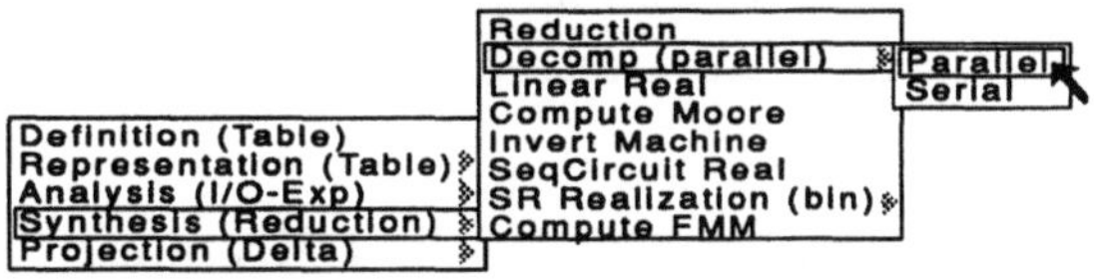

Projection The projection group of commands allows to break down the selected system into its parts. Currently, only one part of the selected system is shown exclusively, which is not conform to the idea of realization trees, which has to be fixed in the future.

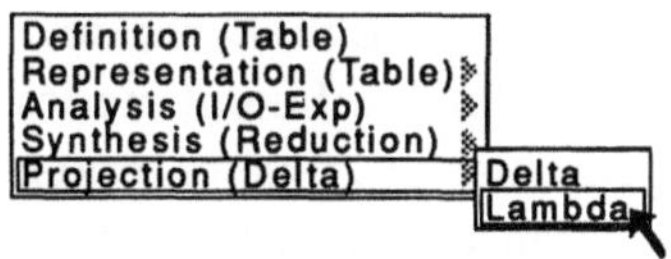

Miscellaneous For some system types, other commands may be available.

The ***middle button menu*** is the same for all system types and allows switching between different realization subtrees, destroying of computed systems together with their subtrees, and low level access to the LOOPS objects via the InterLISP-D inspector and the DEdit LISP structure editor.

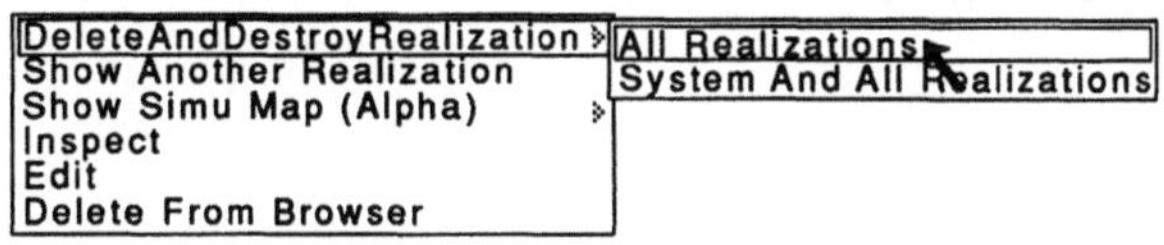

A.4.3 Defining a System

Dependent on the type of the object to be created, a window will appear in which the new system can be edited conveniently. This function is called "define" in CAST.FSM. Some system types have no "define" method yet. In this case, if one wishes to create the system anyway, one has to select the "Inspect" command from the middle button menu in the instance browser on the name of the object. In the InterLISP-D inspector all relevant values have to be set with great care, because in this case there are *no consistence checks* made at all.

A.4.4 Petri Nets Make the Difference

Because the net theory branch of CAST.FSM was developed in parallel to the first version of the original CAST system, systems of a Petri net type are handled a bit different in most situations. Instance creation and definition via the CAST class browser brings up a small "TTYInEdit"-window (refer to the InterLISP-D documentation) in which the user may edit the basic sets defining the net in sequence. For convenience, the user is provided with a default value for each set, which can be accepted by hitting the return key or edited arbitrarily. In the instance browser, one should firstly select the command "new picture" from the "Representation" group of commands. This will show the net graphically in a new window called "Petri net instance browser". All further editing, analysis and synthesis operations are performed in that special Petri net window. On the border and title bar the right button menu is the usual InterLISP-D window menu. Inside, the right button menu is used to switch between three different modes: The first is for modifying the layout of the graphical representation, where nodes may be moved with the mouse to change the appearance of the representation. The second is for graphically editing the structure of the net (adding or deleting nodes or arcs), and the third mode is for playing the token game. Most operations in these modes are performed using the left mouse button. For each mode, a short instruction (in German) is displayed in the InterLISP prompt window.

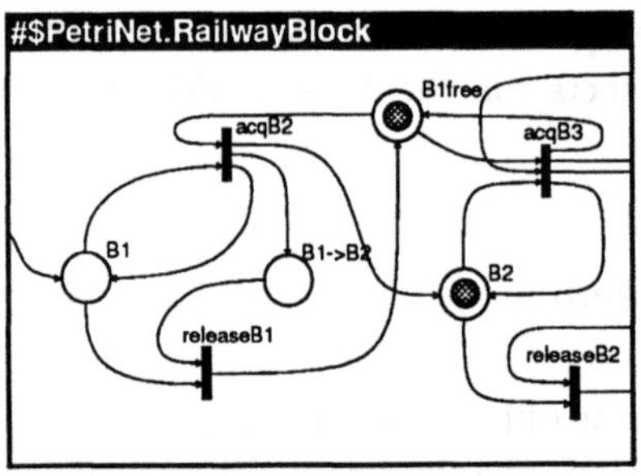

Figure A.6 The Petri Net Instance Browser.

A.5 TABLE EDITING

The systems of the most important system types are edited in a table representation. Usually a system type is mathematically defined as a tupel consisting of some sets and some functions or relations. In the table definition window one firstly has to select the "define set" commands for those sets from the right button menu. The window then contains a default LISP expression (QUOTE NIL) and the caret in the corresponding line. (In InterLISP the blinking cursor, which shows where typed characters will go, is named caret, while the pointer moved by the mouse is named cursor.) Normally one would select the NIL and replace it with a list of elements. In many cases it is more convenient to select and replace the whole expression by typing in an arbitrary InterLISP form (name of a variable or a function call) yielding the desired set when evaluated. Two useful functions supplied in CAST for this purpose are GF and GF2 for computing the elements of a Galois field. These functions are implemented as follows:

GF prim n *[function]*

returns the list of elements of the Galois field of order $prim^n$ in list notation.

GF2 n *[function]*

returns the list of elements of the Galois field of order 2^n in list notation.

Examples:

(GF 2 3) → ((0 0 0) (0 0 1) (0 1 0) (0 1 1) (1 0 0) (1 0 1) (1 1 0) (1 1 1))

(GF 2 3)→ ((0 0 0) (0 0 1) (0 1 0) (0 1 1) (1 0 0) (1 0 1) (1 1 0) (1 1 1))

(GF 3 2)→ ((0 0) (0 1) (0 2) (1 0) (1 1) (1 2) (2 0) (2 1) (2 2))

When all sets are defined in this way, the function tables or matrices can be filled up. They can be initialized by right button commands if desirable. Then the left button pressed over a table or matrix element pops up a menu of admissible values. Rather big tables are filled in this way quickly and conveniently. As continuous scrolling is possible but very slow with long tables, one should make use of the very sophisticated scrolling capabilities of InterLISP-D, such as thumbing and especially by-part-of-page-scrolling.

When the fill-up is finished, the "Exit" command (right button menu) must be used to leave the system editor. Using any other method to shrink or close the window or let it open will leave the system undefined.

A.6 ANALYSIS OPERATIONS

A.6.1 I/O-Experiment

One analytical command common to most system types is "I/O-Experiment" which creates a window where the user can supply inputs for the system and see the state changes and outputs displayed. In these I/O-experiment windows, the right mouse button is the "command button" for selecting the available commands step, init, and undo. Step will ask for an input via a pop-up menu and show the result. Careful interpretation with respect to *Mealy/Moore* type system differences is recommended. The last steps can be undone with successive undo commands. The init command will clear the window, and initialize the system if necessary, for the next experiment.

A.6.2 Lattice

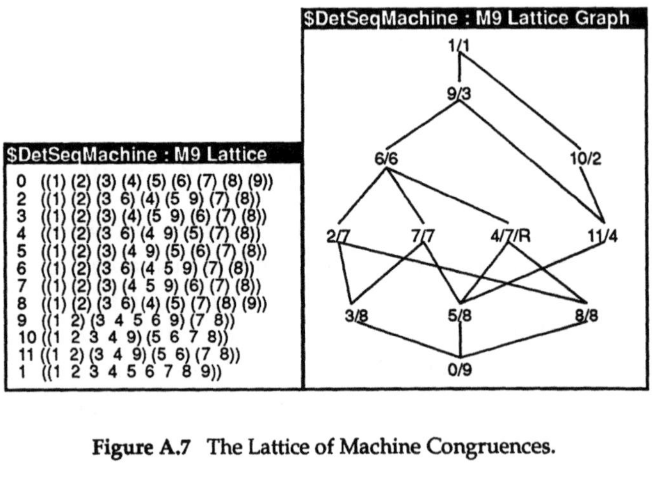

Figure A.7 The Lattice of Machine Congruences.

For several system types, the lattice of congruence relations may be computed. The algorithm is computational expensive, so only practical for small systems. The result is displayed in two windows, one showing the Hasse diagram of the lattice, the other displays the correspondence between congruence relations and the labels used in the diagram. The relations are labelled in the following way: The first item is 1 for the top of the lattice and 0 for its bottom, else a serial number. The second item is the number of classes of the relation, indicating the number of states used by the corresponding quotient machine. One relation has a third element "R", indicating that it is the reduction congruence, which is the greatest output-consistent relation. If this is the "0" relation, the system is reduced already.

A.6.3 Others

There are several other analysis methods. When selected, they will present their results in own windows, together with a textual explanation.

A.7 SYNTHESIS, SYSTEM TRANSFORMATIONS, AND THE REALIZATION TREE

Nodes of the realization tree may be expanded with synthesis operations. The new realization of a selected system will be displayed as a child node in the diagram. Some operations produce more than one child node, i.e., a new subtree. For example, the method parallel-decomposition of a DetSeqMachine named M1 will add a child node named M1.PN1 (parallel network 1) with three additional subnodes, named M1.PN1.1 for the first machine component, M1.PN1.2 for the second machine component, and M1.PN1.LF for the output coding function. The two machine nodes are

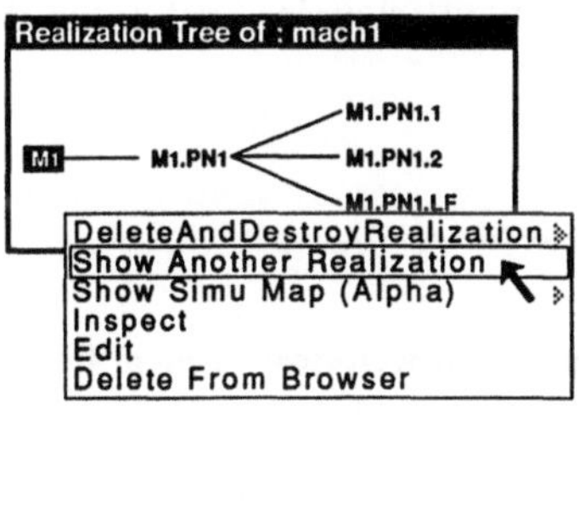

Figure A.8 Switching between several realizations of system M1.

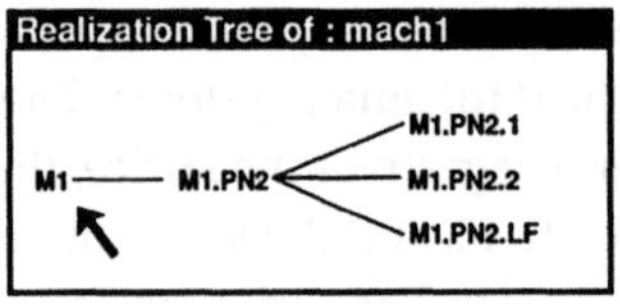

Figure A.9 Another Realization of M1.

of type DetSeqMachine, so they will show the same menus as the original system M1. The output coding subsystem is of type *function* and therefore shows a different left button menu when selected. If the same transformation (possibly with different parameters) is selected for the same node, the name of the resulting nodes will have the serial number incremented, for instance, M1.PN2 for a second parallel decomposition of M1.

Selecting a synthesis transformation on the same node a second time (the same transformation or a different one) will hide the result of the first synthesis operation and show the new realization. So it is possible to expand any node of the current tree anytime, but only the newest version will be displayed. If one wants to return to an earlier version of realization of node x, in the middle button menu of node x the command "Show Another Realization" has to be selected. A pop-up menu shows the different realizations by name. With this command, experimenting with different sequences of system transformations is possible. In this way, CAST.FSM may be called a rapid prototyping tool for *systems algorithms design*.

A.8 LIVING WITH A PROTOTYPE: ERRORS AND BREAKS

CAST.FSM is not a product. In contrast, it is an unfinished prototype. The user of the current CAST.FSM release has to live with many insufficiencies and some design flaws, as well as some program errors.

A.8.1 Beginners and casual CAST.FSM Users

In any way, the user needs the following background knowledge to be able to work with CAST.FSM:

- the dandelion user's guide,
- a little experience with LISP functions and lists,
- the InterLISP-D window system, mouse, and inline editing,
- the InterLISP-D file package,
- general knowledge about LOOPS concepts (Chapter 3 and 4 of the LOOPS manual).

This CAST user's guide in addition to these topics should suffice for using CAST at a demo level. The following hints should help in the case of program errors.

In case of an error, InterLISP-D brings up a new window, called *break window*. The caret (text insertion cursor) in this window indicates that the LISP interpreter waits for user typed commands. The user should use the mouse instead and press the middle button in this window to select the break command BT (for backtrace). Two additional windows will be displayed. One high but narrow window is, in fact, a (scrollable) menu of the current stack frames, shown in calling sequence, newest at top and oldest at bottom. The topmost frame is the one in which the error was detected. The second window is an InterLISP-D inspector, showing the selected stack frame in detail. In some cases it is necessary to select the stack frame with the middle mouse button instead of the normally used left mouse button. The user should try to find the uppermost stack frame, which is not an InterLISP or LOOPS system function. A hardcopy of the screen in this situation will be helpful for debugging the error. If in doubt, one should print several screens with interesting frames selected.

A.8.2 Power Users

Power users of CAST.FSM are required to have deeper knowledge and experience of InterLISP-D [Teitelman et al. 83] and LOOPS [Bobrow, Stefik 83]. As suggested in the book CAST by Pichler and Schwärtzel [Pichler, Schwärtzel 90], any system transformations which are not included in the basic CAST system have to be added by the user. This is possible because of the rapid prototyping nature of the underlying InterLISP development system, but nonetheless it requires programming knowledge in LISP and object-oriented concepts.

A.9 SAVING REALIZATIONS ON FILES

Complete realization trees with all related information such as already computed lattices can be saved on InterLISP files by following these steps:

- Determine the system type of the root of the to be saved realization tree.
- Select this system type in the CAST class browser with the left button.
- Choose the command "Add some instances to a file" from the left button menu.
- Specify a file name *filename* in the browser's prompt window.
- Select the name of the root object from the pop up menu.
- Additional objects may be selected.
- Quit selecting by clicking outside of the menu.
- Type (MAKEFILE '*filename*) into the TTY window.

For restoring saved realizations from a file filename, type (LOAD '*filename*) into the TTY window. To get the realization tree displayed, select "Browse Instance" from the left button menu on the appropriate system type in the CAST class browser.

REFERENCES – APPENDIX

[Bobrov, Stefik 83]
D. G. Bobrow, M Stefik: *The LOOPS Manual*, Xerox PARC, December 1983.

[Teitelman et al. 83]
W. Teitelman, et al.: *InterLISP Reference Manual*, Xerox PARC October 1983.

[Pichler, Schwärtzel 90]
F. Pichler, H. Schwärtzel, *CAST: Computerunterstützte Systemtheorie*, Springer Verlag, Berlin, 1990.

[Praehofer 86]
Herbert Praehofer: *LOOPS-Implementierung Automaten-theoretischer Methoden*, Master Thesis (in German), Johannes Kepler University, Dept. of Systems Science, 1986.

INDEX

MIX
Papier aus verantwortungsvollen Quellen
Paper from responsible sources
FSC® C105338

If you have any concerns about our products,
you can contact us on
ProductSafety@springernature.com

In case Publisher is established outside the EU,
the EU authorized representative is:
Springer Nature Customer Service Center GmbH
Europaplatz 3, 69115 Heidelberg, Germany

Printed by Libri Plureos GmbH
in Hamburg, Germany